The Vital Vastness
—Volume Two

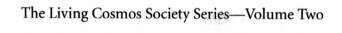
The Living Cosmos Society Series—Volume Two

The Vital Vastness
—Volume Two

The Living Cosmos

Richard Michael Pasichnyk

Writer's Showcase
San Jose New York Lincoln Shanghai

The Vital Vastness—Volume Two
The Living Cosmos

Writer's Showcase
an imprint of iUniverse, Inc.

For information address:
iUniverse, Inc.
5220 S. 16th St., Suite 200
Lincoln, NE 68512
www.iuniverse.com

ISBN: 0-595-21086-4

Printed in the United States of America

CONTENTS

Tome Four
In and Beyond the Milky Way:
An Exploration of the Universe
(The Structural Dynamics of Celestial Objects)

Tome Five
The Unity of the Sun, Earth and Moon
The Solar-Lunar-Terrestrial Linkage In the Historical Process

Conclusions
It's the End of the World as We Know It!

TOME FOUR

In and Beyond the Milky Way: An Exploration of the Universe (*The Structural Dynamics of Celestial Objects*)

"Exact observation of reality shows that spatial relations
—the phenomena of symmetry—lie at the basis of all phenomena
we have studied." Vladimir Ivanovich Vernadskii
(In Zh A Mededev, *The Rise and
Fall of T. D. Lysenko*, NY, Columbia Univ Press, 1969)

"Natural science [is] a science of processes of the origin and development
of these things and of the interconnectedness which binds all these
processes into a great whole."
Friedrich Engels (Ludwig Feuerbach, 1886)

"Indeed, if the history of science is any guide, theory and
observations have never exactly been in phase."[546]; p.244

CHAPTER 1

Earth's Family
(The Structural Dynamics of the Planets)

Recollecting the new model of the Earth, referred to as the Field-dynamical Earth Model (FEM), revealed in the Volume One, will help to uncover an understanding of other objects in the Cosmos. Various phenomena were noted to occur along the 30º to 40º latitudes, and at the poles of FEM. Observed along the 30º to 40º latitudes were weather centers, particle flow, geophysical anomalies, and various other phenomena. Around the equatorial region a ring was observed consisting of protons and electrons flowing in opposite directions, labeled the equatorial ring current and equatorial electrojet. At the poles particle flow was noted from both the solar (inflowing) and polar (outflowing) winds, which produce the auroras. The Earth has a north-south or dipole magnetic field, and a magnetosphere. Also, a belt of radiation grows and shrinks in response to particle flow in the solar wind. Many of the features observed are time-varying phenomena, because the Field-dynamical Earth Model includes a system of time-varying particle accelerators. The following sections of this volume will reveal that each of the planets, stars, comets, galaxies and other objects in the cosmos display a similar structure, in spite of their varied appearances.

1.1 Mercury

The discovery of a magnetic field on Mercury came as a big surprise. A substantial north-south magnetic field of internal origin is observed (*intrinsic dipole*). The inner core region of Mercury should have solidified or froze out early in its history, leaving only a thin shell-like region.[156,190,477,518,551,578,693] This immediately offers a problem for dynamo theory, because it is

> "not considered by most dynamo theorists as a probable source region for traditional planetary dynamo processes to be active. Unfortunately, there are no adequate dynamo theories which predict either the existence or the characteristics of a planetary magnetic field. The existence of an active dynamo places certain constraints on the present-day internal structure and temperature. The most basic constraint is that there must now exist an electrically conducting region in the interior in which electrical currents flow, driven by some energy source." [190]; p.513

Radio emissions have been known to disappear during certain phases, indicating a time-varying component.[255] Mercury's slow rotation on its axis, temperature profiles, and radio emissions do not match the dynamo model at all.[156] Yet, the facts clearly match what could be expected of the Field-dynamical Model.

The magnetosphere strongly resembles a miniature of the Earth's with everything happening more quickly and repeating more often, because it is closer to the Sun.[627] It also contains sodium, which may help generate electricity.[386] This may be the reason that the electric field in the magnetosphere is much greater than that of the Earth; another completely unsuspected observation.[314] In addition, trapped particles are too dense for what is expected of the Mercurian ionosphere. All of these observations indicate that Mercury, like the Earth, has a dynamical mechanism that generates charged particles, electricity and magnetism.

Likewise, six-second fluctuations of large fluxes of electrons and protons were observed, and came as a big surprise. "Such well-defined structure was certainly not expected in the streams of high-energy particles, and it will probably be quite difficult to find a physical mechanism that will accelerate particles with opposite charges simultaneously." [518] The Fields of the Field-dynamical Model are adept at doing exactly this, which demonstrates that Mercury's dynamics result from such a model.

Another surprise is that auroras also occur on Mercury.[156,314] The polar cusps display high levels of helium whose abundance is maintained at a steady state. This observation is unexplained, because a "sink for [hydrogen]" is needed to explain the fact that hydrogen is relatively low compared to helium.[314] However, this is something that could be predicted from the hydrogen-fusion model with hydrogen plasma accelerated into the core, and the subsequent production of helium. The aurora is also time-varying, but being so close to the Sun the auroras should be continuous, according to present models.

Latitude restrictions for phenomena also exist on Mercury. An "aurora" has also been observed on the side away from the Sun (midnight) around the 25º to 35º latitudes.[156] At mid-latitudes there are hazes, and a luminous spot, which were seen in advance of the planet when it is farthest from the Sun (*aphelion*), and trailing when closest to the Sun (*perihelion*). The spot was never at the center, but south of it, and occasionally there were two spots. Issuing from a crater (31), the spacecraft had caught a glimpse of the time-varying Field when it revealed the hydrogen fusion by-products as they were released. Other luminescent effects have been noted at other times.[197,207] Dark, nebulous rings have also been observed that occasionally display a violet tinge.[392]

Bright emissions (*at short UV wavelengths*) were noted on the dark side one day, then disappeared and again reappeared, again indicating a time-varying component. The wavelength noted is typical only of a star—hydrogen fusion—or a moon, but Mercury has no moon. As a result the author comments that it had "no right to be there. What the original emissions were from, the ones spotted on the approach to the planet, remains a mystery."[55]

Plate tectonics on Mercury also display latitude constraints. Polar (*normal fault*), mid-latitude (*strike-slip fault*) and equatorial (*thrust fault*) provinces are noted. The mid-latitude province is centered on the 30° latitudes, and exhibits a nonrandom arrangement. Certain features (*lobate scarps*) lie along the northwest-southeast and northeast-southwest grid directions. This regular arrangement immediately became controversial with claims that it was the result of the effects produced by lighting. Others see this orderly distribution as the result of the far field effects of the Caloris Basin, which is centered on the 30° latitude. In the polar regions, the directions are not as defined, but an east-west orientation is common. Or, in other words, the directions are more or less concentric around the poles.[514,711]

A group addresses the enigma of the nonrandom distribution. "The global map shows that the different maxima are not randomly oriented, but are more or less constant for the entire planet [and there was] an unknown process(es) that produced an important [North] 20° [East] trend."[711]; pp.406 & 428. As on Earth the Fields have also produced nonrandom tectonic features.

Other observations support a Field-dynamical Model for Mercury. The atmosphere is mostly helium, hydrogen and oxygen, but these gases should have escaped along with the solar wind, indicating a local source. Planetary hydrogen fusion and its by-products are an ideal source for replenishing and neutralizing these atmospheric components, while Field dynamics prevent their depletion by deflecting the solar wind. Inert gases, such as neon and argon, are absent, but should be abundant. Some "unknown" energetic process is needed to eject the gas atoms into interplanetary space.[255,314,378,386] The acceleration mechanisms of the Field-dynamical Model are the energetic processes.

There is a low brightness temperature (*about* 200° *K*) in the hemisphere illuminated by the Sun, which does not show a strong variation with phase. Furthermore, it should be 2.5 times hotter than what was observed (*about* 500° *K*).[255] Such a reduced constant temperature indicates Mercury is far more physically independent of the Sun than we ever imagined.

Mercury is one tenth as massive as the Earth or Venus, but is has an exceptionally high density.[477,693] "The observed density of Mercury is confirmed to be too high to be explainable by straightforward condensation and accretion models. Some additional important mechanism is required."[477]; p.665. This observation departs from the traditional scenario of the condensing protoplanetary nebula (condensation) and an era of bombardment by meteors or planetesimals (accretion). This is strong evidence that our present planetary models are at the very least inadequate.

1.2 Venus

Venus is nearer to the Sun than is the Earth, and it was believed that Venus would have a much stronger magnetic field than the Earth.[599] That is, pressure and temperature conditions are believed to have generated a liquid core, and this depends on the size and composition of the planet. Since Venus is the approximate size of the Earth it should have a strong magnetic field, yet it does not. The magnetic field was observed to be near zero strength for a long period of time.[48] However, at times a strong magnetic field was observed. The Venusian magnetic field is time-varying, as could be predicted if it were a time-varying particle accelerator, such as that of the Field. There is little evidence for a vertical or north-south component, as the magnetic field is only horizontal, and varies from orbit to orbit. Admittedly, scientists again say that these observations have nothing to do with our understanding of internal dynamos.[230,592] Furthermore, large-scale variations in the magnetic field take place in response to the solar wind, indicating a solar linkage.[183,600,631]

In spite of its lack of a strong internal magnetic field, Venus has auroras.[183,281,621] The presence of auroras was completely unexpected without a strong magnetic field. An oxygen nightglow at times of solar flares was also completely unexpected. A scientist comments on this enigma: "All the known mechanisms [are] far too weak."[591] In spite of it all, Venus is like the Earth, "In many ways these wave and field measurements resemble those obtained at Earth above the auroral oval, within the polar cusp and at the boundary of the plasma sheet."[641]; p.185. Auroras occurring with such a weak non-dipole magnetic field defy present theories of planetary physics, but support a Field-dynamical Model.

Another surprise was that the magnetosphere is similar to the well-known peculiarities of the Earth's. Ions show acceleration processes occur at great distances from Venus (*in the particulate shadow or in the wake*). In addition, the density of the solar wind influences the level of electrons in the Venusian ionosphere, which again discloses a Venus-solar linkage. A part of the magnetosphere (*magnetospheath postshock flow*) is similar to that of Mercury, Earth, Jupiter, Saturn and the other planets.[48,130,332]

Part of the reason that the atmosphere of Venus is so hot, about 500° Celsius (more than 900° F), is that the magnetosphere bows close to the planet in response to the solar wind, heating the planet's atmosphere. Another reason is that the atmosphere, consisting of ammonia, methane and sulfuric acid, traps heat in a "runaway greenhouse effect."

A number of unexpected, and hence, unexplained phenomena were observed at the poles. For one, the polar regions are actually hotter than the equator. A scientist makes note of this enigma: "This is an unexpected direction of variation, since normally the equatorial regions would be expected to be warmer because of the greater solar heating there. The reasons for this observed behavior are not yet clear."[655]; p.251

The reason for this observation is that hydrogen fusion is taking place in the Venusian interior and heated ions are being accelerated out of and into the poles. Another indication of this internal, dynamic energy source is that Venus radiates 15% more energy into space than it receives from the Sun. The planet is entirely cloud covered except for the poles, undoubtedly the clearing at the poles is due to particle flow. Surrounding the poles is the "polar collar cloud," supporting this interpretation.[655]

A pair of polar vortices cannot be explained by a planet's rotation affecting air masses. Similar vortex structures have been noted in mid-winter on Earth when the polar wind is more active (as commented on in Tome One of Volume One). On Venus there is also strong subsidence at the poles and cold bands that surround the hot spots, as could be predicted from particle flow creating a vacuum.[423,484,655,706,707]

The absence of atmospheric cells and poleward movement of clouds is not explainable by solid-planet physics, either.[114] However, this structure could be expected if an accelerator—the Field—were accelerating two oppositely charged populations of ions (a helical flow). Polar flattening could be predicted from the new model and has been observed (*i.e., due to ferromagnetic constraints*). All of these observations indicate that Venus has the same internal structure as that of the Earth: a core of hydrogen fusion maintained by particle accelerators and magnetic confinement. Hence we see a multitude of comments like the following, "No models exist yet to show plausible mechanisms for generating and sustaining the polar collar or the dipole nature of the polar vortex."[707]; p.679

Other evidence that the same type of planetary model as that of the Earth operates at Venus are phenomena associated with mid-latitudes. For one, there are "holes" in the ionosphere at approximately 33° North and 24° South Latitudes, which are more active at night when oriented away from the Sun (see Figure 4.1). A scientist expresses the uncertainty associated with present models: "the causes of the holes and horizontal stratification, and the generally greater variability of the night side, are not yet entirely clear."[131]; p.834

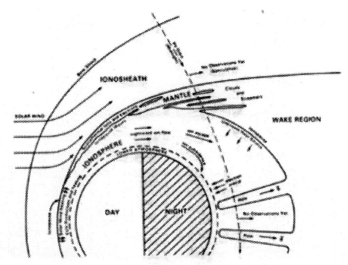

Figure 4.1. Venus Ionosphere. This sketch summarizes the findings of ionospheric experiments conducted on Venus. Note the holes around the mid-latitudes and the fact that an unknown heat source exists on the night-time side of Venus. This is evidently due to particle flow that was unexpected. [from reference 131]

Magnetic fields within the holes display vertical components and are stronger than adjacent regions. Electron temperatures are lower (*about* 2,000° *K*) than surrounding regions of the ionosphere, except in the regions with the least particles (*lowest density*) where there are highly elevated temperatures (*about* 20,000° *K*). Particle acceleration causes heating and a partial vacuum, indicating that this observation is evidence of Fields accelerating particles. Likewise, there is an enhancement of hydrogen ions (H^+ *over* O^+) inside the holes. The plasma depletions inside the holes are known to fluctuate along with electric fields and upward ion flow.[131,341] As could be predicted from a Field-dynamical

Model of Venus, there are discrete mid-latitude Fields along which particles flow, producing various phenomena.

Further support for particle acceleration at mid-latitudes has been reported. In the magnetosphere there is a "particulate shadow" or "cavity" (*in the magnetotail or wake*) that displays an absence of regular ion flow. However, electron fluxes are extremely variable, while ions display acceleration processes that occur in the "cavity" at great distances.[332] Observations of ion accelerations (*magnetotail flux*) were interpreted to be the result of an internal (*intrinsic*) magnetic field, but according to present models it is "not large enough to supply magnetic flux in such a tail."[631]; p.910. Observations indicate the existence of Field dynamics beyond what is called for in present theory.

Gravity or planetary waves, such as occur on Earth in the winter hemisphere, were observed in the middle atmosphere displaying a 4- to 5-day period. Heat (*Kelvin*) waves show the same period and radiate symmetrically about the equator. Both waves are synchronously coupled, indicating a common source, but present models bring about the conclusion that "there is no distinct excitation source."[655]; p.254

Typically it is believed that the solar wind causes the acceleration (*magnetohydrodynamic processes*), but observations indicate that "the differences imply that something else is occurring at Venus in addition."[631]; p.938. Protons and oxygen (O^+) ions appear to have the same speed at some times and different at others. Again observations depart from the traditional: "Our observations provide evidence that the process is taking place, but provide few details of how."[630]; p.183. As could be predicted by the Field-dynamical Model, strong and large-scale magnetic fields are noted during periods of high solar wind pressure, there are large day-to-day variations of neutral hydrogen, and hot, accelerated hydrogen is greatest at the poles.[105,252,484,540-542,708]

Other phenomena lend further support. Usual on the dayside, airglow or an ashen light was instead seen on Venus' darkside.[58] A bright limb band displayed a maximum at the equator along with a ring or particle density.[131,206] A bright spot or so-called "star-spot" of a minute star-like flashing white light was seen on the darkside, and at times the darkside was mottled over with nebulous granules, though this is rare and fluctuates with the Sun.[96] At times brushes of light, roughly fan-shaped in a sunward direction (*maedler phenomenon*), made Venus look like a multi-tailed comet.[97] These phenomena are observations of particle flow.

A spacecraft orbiting Venus in 1982 discovered a marked increase in helium ions, a product of hydrogen fusion.[383,628,629] Meanwhile, a day later corresponding helium and magnetic field fluctuations were also observed near Earth.[628,629] No satisfactory explanation has been proposed to account for either observation.[383,628,629] Yet, the answer is clear when considering the Field-dynamical Model with its production of helium, and its solar linkage.

Venus is surrounded by a radiant glow (*corona*) of hot atomic hydrogen and an enhancement of a radioactive form of hydrogen (*deuterium to hydrogen ratio*). Neither observation is explained by current models, but can be predicted of the Field-dynamical Model. Also, brightness emissions along the equator of vibrationally excited oxygen (O_2) are unexplained, yet it is very similar to the Earth's equatorial electrojet.[281] A ring, particle density and brightness emission along the equator indicate a form of planetary ring.[131,281]

The ionosphere can reach several thousand kilometers to scarcely any ions above 200 kilometers (125 miles) 24 hours later.[230] Again this is evidence of an ion-generating source and accelerators that are time-varying. The large-scale magnetic fields observed in the ionosphere, when the solar wind pressure is high, persist longer than expected from a weak non-dipole magnetic field.[202]

At certain wavelengths (*infrared*) the darkside shows patchy bright and dark cloud patterns about temperate regions that cannot be understood by present models either. Two probes showed an unexpected glow that increased near the surface possibly akin to the glow or lighted displays noted in the Field areas on Earth, such as Saint Elmo's fire in the North Atlantic Field (*only thermoluminescent*). Infrared and ultraviolet wave-length analyzers also disclosed unexplained voltage changes (*gradient drops*).[591] Similar voltage gradients are utilized in physics to accelerate particles. All of these observations would have been explained and even expected if the Field-dynamical Model was known.

Lightning on Venus also gave us a few surprises, and confirms this model. Strong and impulsive (time-varying) low-frequency noise bursts were noted when the local magnetic field was strong and steady. Lightning clustered near Beta and Atla on the eastern edge of Aphrodite Terra, which is a land form with volcanic peaks at about the 30° South Latitude. The surprise was that the lightning came *up* from the planet, and it is not volcanically triggered. Lightning on Earth takes place along with precipitation, but Venus displayed no precipitation.[257,451,640,710] Furthermore, bursts of radio noise, typical of particles flowing along a magnetic field, were noted, but *below* the cloud level.[452,461,640] A region at the 30° North Latitude, with volcanic peaks, is another source of lightning discharges. Lightning is observed on both the day and night sides with very low-frequency waves (*whistlers*) detected deep in the ionosphere.[230,640]

The atmosphere clearly displays the effects of polar and mid-latitude accelerators, such as observed in mid-latitude jets, which are analogous to the jet streams on Earth. Jets above the clouds were observed at 15°, 50° to 55° and 70° to 75° latitudes.[423,555] This demonstrates a particle flow that spreads, like the funnel shape of the Field, about 15° on either side. Hence 30° to 40° latitude fields create jets at 15° and 50° to 55°, while the polar field at 90° latitude creates 70° to 75° latitude jets forming a cold "polar collar." Furthermore, scientists admit that the effect of a planet's rotation cannot explain these jets.

There is a smooth transition from one layer (*troposphere*) to another (*mesosphere*) at latitudes below 45°. However, one layer (*troposphere*) is found at about 56 kilometers (35 miles), while above 45° latitudes it rises to about 62 kilometers (39 miles).[423] Again, the accelerators spread with altitude in the funnel shape, hence there are higher layers above 45° latitudes due to ionizing radiation creating a partial vacuum below this latitude. The satellites Mariner 10 observed the presence of, and Pioneer the absence of, the mid-latitude jets.[481,558] Again, this is predictable from time-varying mid-latitude particle accelerators, which operate at times and not at others.

Vertical wind shear is low between about 30° to 48° latitudes, and high from 0° to 30° and 48° to 60° latitudes.[481] The center of the accelerators is more stable and the outer "rim" of the Field, where particles flow, creates a vacuum that other parts of the atmosphere move into. Hence, high vertical wind shear on either side of the Fields—0° to 30° and 48° to 60° latitudes—would occur. Within the Field latitudes (about 30°) the high wind-speeds "appear unrealistic," because of particle flow and the vacuum created, which are not yet known. The jets have continuous large features which scientists insist "must retain their coherence through some guiding mechanism."[481]

Another indication of time-varying accelerators involves observations of changes in carbon dioxide. Strengths of carbon dioxide vary in a four-day cycle, which is not exactly periodic, but is built-up on successive cycles and then "collapses." In order to produce this effect the cloud-deck must move up and down as much as one kilometer (0.62 miles) simultaneously over the entire planet! This requires a high input of mechanical energy that is extremely difficult to account for in a slowly rotating planet. One author discusses this in an article titled, "Venus Breathes in Steady Fashion": "Therefore the cycle variations point to some unexplained deep-seated property of the atmospheric dynamics."[53]

More evidence comes in observations of Venus' superrotating atmosphere. Winds at high altitude sweep around the planet in only four days, while the rotation of Venus takes 243 days.[376, 499,650] What makes it even more impossible to reconcile with current planetary models is that the clouds rotate in the opposite direction of the planet (*i.e., retrograde motion*).[376,499] This superrotation consists of maximums of hydrogen and helium, and (*ultraviolet*) airglow after midnight, as well as a daily temperature variations.[650] As discussed in Tome One (Volume One) and Tome Five, the mid-latitude Fields become active away from the Sun, accelerating the by-products of hydrogen fusion—protons, hydrogen and helium nuclei and atoms—causing temperature changes.

Venus has the largest ratio between atmospheric and planetary rotation of any body in the Solar System, since the clouds rotate about 60 times faster than the planet's rigid shell.[376,499] However, Venus is not the only planet with a superrotation of the atmosphere, because it has been observed on Earth, Mars, Jupiter, Saturn, Neptune, and Uranus, as well as Titan.[499]

A "Y" on its side and other shapes appear as dark features (*particularly in UV photos*) that are centered along the equator. The "Y" feature persists for weeks reappearing in four-day intervals, and persists at the same longitude for at least decades.[16,257,376] These features, again, show the predicted mid-latitude restrictions: "the large scale markings ([about] 1,000 km. or 600 mi.) situated between the 45° latitudes have lifetimes greater than 4 [days] and move with an apparent angular motion at the equator."[376] (see Plate 4.1)

Temperature and pressure variations also disclose the importance of latitude in accordance with the Field-dynamical Model. Not only are the poles hot, but the deep atmosphere is also cooler near the equator than at 30° latitudes below 40 kilometers (25 miles). One probe's data "show the equatorial region to be (surprisingly) cooler than mid-latitudes below the clouds."[655] This was unexpected and is not understood, because more sunlight strikes the equatorial region.[690] Meanwhile, mid-latitude fields ionizing the atmosphere, producing a partial vacuum, would cause equatorial air to flow away from the equator and into the mid-latitude Fields (*damming*), producing these observations.

Likewise, at the poles the coldest and hottest temperatures exist side by side, the hottest over the pole, and the coldest around the pole along the polar collar.[423] Atmospheric pressure is highest at about the 30º latitudes, with high values at the equator, and near the pole, at about the 60º latitude. The lowest pressure is directly over the pole.[709] Furthermore, the temperature and pressure variations cannot be explained by the rotation of Venus (*i.e., cyclostrophic balance*).[655] Yet, particle flow can explain all of these observations with high pressure occurring in those regions that the atmosphere flows into when balancing the partial vacuum that was created by the ionization of the atmosphere.

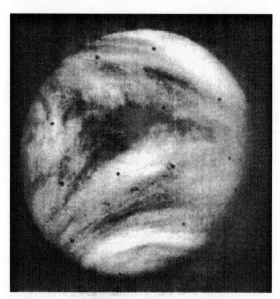

Plate 4.1. Venus' Atmosphere. The super-rotation of the upper atmosphere of Venus is not understood by atmospheric scientists. Wind speeds between 50 and 70 kilometers (31 and 43 miles) are an astounding 360 kilometers per hour (225 miles per hour). Wind speeds are also strong at lower altitudes, but change abruptly at the 50 kilometer height. This could be expected from the funnel shape of the Fields,

which widen at a certain altitude. A dynamical component is also suggested by the constant appearance of the large-scale pattern that forms a horizontal 'Y' repeatedly. [NASA]

Understanding Venus encourages some intriguing questions when other facts are considered. Hot spots in tectonics occur along the equator with elevated circular regions, gravity anomalies, swells in the rock surface (*lithosphere*), rifting and volcanic activity, implying that it is dynamically supported like the Earth.[435,506] The rock surface (*lithosphere*) is thin and hot, not dense enough to allow deformation of the surface or subduction as on Earth. Therefore, it was believed that no plate tectonics could exist, but observations show that plate tectonics do exist.[505,506] As a result, "most topographic features on Venus remain enigmatic."[506]; p.70.

A vast canyon dwarfs the Grand Canyon and is several kilometers (miles) deep situated along the equator. From rim to rim it is about 240 to 320 kilometers (150 to 200 miles), and at least 1,450 kilometers (900 miles) long. In contrast, the Grand Canyon is about 1.6 kilometers (more than a mile) deep in some places, 6 to 29 kilometers (4 to 8 miles) wide, and around 349 kilometers (217 miles) long.

There are also something like the rift valleys of the mid-ocean ridges on Earth, but mostly along the 20° latitudes. This trough-like depression is interrupted every 50 to 95 kilometers (30 to 60 miles) by "right-angle" fractures typical of some (*transform*) faults on Earth. Meanwhile, the terrain of Venus should be flattened due to the high temperatures. Yet, the positive and negative land-forms indicate that it releases the extra heat somehow, and there is a dynamical component involved.[457]

Typical of minerals being aligned with electromagnetic (and/or electrostatic) fields (*ferromagnetism, etc.*) there is polar flattening, an equatorial disc, and large equatorial flattening.[157] Gravity anomalies are most pronounced at about 28° North Latitude/290° longitude, and along the equator at 205º longitude. They do not correspond to land-forms in a strict sense, but elevated regions are often observed.[157,506,536]

Clouds display three distinct layers: an upper layer from 60 to 57 kilometers (48 to 35 miles), a middle layer from 57 to 51 kilometers (35 to 32 miles), and a lower, very narrow layer only about 100 meters (300 feet) deep. The upper layer is an admixture of an iron compound (*about 1% FeCl₃*) in drops of sulfuric acid.[256,257,428,438] It may generate electricity, and along with methane may be superconducting (organic molecules, etc.). Though the high temperatures make this doubtful, the inflowing frigid air may do the trick. The high neon ratios in the atmosphere are like the primitive solar wind, which makes it very difficult to explain the origin of the atmosphere without the Field-dynamical Model. Also unexplained by present theories on the formation of the Solar System is the abundance of carbon, nitrogen, and other elements (*Ar, Kr and Xe*), which are similar to the Earth.[536]

There is evidence that Venus was once not so hot, and had water and carbon dioxide, but the Venusian clouds now contain sulfuric acid causing a runaway Greenhouse Effect that traps the heat.[256,431,438,734] Furthermore, the magnetosphere is pushed close to Venus by the solar wind, thereby heating and ionizing the atmosphere. Could the present high level of solar activity have made the clouds as such (*photoionization and ionizing radiation from hydrogen fusion*) and heated the atmosphere? It is believed that Venus may have once had life, and may presently have some form of life in its clouds.

1.3 Mars

Like most of the inner or terrestrial planets, Mars displays differences between what was expected of dynamo theory, and what has been observed, when considering its magnetic field. Strong controversial differences exist between the strength of the internal (*intrinsic*) magnetic field and its ability to deflect the solar wind above the ionosphere.[493]

The evidence demands a stronger magnetic field (*mass flux is two orders greater*). Yet, all the observations indicate that Mars does not have a magnetic field, though it may have had one in its early history, or it is time-varying, and has not yet been observed.[406,627]

The magnetosphere away from the Sun displays a flow with velocities much larger than the planet's escape velocity.[584] This indicates some type of acceleration mechanism exists that is not explained by present physical theory. Unexpectedly, Mars also has an aurora (*UV dayglow and CO*) indicative of magnetic and/or electric lines of force.[280]

It becomes immediately apparent that something unique is affecting the polar regions of Mars when it is noted that circular basins exist at each pole. Each basin contains stacked plate units that resemble a pinwheel (see Plate 4.2). It also resembles the patterns noted on the polar regions of the Sun, comets, galaxies and other celestial objects, as will be discussed. Polar stress patterns cannot be explained by simple convection models, but require episodes of flow descending towards the interior near the poles. This is an accurate way of describing a time-varying particle accelerator that accelerates particles towards the interior operating since the planet's formation. The polar basins cannot be explained, as once attempted, by two impacts of the same energy striking each pole.

Trenches surround the South Pole at 80° latitude, while the polar ice cap covers one half. The North Pole has a large valley at 85° latitude that nearly cuts the northern cap in two.[751] In winter, the North Pole's cap spreads to about the 40° latitudes and begins to dissipate in spring, following the Martian equinox.[374]

Spiral clouds with a narrow jet occur near the North Pole during the season when winds are at their weakest. It is also one of the most active weather centers, especially in winter, and the clouds that form should be, according to physical theory, at altitudes of 6 to 7 kilometers (10.7 to 11.3 miles), but instead reach 50 kilometers (80 miles).[374] These spiraling cloud patterns are very much like cyclones on Earth. Ozone appears nowhere else on Mars but over the poles, and displays great seasonal change; in spring it decreases and reappears in summer.[430,751]

The caps are mostly dry ice and water, and like the polar regions, they are terraced.[191] The growing and shrinking of the ice caps and their composition indicate that they generate electricity (*grounded ice caps produce electrostatic and VHF fields*). These facts alone suggest that there is a dynamic mechanism responsible for the terraced caps and their season shifts.

Mid-latitudes are very active with regard to producing weather phenomena. Between the 30° latitudes there is the Tharsis Region, with its four giant volcanoes, which is one of the most meteorologically active areas on Mars. In spring and early summer the clouds form a system of waves that are parallel and evenly spaced, which indicates a time-varying mechanism in their formation. There are also gradients of heat and gravity waves. In both hemispheres, between 25° to 55° there are lee waves and wave clouds, which is approximately a 15° spread from the 40° latitudes.[374]

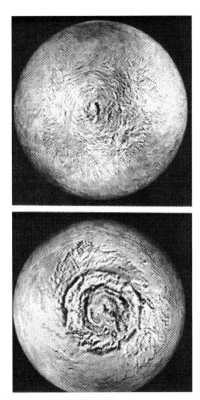

Plate 4.2. Mars Poles. These photos show the spiral of stacked plates at the Martian poles. Here we see an excellent example of the result of an applied field, the Field-dynamical Model's dipole field, influencing the arrangement of minerals (i.e., ferromagnetism, etc.; see discussion on planet formation, Chapter10). Compare this with Figure 3.4 of Volume One, and Figure 5.1, and Plate 4.6 in this volume. [NASA]

White, and especially yellow, clouds increase around the closest approach to the Sun (*perihelion*) between the equinoxes. For the Northern Hemisphere in autumn to spring, and especially in winter, clouds become abundant (the Earth and other planets also have more deep-seated changes in winter).[751] Blue clouds are greatest between the

25° to 30° latitudes, a secondary peak occurs along the equator and the lowest is at 30° South Latitude. Furthermore, "all 3 cloud types reach peaks in the zone between 0°-30° [North Latitude]."[751]; p.111. A drop in water vapor is noted for latitudes higher than 30° with the northern summer showing a peak in water levels and white clouds.[751] Dust Devils, which are dusty, miniature tornado-like winds, tower 6 kilometers (3.5 miles) above the terrain mostly at 33° to 43° North Latitude around Arcadia Plantia (142°-155° West Longitude), and a few at 37° North Latitude around Utopia Plantia (192° West Longitude).[239] The association of physical structures and weather phenomena between the 30° and 40° demonstrates the influences of the Field-dynamical Model.

All of these observations of weather phenomena are consistent with particle accelerators along the mid-latitudes, which become more active when a hemisphere is pointed away from the Sun, and when Mars is closest to the Sun—just as occurs on Earth. That is, parallel clouds, and lee, heating and gravity waves are produced by the time-varying character of the Fields. The seasonal changes are the result of the interactions with the Interplanetary Magnetic Field (IMF), which constitutes a solar-Mars linkage.

Dust storms and a darkening wave also display these latitude restrictions, including phenomena associated with the polar regions. Global and semi-global dust storms originate from Solis-Planum Thaumasia Region (30° South), Hellespontus (30° to 50° South) and Isidis (20° to 30° North) latitudes. Short-lived bands occur near the South Pole, as well as the southern low latitudes in spring. Along mid-latitudes in the Northern Hemisphere storms occur in autumn and winter, while in high latitudes they take place in summer. These transitions take place rapidly. Atmospheric physicists state the failure of present theory: "There is no adequate theory for whatever instability in the Martian atmosphere triggers such abrupt transitions."[374]; p.97

A darkening wave issues from either polar area and moves toward the equator, crosses it and then fades away at the 15° to 20° latitude of the other hemisphere. It is said to be vaporized carbon dioxide, but the exact same amount that issues from the polar region condensates in the other hemisphere. This phenomenon remains unexplained. A physicist studying Mars explains the enigmatic character of these waves: "No hypothesis ever advanced for the explanation of the seasonal darkening wave has been able to explain this one point!"[751]; p.290

Since the Fields are at 30° latitudes and spread 15° due to their funnel or cone shape, it would follow that the carbon dioxide would condense at 15° latitudes. This is especially evident when it is noted that the darkening wave begins at summer solstice at the pole. It then moves into the winter hemisphere at a time when particle flow creates a vacuum allowing the upper, colder layers of the atmosphere to condense vapors in the atmosphere. This is also suggested by the height of the north to south dust storms coinciding with the maximum of the darkening wave, which moves in the opposite direction (south to north) or vice versa. That is, the dust moves north to south in the southern summer solstice, while the darkening wave moves in the opposite direction of south to north.[372,751] Furthermore, there are daily and seasonal fluctuations that appear to be consistent with the freeze-thaw cycle.[372] Dust storms and darkening waves display the predictable limitations of the Field-dynamical Model.

Atmospheric circulation also demonstrates the latitude restrictions of the new model. The lower level of atmospheric cells joins the upper level at the 30° latitude in each hemisphere. The 30° latitudes clearly show stronger winds. Again a relationship with the Sun is evident with zonal winds at solstice being intense westerly jets in the upper mid-latitudes of the winter hemisphere. The summer hemisphere displays a weaker easterly at this time, but a fairly strong current near the equator where the mid-latitude Fields meet. Furthermore, during the equinox there are standing waves in both hemispheres in the upper mid-latitudes. A maximum (*amplitude*) in daily tidal pressure, in the Southern Hemisphere, is near the 8° latitude during the solstice, and 15° latitude during the equinox. Pressure lows were noted in Syria-Tharis at 5° South Latitude and Tempe at 40° North Latitude, while highs were in Hellas at 50° South Latitude, and a trough around the pole at 60° latitude.[751] A planetary scale wave occurs in the late winter at 23°, 48° and 45° to 65° North Latitudes. Temperature and pressure gradients fluctuate more between the 30° and 45° North Latitudes.[191] A similar latitude relationship, as occurs with clouds, dust storms and darkening waves, also shows up with winds and pressure.

Bright and dark spots have been noted along the mid-latitudes from the Late 19th to the middle of the 20th Century.[197] White, bright spots were seen at the 40° North Latitude on 5 July 1890, the 30° South Latitude on 10 June 1892, the 50° South Latitude 3 July 1892, and at 25° South Latitude from 10 to 17 July 1892 (only one spot). The most prominent at this time was at the 47° South Latitude from 11 to 13 July 1892.[36,43] Others were observed from 5° to 32° South Latitude, and 18° to 27° South Latitude between the 250° to 290° longitudes, and were reported in 1924.[726] White flashes and a string of spots and flashes were observed around the South Pole at about the 60° to 65° latitude, as reported in 1937.[764] Luminous displays were also noted at Sithonius Lacus in June of 1937, Tithonius Lacus in December of 1951, and Sinus Meridiani (0° to 10° North) and Sinus Sabaeus (5° to 15° North) on 1 July 1954.[83] A beam of bright light was observed in December of 1900.[39] Dark spots were seen near Mare Tyrrhenum on 1 May 1952; Hellspontus Mons (35° to 50° North) on 3 July 1952; Castorus Lacus on 7 July 1954, and Ascuris Lacus on 23 September 1954.[83] These luminous displays (masses of ionized particles) and dark spots (neutrons) suggest hydrogen fusion by-products and their locations are typical of the new planetary model (similar observations were noted for the Earth in Tome One).

Evidence of massive floods also indicate the operation of the new model (see Plate 4.3). Floods carved great gullies near the equator that reach 2,900 kilometers (1,800 miles) in length, and up to 200 kilometers

(124 miles) in width. One deep canyon stretched a length comparable to the distance from Boston to San Francisco, and its depth is 5.6 kilometers (3.5 miles). Compared to the Grand Canyon's length of 349 kilometers (217 miles) and depth of 1.6 kilometers (1.0 miles), the Grand Canyon is dwarfed again. The water is not atmospheric, but internal, and the amounts and timing are associated with volcanic activity.[323,663] Huge catastrophic floods took place in Mangala (4° to 10° South), Areas (2° to 10° North) and Kasea (21° North) Valles.

Erosion channels, known as arroyos, concentrate in areas equatorward of 40° North and 60° South Latitudes.[663] Arroyos are also observed in Tiu (10° to 18° North), Simud (0° to 14° North) and Shalbatana (1° to 15° North) Valles. Craters associated with the arroyos show a number of water flows separated in time.[342] The temperature and atmospheric pressure would have had to have been much greater during a number of periods, or some other condition(s) allowed for water's presence. A physicist comments on this: "A mechanism for generating or releasing water at a rate rarely if ever matched on Earth operated repeatedly."[528] Some evidence suggests that ice streams have carved Martian outflow channels, as well.[488] The source of the water is the by-products of hydrogen fusion, hydrogen and oxygen combined, to produce water.

Other physical features of Mars confirm the Field-dynamical Model is responsible for other physical characteristics. The Southern Hemisphere is more heavily cratered than the Northern Hemisphere. Mars has about twice as many craters as our moon.[519] What were once thought to be canals are today known to be aligned craters. A physicist who is a specialist on Mars describes this enigma: "it remains debatable whether these crater alignments over thousands of kilometers resulted by impacting bodies or whether they were internally produced"[751]; p.437. There is evidence that many craters in the Solar System are not caused by impacting bodies, but are internally produced (see Chapters 2 and 3 of this volume).

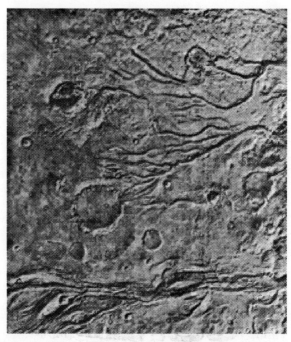

Plate 4.3. Water on Mars. Many large outflow channels on Mars show how this plateau, *Chryse Plantia*, was eroded by water. It is among a myriad of other evidence that scientists believe life was or is possible on Mars. [NASA]

The major features of plate tectonics on Mars are due to clockwise or counterclockwise rotations of features between the 40° latitudes. Tempe-Arcadia, Ascraeus and Pavonis Mons rotated 11.5°; Syria-Tharsis, and Vallis Marineris Rift 13.0°; Memnoia Rills 17.0°; Labyrinthus Noctis 37.0°; Arsia Mons 45.5°; and Olympus Mons 15.0°; together they created the major feature known as Petharia. Hesperia-East Rift rotated 25.0°; Tyrrhenum-West Rift 9.0°; Elysium 63.0°; and Syrtis Major-Arabia 89.5° to produce the major feature called Rhelysia. An expert on Mars states, "In fact, the pattern of crustal stresses near the equator and extending into the middle latitudes can also be described as rotations about various poles along the equator."[751]; p.515. Like the Earth's plate movements, such as the Gibraltar pivot, plates rotate on Mars.

Four major shield volcanoes called Arsia, Pavonis, Ascraeus and Olympus Mons are located between 20° South to 30° North Latitudes. The same expert comments that, "all originated in a line at the equator (not at the same time) in the rift."[751]; p.518. Fractures in the crust on a planet-wide scale (*lineament grids*) appear to be the result of internal stresses with a pole of ascending currents at 40° South Latitude at Alba Patera and descending currents at 40° South Latitude at Hellas Basin.[751]

Other observations more thoroughly reveal the new model. Brightness, the darkening wave, brightening of Lunae Pallis at about 35° North Latitude, and a blue haze noted on Mars fluctuate with solar activity.[621,751] The ionosphere shows ion species typical of hydrogen fusion and ionizing radiation (O_2, CO_2, O^+, H^+, He^+, NO^+, C^+, N_2, COH^+ *and electrons*).[576] Very high levels of iron, as well as sulfur and chlorine make the Martian soil a very likely conductor, as well as generator of electricity.[15,177,750]

Experiments performed on the Martian soil indicate a complex chemistry that is very much like life processes, only no organic molecules were detected. However, low levels of organic matter may have gone undetected by the instrumentation, or the organic molecules may have been absorbed by the soil and oxidized in life-like processes.[5,106,501] Meanwhile, organic material was identified on a meteorite that appears to have come from Mars.[21] Another possibility is that life exists under the surface of Mars.

The ice caps could support plant life and a life system could be developed over the entire surface of the planet. Mars was a warm planet with a much denser atmosphere and more moisture than it has today, and could have supported some form of life then.[14] Now the temperature fluctuates wildly because of its thin atmosphere, and falls down to 105° Celsius (220° F) below zero.

1.4 Jupiter

Jupiter's magnetic field is about four times stronger than that of the Earth's. As a result, the magnetosphere extends beyond its moons to about one hundred times Jupiter's radius.[680] Radio bursts, caused by particle flow along the magnetic lines of force, are almost entirely dependent on the phase of Io, Jupiter's largest moon.[10,234] There is also a correlation between solar wind boundaries and (*decimeter*) radio emission.[124] Particle flow along the field system can be witnessed in that compression of the magnetosphere coincides with the activation of the equatorial current.[680] These observations alone indicate that a Field-dynamical Model also operates on Jupiter.

Jupiter's aurora is the largest ever viewed by humankind, approximately 29,000 kilometers (18,000 miles) long at the North Pole (as observed 5 March 1979). It is much stronger than expected by present models, and due to electrostatic forces it is attended by lightning.[335,674,680] Another contradiction to the present model is that those particles which precipitate along field lines in one hemisphere are inhibited from reaching the other hemisphere.[620,675] Yet, this phenomena can be understood as particle acceleration along field lines and into the Jovian interior. Likewise, the observation of X-rays, typical of particle acceleration, were noted at both poles.[522] Auroras display greater acceleration at about the 180° longitude, which is unexplained by present theory, but is associated with a magnetic anomaly.[335,674] The moons, Io and Europa, are known to produce auroras by triggering particle flow.[225,280] Jovian auroras are not rare and have been observed repeatedly.[75,140,180,186,195, 225,263,280,674]

The bright polar regions of Jupiter are clear of atmospheric gases evidently due to particle flow.[684] Typical of the Field-dynamical Model, hydrogen (*H and H_2*) enhancements at the poles are time-varying.[280] There is also a polar wind of hydrogen, and enhanced methane emission bands surround the poles at about the 60° latitudes.[280,543] An unusually large abundance of hydrogen is associated with the auroras, but it cannot be explained by present-day models of Jupiter, nor by solar wind sources.[186,280]

Latitude restrictions of various phenomena are conspicuous in observations of the atmosphere, and those clouds that make up the bands and spots of vivid color. For one, the colors in the banded structures are parallel to the equator.[159] Blue light was noted at latitudes below 40°, and blue and red bands increase above the 40° North and 48° South Latitudes.[685] The bands of color display distinct changes at about the 30° latitudes.[568] Sharp boundaries exist in the cloud systems, which have time-varying characteristics.

The persistence of the dark brown and great red spots departs from the traditional model of Jupiter's atmospheric dynamics. An isolated dark brown feature is noted at 30° South Latitude (330° longitude) that has persisted for at least decades.[568] Likewise, the Great Red Spot, which is about 50,000 kilometers (31,000 miles) in length and 11,000 kilometers (6,820 miles) in width, was first observed in the 17th Century and continuously from 1878.[621] To consider the Great Red Spot as a three-century-long storm system is utter nonsense. There must be a dynamical component that stabilizes its position. Furthermore, typical of the new model's solar linkage, the visibility of the Red Spot fluctuates with solar activity.[94] It appears at the 20° to 30° South Latitudes, has had flare-type discharges (particle release), and it is probably organic (superconducting) molecules that make it appear red.[476,568]

The Field-dynamical Model is evident in that the Great Red Spot's and other atmospheric features' brightness are correlated with solar activity.[322,476,621] This is particularly evident when their brightness reaches a peak during sunspot minimum, which is the opposite of what would be expected from less solar wind plasma (*i.e., solar forcing*). In contrast, the bands take longer to rotate, particularly the north equatorial current, when there is a higher level of solar activity.[621] These bands display a differential rotation, as occurs with the Sun's surface features.

Heat distribution also reveals latitude restrictions. Warm features are evident at about the 35° latitudes.[568] In contradiction to solar heating, there is more heat above the 45° latitudes.[598] More heat should exist along the equator, but here we see the 15° spread of the Fields, and the heat produced by particle flow. When considering the vertical cloud-structure that exists, physical theory works well near the equator, but for latitudes of 40° or greater an "unknown" mechanism is required (*mass production rate needs to be a magnitude larger*).[684]

Particle flow was observed along the mid-latitudes of the magnetosphere (*field-aligned streaming away from the near-equatorial current sheet*). Because the Field-dynamical Model is not considered, comments surfaced that it is the "most prominent" and "puzzling" feature. It requires mechanisms unaccounted for by present theory, such as a planetary wind (*i.e., perpendicular electric field, intensity gradient at the equator and strong diffusion*).[91,503]

A two-component plasma, consisting of hot and cold plasma, discloses an acceleration mechanism that is time-varying.[100] A field-aligned component at about the 30° latitude deflects back toward Jupiter, and it is time-varying.[509] Ten times more energetic (*O and S*) ions flow outward into the magnetosphere, which indicates an internal energy source.[208.297]

A number of other features fail to be explained by the present models of Jupiter. Pioneer 10 observed changes in the atmosphere that were much too great to be resolved by present theory.[280] The power for the aurora, heating of the atmosphere, and a sink for major ions at high latitudes are also unexplained.[80] Temperature variations in certain features, such as the major white ovals and the Great Red Spot, are colder in some regions closer to Jupiter's surface (*upper troposphere and tropopause*), and relatively warm in another region above this (*upper stratosphere*).[192,507] "At the present time, no satisfactory explanation for this behavior appears to exist."[192]; p.8769. This enigmatic nature of its workings is because the Field-dynamical Model is not being considered, and are due to particle flow.

Further support for the new model are the wind peaks that occur at the 20° to 30° latitudes, and less so at 10° latitudes.[677,680,681] Likewise, flare-like discharges and lightning take place, and there is a high concentration of hydrogen (*lyman-alpha*) in low latitudes of the upper atmosphere.[280,476] Possibly the greatest indication of a hydrogen fusion core is the energy it receives from the Sun is only a small portion of what is radiated by Jupiter.[51] Likewise, Jupiter has often been referred to as the "sun that never made it."

According to widely held theories, Jupiter should not have rings, but it does. Something must constantly replenish the rings, otherwise they would not be there. The mid-latitude Fields are ideal for this purpose, and bursts of hydrogen have been noted in the vicinity of the rings, as well as, time-varying glowing emissions (*electroglow*).[280] Three major components to the rings (water, ammonia and methane) are hot at times and cold at others. Wedge-shaped, the inner region is thinner (*0.4 R_J*) than the outer region (*1.4 R_J*).[395,554,597] Rings will be discussed more in detail in a later section, and are not the result of the gravitational interaction of the moons.

Like the rings, numerous observations reveal Field contours and particle flow along field lines. The blue-gray hot spots in the equatorial region have emissions of radiation that are time-varying.[568] The magnetic field has a high curvature along the equator and charged particles spiral along field lines into the equatorial region.[113] The banded features are oriented parallel to the equator.[159] Also in the equatorial region are ultra-high frequency and super-high frequency waves (*whistlers*) generated by lightning.[659] Ions are repelled by the moons, and there is a current flowing away from the equator that indicates a parallel electric field.[311,349] The differential rotation of Jupiter is like the Sun, except that there is a jump in speed between the 15° latitudes.[621]

Also, ion decreases observed in the vicinity of the rings cannot be explained by their absorption by the moons Io and Amalthea.[172,609]

A similarity that can only arise from Field-dynamical Models is that the east to west airflow in the Earth's upper atmosphere (*troposphere*) is correlated with changes in the Great Red Spot. One scientist describes the similarity between the Earth and Jupiter: "the result obtained is evidence that zonal circulation experiences parallel changes both on the Earth and on Jupiter"[621]; p.349. The parallel changes are the result of the Field-dynamical Model in its solar-Earth and solar-Jupiter linkages. Likewise, variations in color changes, brightness and turbulent processes in the atmosphere occur with changes in solar activity. Hydrogen production also fluctuates with solar activity (sunspot number) and the large maximum of March 1979, far beyond solar input, remains unexplained.[81] That is, until now.

Other evidence supports the understanding that Jupiter's workings involve a Field-dynamical Model. Its composition is mostly liquid metallic hydrogen due to the high pressure and low temperatures.[408] The atmosphere is mostly hydrogen with helium and some ammonia, methane and water. The organic (*unsaturated*) compounds (*HCN and chromophores*) in the atmosphere are unexplainable due to the highly reducing atmosphere (*only 8 H_2*), and require a source of hydrogen fusion below the protective shield of the magnetosphere.[271] Some form of life may very well exist in the atmosphere of Jupiter, or one or more of its moons.[408]

1.5 Saturn

Saturn has radio signals similar to those of Jupiter along with a magnetosphere and radiation belts.[275] The leading edge of the magnetosphere (*bowshock*) and proton levels are responsive to the solar wind.[565,768] Electrons and protons are accelerated within the magnetosphere.

As with the Moon and FEM, radiation is depleted by the moons due to particle flow along field lines (*i.e., plasma torus, bow shock, electrostatic repulsion, etc.*).[492,672] Radio discharges are impulsive, as could be expected of a time-varying particle accelerator.[260] Electrons peak at 90° with respect to the local dipole magnetic field.[2,441,442,725] The "most remarkable and enigmatic aspect" is the high agreement (*axisymmetry*) of the magnetic field and the rotation axis, which may be exactly the same (less than *1°*). This close conformity is not understood within the framework of dynamo theory.[189] Indicative of an internal energy source, Saturn radiates 2.4 times the energy it receives from the Sun. These facts alone suggest that the Field-dynamical Model operates on Saturn, but there is more.

Typical phenomena of the Field-dynamical Model take place at the polar regions of Saturn. Aurora occur at approximately the 80° latitudes around the poles, and they also involve a polar wind.[82, 582,635] Typical of particle accelerators, X-rays are noted during auroras.[303,424] Helium, a product of hydrogen fusion, is emitted at the poles, and hydrogen emissions fluctuate with solar activity.[280,401,664] Polar brightening is evidently the result of particle flow, and polar phenomena are time-varying.[159,310]

A radio emission known as the Saturn Kilometric Radiation or SKR was observed. In some "unknown relationship" the SKR is highly correlated with the solar wind.[189] SKR is also fixed in relation to the Sun (i.e., sunward) as would be expected of a solar-Saturn linkage. In the Northern Hemisphere it takes place at the 70° to 80° latitudes, and in the Southern Hemisphere it occurs between the 60° to 85° latitudes.[189,402] This fix in relation to the Sun and the wider latitude spread in the Southern Hemisphere can be explained by interaction between the IMF and Saturn's magnetic field, which is tilted 15° with the North Pole pointing towards the Sun.

Latitude restrictions of phenomena are also evident on Saturn. There is a differential rotation like that of Jupiter and the Sun.[621,677] The equatorial belt extends to 20° latitudes and the temperate belt is at 40° latitudes.[677] High-altitude clouds occur in the equatorial zones, and at 45° to 50° latitudes, which are also those areas with the highest wind speeds.

Between these two regions we observe very low altitude clouds and the greatest cloud top pressure.[712] This is predictable with Fields at the 30° to 40° latitudes with a 15° latitude spread along which particle flow creates a vacuum producing low clouds, and the inflowing atmosphere producing high pressure at the cloud tops (damming). Temperature is lower along the equator than about the 5° to 20° latitudes, revealing the effects of particle flow.

Wind speeds are highest along the equator and are superrotating, which is unexplained by present theory. Low wind speeds occur between 30° to 45° latitudes, and a jet is noted at the 40° to 50° and 60° latitudes. The small jolt in wind speed at 30° latitudes and the westward jets at 40°, 55° and 70° latitudes resemble features on Jupiter. Clouds of various types are also like Jupiter. Wispy cloud features occur at 10°, and brightest at about 7° latitudes. Ovals occur at 30° latitudes, and a brown spot at 42° North Latitude. The large Red Spot, called Anne's Spot, is visible at about 55° South Latitude (a 15° spread from the 40° South Latitude). A ribbon like feature is evident at 46° North Latitude near large temperature changes (gradients).[381,677]

Like Jupiter, the banded structure is parallel to the equator.[159,381,621] Mid-latitude emissions are triggered by sunlight (*photoelectric*) or particle flow (*electroglow*).[280] Large scale holes exist in a (*F*) region along mid-latitudes. An unexplained acceleration of hot hydrogen at mid-latitudes and high altitudes also exists.[458,636] Plasma displays two-components, hot and cold, indicating a time-varying acceleration mechanism.[673] All of these phenomena could be predicted for a Field-dynamical Model of Saturn.

Along the equator Saturn has rings and what is called the Saturn Electrostatic Discharge or SED. A ring current, bursts of hydrogen (*lyman-alpha*), and a hydrogen glow near the rings, especially the B-ring, are observed.[179,188,401] Also, heavy ions, electrons, and the rings are strong sources of radio emission, indicative of particle flow along magnetic field lines.[262,673] Likewise, SED has been associated with low-energy charged particle fluxes.[188] SED revolves or rotates like a search light beacon, and is confined to a narrow band along the equator (*within 5 minutes*). The spokes of the B-ring, polar aurora, brightening and SED are all time-varying. All, except the SED, are associated with a possible magnetic anomaly. More about the rings and SED will be discussed in the section 1.10, on planetary rings.

Other "mysteries" only serve to confirm the new model. Saturn's brightness fluctuates with solar activity, though irregularly (i.e., time-varying).[621] The power for the aurora, heating of the atmosphere, and electron concentration in the ionosphere are without explanation according to present hypotheses.[82] Likewise, the excess of hydrogen is not explained.[336] Indicative of an internal heat source is the fact that Saturn radiates 2.4 times more heat than it absorbs from the Sun. Yet, all of these observations can be predicted from an understanding of the new model.

1.6 Uranus

The magnetic field of Uranus was one of the many surprises of the Voyager mission. Originally it was believed that Uranus had no magnetic field or magnetosphere.[665] Yet, it is very complex, revealing more of the Field-dynamical Model than most other Voyager observations of the other planets. There is a magnetosphere inside a magnetosphere with lines that are known to dive back below the surface before reaching the equator.[415] The magnetosphere displays convection like that of the Earth, while the whole system corotates with the planet.[387,443] The rotational axis and the magnetic field are "offset" by 55º to 60º.[24,529,552] Another way of looking at this phenomena is to interpret this offset as actually the observation of a 30º latitude field, which is being assumed to be a displaced dipole field, because the phenomena observed are typical of a polar region. A dipole is speculated in order to preserve the opinion that magnetic fields are always generated by dynamo processes.

Meanwhile, the evidence is there for the Field-dynamical Model. Radio emission at the weak "North Pole" is time-varying. It is an impulsive solar-wind driven source with at least six separate components.[224,404] Auroral zones are much closer to what has been called the "equator" than to the rotational poles.[552] The emissions are like those at Earth, Jupiter and Saturn, and are unexplained (*including a dichotomy between narrow band bursty and smooth emissions*).[224,473] Auroras are "self-excited," or better stated, it is driven internally, not solar-wind driven, and disturbs the upper atmosphere.[181,280] In addition to the aurora, there is also airglow and electroglow in the form of a double helix in the tail of the field.[242,387] The electroglow has been so-named because it is believed to be due to electrons exciting hydrogen in the upper atmosphere. However, the source of the electrons' energy is not known according to present models.[24,380] The hydrogen column (*Lyman-alpha*) is unexpectedly strong and time-varying (*by a factor of two in 24 hours*), which also indicates a magnetic field.[235] Aurora are observed at 44.2° North Latitude and spread about 15° in diameter typical of the Fields. The level of electrons that feed into such a diameter are insufficient to account for the observed intensity, requiring a dynamic component.[529]

The energy sources of the atmospheric phenomena also disclose the Field-dynamical Model. The upper atmosphere reaches a temperature (750° K) that causes it to balloon 6,000 kilometers (3,725 miles) above the clouds. About 30% of the heat radiated comes directly from Uranus. Solar input alone cannot account for the high temperature, and therefore, another source is required.[380] The source is the hydrogen fusion core of the Field-dynamical Model.

Latitude restrictions of atmospheric phenomena also disclose the model. Solar input alone should produce warmer temperatures in the South Polar region than at the equator, but temperatures are nearly uniform, even for the dark North Pole, and the illuminated South Pole. The equator has the warmest and the 30° latitudes have the lowest temperatures at both 15 and 60 kilometer (9 and 37 mile) altitudes.[529] Typical of the Fields, ionization has produced a partial vacuum with the upper and colder atmosphere flowing in and producing these observations.

Two convection cells in the atmosphere are displaced on either side of the 30° latitudes. There is strong mixing of the clouds with the atmosphere, and they are formed in an area of strong vertical wind shear. Sources that drive the winds are not understood in light of present models, because they should typically be horizontal. Like Earth, equatorial winds flow against (retrograde) and mid-latitude winds flow with (prograde) the rotation of Uranus. Meanwhile, solar heating should cause expansion toward the equator, and the Coriolis Force should result in retrograde winds, yet observations show the opposite.[24,353,529,694] This is unexpected, but a scientist insightfully relates that it could be due to "internal dynamics."[24] Likewise, in the downstream magnetosphere there are repeated enhancements of proton and electron flows.[529] These time-varying internal dynamics are expected of the Field-dynamical Model.

Other observations reveal the anticipated. Ammonia concentrations are greater from 15° to 45° latitudes. That is, a 15° spread from the 30° latitudes. Concentrations are strongest in the 30° to 40° South Latitudes and depleted in the South Pole.[353] In contrast, and expected of particle flow, there is a "gap" in methane concentrations at the 30° South Latitude.[337,529] Thermochemical equilibrium cannot explain the gradients and concentrations of ammonia, sulfur and other chemical species. Likewise, latitude differences require pole to equator variations in temperature-pressure profiles and/or latitude variations of an internal heat source.[221] All of these observations are a very adequate description of the new model.

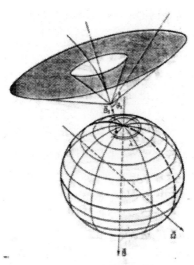

Figure 4.2. Uranian Kilometric Radiation (UKR). This schematic sketch of the cone shape of the UKR reveals the cone shape of the Field. [from Zarka, P., Lecacheux, A. (1987) Beaming of Uranian Nightside Kilometric Radio Emission and Inferred Source Location. Journal of Geophysical Research 92A:15177-15187, copyright American Geophysical Union, 1987].

Uranus has a difference between polar and equatorial radii that produces a polar flattening, which is larger than predicted for the rotational period noted.[529] This calls for new interior models and is explainable by Field dynamics controlling overall geometry. Polar flattening is found on most, if not all, of the planets.

Lightning discharges in the atmosphere generate an emission known as the Uranian Electrostatic Discharge.[529] Uranian radio source footpoints are at the 40° to 62° South Latitudes.[472,779] Much of this Uranian Kilometric Radiation is centered around 45° South Latitude. There are symmetrical emissions in the cone shape of the Field evident at the 45° latitude (about 235° longitude). On the dark side of the Southern Hemisphere, it also has a quiet region in the center.[266,403,779] This is an accurate description of Field structure, as pictured for the Earth in Tome One (see Figure 4.2). Like the Sun, Jupiter and Saturn there is a differential rotation of the clouds that varies with latitude.[242]

The rings of Uranus came as a surprise and are constrained to the equator. As with the rings of other planets, these rings appear to require an ongoing process to maintain, and appear to be much younger than the planet. Furthermore, the rings cannot be explained by the effects of the Uranian moons (*sheparding*).[24,529] Forces should spread the rings out into wider and wider sheets. However, particles were observed to spread out, but where prevented from going further, as if there were "an impenetrable wall."[529] More will be discussed on rings in a later section.

1.7 Neptune

When it was discovered that Neptune had an odd magnetic field too, a comment that seemed essential long ago was finally made. The observation "should force theorists to rethink their ideas about how magnetic fields are generated within the interiors of planets."[418] Planetary geologists tried to explain the offset magnetic field of Uranus by claiming that it was undergoing a polar reversal. But, when it was discovered that Neptune appeared similar it certainly left that explanation open to question. The odds that both planets would be reversing at the same time that Voyager passed by is ridiculously small. The Neptunian magnetic field is "tilted" nearly 50° away from the rotational pole, or it could be said to be at about the 40° latitude if it were not insisted that it is the pole.[553,696,746]

Observations revealed that there are relatively large non-dipole contributions to the magnetic field. These contributions are similar to the Earth's surface and life contributing to the geomagnetic field. Like Uranus, the offset field of Neptune is not typical of dynamo theory, and is, therefore, "unexplained."[553] These facts alone suggest that a Field-dynamical Model is responsible for the observations.

Other phenomena noted also confirm this conclusion. Very intense short radio bursts (*strongly polarized*) from Neptune "presumably" originated in the South Polar region. The bursts were interpreted as a rotation period of about 16 hours.[696,746] This may be the case, but the bursts could simply be due to the time-varying aspects of the Field in this region.

Encouraging the idea that it has an internal energy source, Neptune emits 2.7 times the energy it receives from the Sun.[696] Neptune's spin rate was believed to be about 17 hours before the Voyager observations. This spin rate was needed to explain why Neptune radiates much more heat than Uranus. That is, the spin rate is theorized to be related to the mixing of the planet's interior, which affects the rate at which heat reaches the surface where it is radiated away. However, with the 16-hour spin rate, this mixing theory is in trouble. These facts encourage the idea that there is an internal heat source.

Though it is furthest from the Sun of the other gaseous planets it has a well-developed ionosphere and atmosphere whose composition is distinctive and complex.[718] Weak auroral emissions on the night side are dominated by ultraviolet, and so, appear invisible to the naked eye. The plasma has two components consisting of light and heavy hydrogen ions (H, H_2 and H_4). The location of the auroral cusp region, and the reconnection of the magnetic and interplanetary magnetic fields change considerably over the 16-hour rotation. Once every 16 hours the South Pole points directly into the solar wind, and the North Pole points directly down the magnetotail. A half rotation later there is the escape of protons and the upward acceleration by electric fields.[101,141,444] Hydrogen fusion is suggested by the fact that protons are in significantly lower fluxes than electrons throughout the magnetosphere.[695] These observations indicate hydrogen fusion and the release of fusion by-products in a polar wind.

Several events in the magnetosphere appear similar to the lightning-generated high-frequency waves, known as whistlers. However, there is so large of a dispersion in the signals that it seems that it could not be due to lightning. Observations require path lengths and plasma densities "that are much larger than anything plausible in the Neptunian magnetosphere." As a result their origin is "unknown."[334] Meanwhile, Field dynamics are responsible for the phenomena, but are not considered by theorists.

Phenomena restricted to certain latitudes are conspicuous on Neptune.[193,338,679,696] For one, generally speaking, the poles and equator are warmer than the mid-latitudes even though the solar (*obliquities*) and internal heat fluxes are different. Like the other planets, this is due to particle flow generating a partial vacuum in mid-latitudes. The Great Dark Spot (S22), similar to Jupiter's Red Spot, is located at 22° South Latitude, and a smaller dark spot is noted at 55° South Latitude. These features are the product of particle flow along Field lines with a spread of about 15° from a source somewhere between the 35° and 40° South Latitudes. This is why a strong thermal wave-like structure occurs longitudinally, some of which is associated with the Great Dark Spot. Likewise, the Scooter, a bright feature at 42° South Latitude, is the result of similar phenomena. Cloud groups form bands of 6° to 25° North Latitude as the result of mid-latitude Field spread and equatorial particle flow. A broad band of low temperatures occurs at about 45° South Latitude, and there is probably another band in the Northern Hemisphere. The bands that form at 45° to 71° North Latitude are due to mid-latitude and polar Field spreads. All of these and other features are time-varying, and change in size, shape and brightness.[679]

Bright short-lived "streamers" appear at latitudes of the Great Dark Spot and at about 27° North Latitude. Methane band images revealed bright bands within 15° of the South Pole, and a small feature at the south rotational pole.[337,679] The appearance and disappearance of the Great Dark Spot at 22° South Latitude, and another at 25° North

Latitude attest to the time-varying nature of the atmospheric dynamics.[338] Again, the spread of the Field is 15° so the sources are at about 40° latitudes and the poles.

Likewise, the brightest cloud features associated with the Great Dark Spot in 1986-87 imaging were located at about 40°, and in 1988 about 30°, South Latitudes.[338] The bright features are due to ionization of the atmosphere. These features tend to appear in the direction away from the Sun, and during the Voyager flyby were observed at 33° South Latitude.[679]

Neptune, like Uranus, has a subrotating atmosphere at low latitudes, and a superrotating atmosphere at high latitudes.[193] The high latitudes are superrotating because of the spread of the Fields with altitude (more surface area is ionized). Assuming that Neptune rotates ever 16 hours, then the winds blow the "wrong way" relative to the interior.[418] Furthermore, the Solar System's fastest winds appear to whip through Neptune's atmosphere at an astounding 600 meters per second (about 1,340 miles per hour)![338] The high wind speeds, the persistence of the large oval "storm" systems, and the hour to hour variability of small-scale features were unexpected in an atmosphere that receives one 20th as much energy from internal heat and absorbed sunlight as Jupiter, and only one 350th of that of the Earth.[679] There is no question that these winds are powered by an internal and powerful energy source.

Typical of the inadequacy of traditional models a group of scientists remark: "The mid-latitude cool bands, reminiscent of Uranus, are not simply explained in terms of radiative forcing. The similarity in the meridional structure of temperature and thermal winds on Neptune and Uranus was unexpected."[193] This is in spite of the differences in solar heating, which is greater in the equatorial region of Neptune and the polar region of Uranus.[696]

Another observation was made using the University of Hawaii's 2.2-meter telescope. It was discovered that Neptune's northern hemisphere is now brighter that its southern hemisphere. For more than a decade the southern hemisphere has been brighter, although the hemispheres were roughly equal from the late 1970s on. Present models of Neptune offer no explanation for the phenomenon, but with a Field-dynamical Model this is no mystery.

Uranus has two narrow and two broad rings. Particles are concentrated in a dense disk about 1,000 kilometers (620 miles) thick centered on the equatorial plane. In addition, low-frequency radio emissions are displaced outward in a disklike beam along the magnetic equator. Again, the short evolutionary timescales for this ring system, like others, is an unsolved puzzle for conventional models, and requires confining mechanisms and the replenishment of ring material. There are also no moons that might keep the rings in place gravitationally.[334,679]

1.8 Pluto

Most of what is known about Pluto is indirect, since it reflects almost no light, and was mostly studied from its transits with other objects (*occultations*). Even recent pictures from the Hubble Space Telescope are very fuzzy images. The Voyagers did not capture an image of Pluto, because of its eccentric orbit which is inclined 16º to the ecliptic plane. In fact, normally the nineth planet, from 1979 to 1999 Pluto was inside Neptune's orbit, making it the eighth planet for 20 years.[204]

In spite of our lack of knowledge about Pluto, there is still some evidence that it too has Field-dynamical Model structure. Pluto has bright polar caps of methane ice that extend to the 45º latitudes. The equatorial region is a dark red band.[110,704] Abundant methane on Pluto should have rapidly darkened (*due to solar insolation and charged particle environment*), but it has a high reflectivity (*albedo*). This means that it must be replenished with fresh frosts, which indicates that there must be an atmosphere.[147,691] This observation was something unexpected for such a small , distant planet (it is about two-thirds the size of Earth's moon). Annual (*volatile*) transport cycles of the ice undoubtedly generate electrical phenomena, and are likely to be super-conducting in such far and frigid regions of outer space.

Other observations encourage the idea that a Field-dynamical Model controls phenomena on Pluto. The 6.4-day brightness variation suggests a time-varying component. Charon, Pluto's moon or companion planet, cannot account for the variation, though it rotates with the same period.[204] This variability has been claimed to be due to our view of Pluto, and the light and dark spots on its surface, or the recession of the polar caps.[147] However, the similar period between Charon's orbit could indicate a particle flow triggered by Charon that through delays produces the brightness variations out of phase with Charon's orbit. Pluto and Charon may be a double planet sharing the atmosphere, ionosphere and magnetosphere.

Pluto also has a weak Venus-like or comet-like interaction with the solar wind. There is a magnetic field, an atmosphere and indications of a magnetosphere (also, an ionopause and ionosphere).[87] These observations were a surprise according to conventional theories, but would be expected of the Field-dynamical Model.

Pluto has a higher density than previously thought, presenting a problem for Solar System formation models. It is rock-ice, not just ice as was originally proposed.[110] If the mass is wrongly estimated then it cannot explain the perturbation of Neptune and Uranus, which were used to discover Pluto in 1930.

1.9 Other Similarities

It is obvious from the above discussion that the planets resemble each other at a number of levels. However, there are additional likenesses between the planets that calls for new models of planetary dynamics. The observations depart from traditional models, as conventional interpretations are at a impasse for explanations.

One such parallel between the planets involves the dynamics of their atmospheres. Solar energy inputs to each planet's atmosphere can vary by a factor of 1,000. Yet, the cloud-top wind speeds are roughly the same for all the planets from Venus to Neptune. A team of scientists discuss this unexpected phenomenon:

> "That the winds are of the order 100 [meters/second] on all the planets is a challenge that any theory of atmospheric dynamics must face. Wind is perhaps the most fundamental variable in dynamic meteorology, yet there is no unified theory that accounts for the wind speeds observed in all planetary atmospheres. Over the years Voyager wind observations revealed many surprises, including the high-velocity, localized winds in the midst of Jupiter's long-lived features, the high wind speed at Saturn's equator relative to the interior, and the dominance of east-west winds at Uranus despite that planet's peculiar orientation. The resemblance to the other giant planets is probably just the first of Neptune's surprises. The only sure lesson in science of comparative planetology is humility."[338]

Wind speeds are similar for all the planets, despite the different atmospheric compositions and distances from the Sun, simply because their dynamics are controlled by the same mechanism: the particle flow of the Field-dynamical Model.

Other similarities were just as surprising. Both Jupiter and Neptune have great spots that cover approximately 15° in latitude by 30° in longitude, occupy stable positions in latitude bands centered near the same latitude, and are associated with bright, active regions.[338] Present theoretical models cannot account for this resemblance and stability, because they are claimed to be storm centers. Storm centers could not be so similar, and so fixed in position for so long.

For example, can you imagine a hurricane on Earth staying fixed in position for even a few days? Jupiter's Red Spot has been in the same position since we trained our telescopes on it almost three centuries ago! It is obvious that an underlying mechanism, dynamically related to the interior is constantly supplying energy to the spot, and the cold depths of outer space make the gases reveal its position, unlike the Earth or other inner planets closer to the Sun.

Another enigma is the equatorial rotation of the planets and the Sun. Venus, Jupiter, Saturn and the Sun are equatorial superrotators, which is unexplained and unexpected. A group of atmospheric scientists comment on what is required for this phenomenon. Their statement practically asks for the Field-dynamical Model: "Maintaining equatorial superrotation requires some pumping of angular momentum into the equator by organized waves or eddies."[679] They also have a large number of alternating zonal currents that are also unexplained.[193] These phenomena are considered enigmas, because present theories are inadequate.

1.10 Planetary Rings

Until little more than a decade ago it was believed that only Saturn had rings. Today every planet that we have been able to study in detail has a ring or rings of some sort. Even the Earth has the equatorial ring-current and equatorial electrojet. Typical of our present focus on gravitational theories, attempts have been made to explain rings gravitationally, but without any real success.

The most widely held theory is known as sheparding. This theory claims that moons and moonlets pull the ring particles along their orbits by gravitational forces. In conjunction with this effect, it is said that the self-gravity of the rings maintain other aspects of ring dynamics. While both of these effects could explain some ring structure and dynamics, they fail to resolve most aspects of ring phenomena.[127,153,223,249,333,525]

Because most moons and moonlets have unique and even eccentric orbits, they create what is referred to as resonances. Proposed resonances, a "vibration" caused by the satellites' motion in orbit, are not strong enough to do the job. Accordingly, rings should only be tens of meters wide, but instead are kilometers or thousands of kilometers wide.[249] One of the major flaws of the theory is that the two types of proposed mechanisms of ring confinement, sheparding and self-gravity of the rings, involve unseen satellites that in all probability do not exist outside imagination and predetermined theory.[127,223,249]

Dust grains that pass through the plasma surrounding the planets will exert a drag on the grain (i.e., plasma drag). This would cause a loss in orbital speed and the grains should fall towards the planets. Obviously this does not happen, or the rings would not be there.[333] An original moon(s) that had broken up by some given situation (tidal break-up or collision) does not fit the observations either.[249] Two scientists comment on an observation that reveals the failure of gravitational theory as the sole explanation: "Even when the two shepards passed each other (called a conjunction), the ring between them refused to unravel. Between the Voyager encounters, gravity had regained the primary role as the shaper of rings—electromagnetic forces had lost their appeal—but conjunctions seemed the most likely time to see gravitational effects."[249]; p.150

Theorists cannot explain how one resonance produces a spiral (*density*) wave, a second a sharp edge, and a third subtle narrow bands or gaps. Satellites produce turning forces (*torques*) that tend to confine a ring, but also transfer so much energy that confinement is impossible. In this regard a group of physicists state the inevitable: "Our assessment of this paradox suggests that the accepted description of the sheparding mechanism is seriously flawed."[127]

A number of features disclose what could be expected of the Field-dynamical Model of the planets, such as observations of magnetospheric plasmas and radiation. Saturn displayed a magnetospheric plasma that was greatly decreased in the vicinity of the rings.[250,384] The first evidence of a Jupiter ring, uncovered by Pioneer 11, showed data (*at 1.8 R_J*) that indicated a very curious, unexplained structure in the intensity of radiation. No radiation was noted in the ring plane, but it did occur on either side of it.[249] The nearly undetectable E-ring of Saturn revealed an equatorial corotating hot plasma. Plasma was enhanced on each side of the equator, while there is significantly less at the equator.[384] This is why plasma drag has not destroyed the rings, because plasma is guided away from the equator, the region of the rings, by the Fields, since the Fields deflect from each other, and back towards the planet, along the equator.

The flow of ionizing radiation would create electrical charges on particles (creating different ions and chemical species), leading to the electrostatic levitation of ring particles. Electric potential developed by a surface in space is theoretically of general astrophysical interest, because it is a widespread phenomenon.[154,464,755] Particles with a charge would be lofted by levitation and maintained by electromagnetic fields. Electrostatic bursting would destroy ring particles resulting in short lifetimes for the rings. Therefore, a process is required that continually replenishes the particles. The entire set of facts indicates that the phenomenon has to be time-varying and not in the ring plane itself, and the Fields must deflect from each other along the ring plane, as expected for the Field-dynamical Model.[154,249,333]

The interaction of charged dust particles and the geometry of the Fields would lead to the rings, and their replenishment and individual characteristics. Whenever there is high-velocity flow it would lead to grain destruction and fragmentation, creating gaps, grooves, sharp edges and ethereal rings. Areas of low-velocity flow (or no flow) would lead to grain growth and generate rings.[333,385] The various charges and atomic species would develop into discrete bands, due to the electrostatic and electromagnetic forces of the Field-dynamical Model. The geometry of the Field-dynamical Model, as well as the particular characteristics of a given planet, lead to the various ring systems (see Figure 4.3).

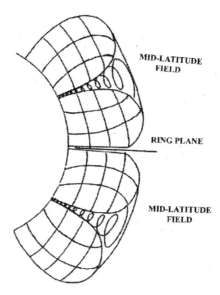

Figure 4.3. Ring Formation. This is a crude schematic of how rings form as a natural outcome of Field interaction involving the Field-dynamical Model. The Fields have been basically cut in half to reveal how they interact along the equator (or another region) to form a ring.

The magnetic field appears to control ring positioning, because they are always oriented in relation to the main magnetic field. This orientation discloses that an interrelationship exists between the Field system, including the main magnetic field, and the rings. A strong interaction between ring particles, radiation near the planet, and the magnetic field have been observed repeatedly.

Evidence of this Field-dynamical Model's control on rings is plentiful. The tilt of Saturn's rings is as much as 30° north or south typical of Field latitudes.[228,229] As could be predicted, the rings are a strong source of radio waves due to particle flow along field lines.[677] The ring particles are partly ice, which when chipped produces an electric field that has long-range force, allowing for the development of extensive ring lengths.[187] Near Saturn's B-ring, one of the densest rings, there exists an enhanced concentration of hydrogen (*the ring cloud is 5.0 R_S along the orbit and 1.5 R_S vertically*).[635] Bursts of hydrogen nuclei (*Lyman-alpha*) have been discovered near the rings, as could be suspected from time-varying fields and hydrogen fusion.[179] Diffraction of radio waves and many unidentified wave-like phenomena in Saturn's rings indicate the same.[496] Furthermore, during the crossing of the ring plane, just at the edge of the G-ring, the plasma-wave experiment of Voyager 2 detected intense impulsive noise.[333] Likewise, theoretical descriptions to account for ring gaps include high fluxes of electrons and protons that destroy ring particles.[95] The ion species, radio waves, and noise are all time-varying.

According to widely held theories Jupiter should not have rings, but it does. Jupiter's ring system is made up of three major components, and is composed of water, ammonia and methane.[121,394] Typical of time-varying fields the rings are hot at times and cold at others, while their radii and thickness vary with longitude and time.[121,597] Jupiter has a dusty magnetosphere above and below the equatorial plane, but not in the ring plane itself. The peculiar behavior of some particle counts inside one of the moons' (Amalthea) orbits appears to be due to field geometric effects.[223,249,525] Likewise, particle movement is closely related to the electrodynamic coupling of the moon, Io, with the magnetosphere acting as a source and sink of plasma. Plasma and neutral production mechanisms in the A-ring, B-ring, and C-ring are partially due to field-aligned ionospheric flow. The thin, faint E-ring involves the confinement of corotating hot plasma. Like Saturn, plasma is enhanced on each side of the equator and diminished within the equatorial plane, along which are the rings.[384]

Seven of the nine Uranian rings are inclined to the equator, and gravity does not explain the observations. The rings precess uniformly under the influence of the magnetic field. Uranus also has radio emissions along the equator, indicating particle acceleration along field lines. They are narrow and dark, not wide and bright as rings "should be." Gravity should spread the rings out wider and wider, but there is no evidence that this has or is taking place.[24,249,404,679,696]

The surprises were many, but are expected of the new model. "Profiles of these strange Uranian rings confirmed that the particles tended to spread outward toward the edges of each ring, but were then prevented from going farther, almost as if they were piling up against an impenetrable wall."[529]; p.258. "Stable, non-transient ring arcs would obviously require a longitudinal confinement mechanism; transient arcs require the continual creation, dispersal, and replenishment of local concentrations of ring material."[679]; p.1431

Neptune's rings stretch along about 45° around the planet, and often appear as ring arcs. Time-varying and solar-linkage characteristics typical of the model can be noted in that these ring arcs display similar brightness variations as the F-ring of Saturn. Theorists indicate that ring arcs could eventually spread to form a complete ring within a few decades.

Again, no moon or moonlet has been found to account for a gravitational origin of these rings. The Field-dynamical Model is evident in the ring arcs, called the "string of sausage," which are in one 35° sector. This comment reveals the need for the new model: "What we've been doing has been racking our brains for a mechanism. At the moment we're stuck. Maybe we're just missing something."[420]

Many enigmas remain for the present theoretical framework. Present theory renders untenable any explanation involving particles in synchronous, planet-aligned rotation.[209,577] Meanwhile, this is what we find exists for all the planets whose rings have been studied.

One of the most puzzling features of the rings is their short lifetime. For the most part, they could only be 1,000 years old or less. A ring of dust would have been swept away by solar radiation, plasma drag and micro-meteoroids.[104,127,249] Electromagnetic forces, their origin "unknown," deposit the ring material, which is unexpectedly very young.[104,127,209] Scientists studying planetary rings comment: "Thus these short timescales remain perhaps the most intriguing puzzle in planetary ring dynamics."[127]; p.732.

The Field-dynamical Model produces, maintains and controls the dynamics of the rings. As a result, scientists come to the inevitable conclusion. "Evidence exists for electromagnetic processes in the rings superimposed on the overall gravitational dynamics."[209]

Observations of Saturn clearly indicate the Field-dynamical Model. Aside from the discussion above there are many ring-related electromagnetic field phenomena. The magnetosphere contains a large corotating inner plasma (*torus*) with oxygen ions, electrons and some protons. An inwardly diffusing feature of the magnetosphere appears to be effectively "absorbed" by the rings in a region with the lowest charged particle density (*at 2.3 R_s*). Absorbed may not be the right word, because the particles are being accelerated along field lines, and are deflected away from or are engaged in reactions that produce the rings. Within the central region (*1.3 to 2.3 R_s*) protons and electrons form from neutron decay.[209] These observations are assumed to be cosmic rays splashing off the ring and satellite material, but are really the Field's contours and interaction, and hydrogen fusion is the source of the so-called "cosmic rays" and the neutrons.

The composition of the ring particles also indicates the new model, which is particularly obvious in the case of Saturn. Very tiny or micron-sized beads of ice would have a very short life span, because they would erode due to blows from micro-meteorites and atomic-scale disintegration. Like evidence on Mars, the Moon and elsewhere, there has to be a source of water, as this quote indicates: "Apparently a million tons of water had somehow been dispersed as micron droplets and quick-frozen in space."[249]; p.156. Likewise, Jupiter has water, ammonia and methane as ring components.[121,394] Hydrogen fusion produces very active species of hydrogen, carbon, nitrogen and oxygen, which could combine to produce water (*or calthrate hydrates*), ammonia, and hydrocarbons, such as methane.

The C-ring and Cassini Division are darker, and less red than the A-ring and B-ring. No conventional explanation can link ring particles separated by the more than 25,000 kilometers (15,500 miles) of the B-ring. Furthermore, the A-ring's edge, B-ring/C-ring boundary, and the inner C-ring have no apparent gravitational or resonance associations.[249] The only possible explanation for such observations is that the different atomic species carry distinct charges that are accelerated into discrete bands by electromagnetic forces and Field contours (see section 10.3 for discussion).

As has been too often the case in science, preconceived ideas block progress. Moreover, proposed theories fail to explain all ring phenomena. A scientist, who argues in favor of electromagnetic and electrostatic forces, discusses this problem in relation to ring phenomena:

> "How is it possible for so many conscientious observer-analysts to encounter so many blocks to progress? Part of the answer to this question seems to be that preconceived ideas have been converted into fixed ideas. Then, when new data are received which do not conform to the fixed ideas, an impediment to progress is experienced.
>
> Notwithstanding the tendency to dispose of untoward data, another part of the answer to the question is that something in or about the data is being overlooked. Oversight unobtrusively is convenient when fixed ideas are being promulgated. However, oversight also can occur because of presumptive expectations that confirmative new findings will be obtained. Important facts have an uncanny tendency to remain obscure."[104]; pp.4-5

Along the lines of obscure facts are the peculiarities of ring geometry. The greatest problem facing theorists is how narrow rings and sharp edges are formed. Jupiter's bright ring has a sharp outer edge. Uranus has surprisingly sharp rings that are not explainable by sheparding, and the proposed moon that would do so is not observed. In fact, the greatest theoretical problems for Uranian rings concerns explaining the dynamics of narrow rings and their sharp edges. Saturn has unexpected narrow ringlets and sharp edges whose dynamics are far more subtle than what would be due to collisions and a few major resonances. The inner-B/inner-A edge, and the structure within four of the five bands in the inner Cassini Division all show sharp, peaked inner edges and gradual somewhat outer edges.[209,249,250,333,384] One suggestion is that a destructive process is responsible for the sharp edges.[209] Ionizing radiation flowing along the geometry of Field lines is the destructive process that produces sharp edges.

Ring thickness is a mere five or ten meters (15 to 30 feet) for rings that are 270,000 kilometers (167,400 miles) across! Scientists describe how this relationship within a portion of the rings compares with Earthly phenomena: "If the depth of the Atlantic Ocean were in the same proportion to its 4,000-kilometer breadth as this thickness is to the breadth of the rings, the abyssal depths of the Atlantic would not quite reach 1 meter; you could never get in over your head."[249; p.159]

Gaps and eccentric rings are also without explanation in present-day convention. Saturn's rings display clear gaps, with or without embedded ringlets, in the A-ring, C-ring and Cassini Division, but not the B-ring. Several, but *not* all, gaps in the C-ring and Cassini Division are associated with strong gravitational resonances. Clear gaps are almost always associated with embedded or adjacent narrow ringlets. B-ring structure is irregular with no association with either strong gravitational resonances or local perturbing moonlets.[209]

The moonlet-gap connection at that still leaves most of the 10,000 ring features unresolved. For example, Jupiter has a single ring disc, but at least sixteen moons.[249] Meanwhile, Jupiter's magnetic field has a high curvature along the equator, and charged particles have been observed to spiral along field lines into the equatorial region.[113]

Likewise, those particles that precipitate along field lines in one hemisphere are inhibited from reaching the other hemisphere.[675] All nine rings of Uranus have significant width perturbations (0.5 to 2.0 km. or 0.3 to 1.25 mi.) that are all larger than orbital radii differences (*except Gamma and Delta*).[282] Meanwhile, ring gaps, as well as eccentricities, can be caused by high fluxes of electrons and protons.[95] The Field-dynamical Model accelerates electrons and protons along Field lines, creating these phenomena.

Other ring phenomena, such as braids, ripples, kinks and grooves, can only be the result of the Fields and particle flow. Gravity cannot be used to justify the existence of the braided F-ring, but fields and electrostatic forces can. Any associations with moons or moonlets simply facilitates in the dynamics of particle flow. One observation of the F-ring showed no braids, but four strands, indicating a space- and time-varying phenomena. The multiple F-ring is also kinked and clumped.[249] The composition and charge of ring particles causes them to react to the Field-system's contours in distinctive ways.

Saturn's rings have waves or ripples that move inward, unlike a pebble tossed into a pond, as would occur with gravitation. A moon, such as Mimas, would cause outward-propagating spiraling (*density*) waves, which would bring particles about 500 meters (165 feet) above and below the ring plane.[249,669] Saturn's rings are quite substantially thinner than this (by a factor of 100 in some cases). Furthermore, these ripples are more like the grooves of a phonograph record. To think that gravity could produce this is like claiming a huge steel ball at a distance could put grooves in a vinyl disc generating a musical record. As most of us know, it requires controlled electromagnetic equipment to produce such recordings. Likewise, controlled electromagnetic forces are required to produce ring grooves, braids and kinks.[104,333]

Luminous phenomena in the vicinity of the rings have been observed repeatedly, and prove that electrostatic forces are at work. From the 18th Century onwards luminous points on the ring edge, luminous objects and the disappearance of one side of Saturn's rings have been observed. One astronomer (Maurice Ainslie) observed a luminous object pass through the rings and devour ring material as it moved. This object was a mass of ionizing radiation, which can destroy ring material. The rings are known to change in brightness, evidently due to time-varying particle flow.[368] White spots where noted on the rings and the planet numerous times.[47,220,696] Bright projections on the rings and "sparkling flocculence" where also observed.[42,44] For example, a white projection was seen near a white spot at the edge of the south equatorial band on 29 September 1910.[41] Voyager 1 took a photo of Saturn and five of its moons, which included an unexpected luminous object. Images of the F-ring show variable luminosity that includes observations of a helical core and emission jets.[104]

One event, in late September 1990, a large white spot appeared on Saturn. Shortly following, this spot developed into a 21,000-kilometer (1,305-mile) long oval. By November, it had developed into a planet-encircling band. Other white spots were observed in 1876, 1903, 1933 and 1960. Saturn's orbital period around the Sun is 29.4 years, which is close to the times of these observations, and may indicate a Saturn-solar linkage is involved.

One of the best examples of ring phenomena that demonstrate the new model of planets is the spokes of Saturn's rings (see Plate 4.4). Spokes were observed at magnetic lines of force that have their feet at latitudes between about 39° and 45°. They are darker than the surrounding regions, have a characteristic wedge shape, and continually reappear at the same locations (time-varying).[209] The spokes occur along a region of possible ionospheric jet streams and high-speed winds at latitudes typical of the Fields (approximately 46° North to 32° South).[165,209] Appearing only in the outer region of the B-ring, the spokes corotate with Saturn's magnetic field.[209,211,249, 254, 676] Some scientists suggest that the spokes become visible because particles are polarized by weak electric fields, while others contend that the fields are strong.[209,333] At the time of observation a pronounced peak occurred around the 115° longitude.[333]

Planetary physicists assert that, "The spokes were not darker particles or areas of fewer particles; they are microscopic particles that are perhaps elevated above the main part of the ring by some magnetic-electrostatic process."[249]; p.142. "However, the question, how relatively weak electromagnetic interactions can exercise such a profound effect has remained a major puzzle."[532] Meanwhile, electric and magnetic fields and electrostatic effects along Field lines in combination are responsible for these features.[1,532,689]

A relatively dense plasma is required near the rings in order to generate a strong enough surface electric field to lift dust particles off the rings.[333] A team of scientists discuss this in the framework of current theories: "It is unlikely that the average plasma density near the rings is large enough to do this. As long as the drifting plasma is dense enough, dust will be elevated marking the radial trail of the plasma."[333]; pp.326-327

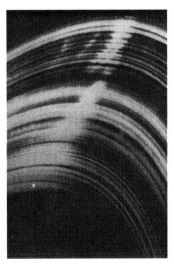

Plate 4.4. Saturn's Ring Spokes. This Voyager 2 picture shows the "spokes" in the rings. The spokes cross over a series of complex rings

and ringlets, and continually reappear in the same locations. Features like this require a dynamic mechanism, or they would not continually reappear in the same locations and cross various unique rings. [NASA/NSSDC]

This trail reveals the Field-dynamical Model, because the greatest spoke activity is spaced at 45° intervals in longitude (*magnetic longitude bins of* 45°). Likewise, the Fields on Earth, as noted in Tome One, are similarly spaced. A funnel shape to the Field would account for the wedge shape of the spokes. Plasma acceleration along field lines is the suggested mechanism (field-aligned currents parallel to the magnetic field).[333,532] Electrons are ejected from parts of Saturn's rings, and possess significant electrical conductivity.[532] In contrast, present theory leaves open the questions of the origin of the dense plasma and the time-varying nature of spoke activity.[333]

Both of these enigmas immediately suggest the new model is at work, and other observations confirm this. The Voyager had detected radio emissions typical of lightning. The phenomenon is referred to as the Saturn Electrostatic Discharge or SED. SED revolves and rotates like a search light with a period close to atmospheric and magnetic field rotational periods.[104,402] Spoke activity, SED and aurora fluctuate together, as could be suspected from the Field-dynamical Model.[188,249]

A more detailed look at SED makes the new model obvious. SED appear like lightning, but a million times more powerful than lightning on Earth. However, this occurs across the rings where there is no atmosphere, and so, SED is invisible to the naked eye, but can be "seen" with radio monitors.[249,333,745] The source is most evident along the B-ring or an extended lightning "storm system" from within Saturn's atmosphere.[333] The source is confined to a very narrow region near the equator where low-energy charged particles are noted.[188,402] Only in the rapidly rotating equatorial regions, near the ring plane, was there any lightning at all. Hydrogen (*Lyman-alpha scattering*) was observed along with SED, which requires a source of hydrogen over the entire ring system in order to preserve the hydrogen atmosphere against losses upon collisions with the ring system. Acceleration of electrons and ions crossing a potential gap is the source of the radio waves. Scientists claim that its physical scale "must be tiny," because it was below the resolution capabilities of the Voyager's instruments.[745] SED is not fixed relative to the Sun as is the Saturn Kilometric Radiation or SKR.[778]

SKR is a long-wave radio emission that Saturn's rotation sweeps across space like a searchlight beacon. SKR is highly correlated with the solar wind in some "unknown relationship." It has a fixed relationship with the Sun taking place in a region near the poles and aurora.[402] SKR is an observation of the solar-Saturn linkage of the Field-dynamical Model.

Maximum spoke activity is associated with the same sector of the magnetic field as the SKR emissions, and spoke activity has the same period as SKR emissions.[249] A statement is made in the face of the inadequacy of present theory: "The mechanism of this linkage between SKR, spokes and aurora remains obscure."[249]; p.149. Both SED and SKR are fixed in relationship to Saturn's magnetic field, and fluctuate along with spoke activity and aurora.

SKR also discloses the Field-dynamical Model. SKR is highly correlated with the solar wind and is fixed in position in relation to the Sun.[188,402] This relationship is due to the fact that the Northern Hemisphere is tilted toward the Sun, and therefore, is more in a position to interact with the Interplanetary Magnetic Field (IMF).

One author gives very convincing evidence that electromagnetic forces are involved in ring formation and dynamics, and discusses the failure of gravitational theories. The dynamics of the rings, such as their flat or planar shape is one example: "Planar shapes can be constructed with electro-potential fields, but only with appropriate field combinations. Hence, such combinations are not likely to occur [and] clearly some new mechanism is called for."[104]; pp.39 & 69. The short timescale for ring systems is also an unsolved problem for conventional theorists. The conclusions are clear: "Each is constrained to an orbit close to the equator of these oblate giants, appears to be much younger than its planet, and requires ongoing processes to maintain its very existence."[529]; p.255

So far, rings appear to be a universal phenomena for the planets in our Solar System. The evidence is particularly obvious for Saturn, Jupiter, Neptune and Uranus. Venus has a ring, particle density, brightness emissions, heat (*Kelvin*) waves, and a structured cloud pattern along its equator.[130,131,280,281,655] Mercury has had dark, nebulous rings with an occasional violent tinge.[392] Mars has many phenomena along the equator, but no ring, per say, has been observed as yet. However, the next mission will undoubtedly discover some form of ring phenomena if nothing but a ring current in the magnetosphere. Pluto is too far away and eccentric in its orbit to distinguish any real details, let alone rings. However, the ice caps do grow only to the equator, and a reddish band exists along the equator—so ring phenomena seem likely. Unified Theory predicts that more rings will be found when sufficient observational data are collected.

Planetary rings and moons' orbits tend to be the narrowest near the planet and thicken away from the planet. For example, Jupiter's ring is a wedge shape with its inner region about 28.5% of the outer region. This is to be expected from the Field geometry. Likewise, Jupiter's moons orbit the planet along the equatorial plane until the outer moons (Leda, Himalia, Elara, Lysithea, Ananke, Carme, Pasiphae and Sinope) are reached who orbit above and below the plane. This is also true of the Saturnian and Neptunian systems.

Another example is our Solar System itself, where all the planets orbit along the ecliptic plane until we reach Uranus, Neptune and Pluto, the three outer planets, who have orbits that are inclined away from the ecliptic. The furthest out, Pluto, has the most eccentric orbit.

The asteroid belt is also a ring. Our Sun has a ring of globular matter and the IMF in the equatorial, and appear ring-like.

Many astrophysical objects have ring-like structures. Comets, such as Halley's Comet, have rings. Likewise, protostars, and protoplanetary and planetary nebulas have rings of matter as stars and planetary systems form. The disks of galaxies have many similarities with planetary rings. Both are spatially thin structures with innumerable discrete objects whose random motions are small compared to their circular speeds, and have considerable internal structure. Galactic clusters and superclusters also display ring-like structures. This type of phenomena is also somewhat true of the electrons orbiting a nucleus, particularly those with a higher atomic number and more electrons (the electrons do not orbit around the poles of the nucleus). All of these phenomena will be discussed later. So far the evidence points toward a Unified Theory of the Universe.

1.11 <u>Earth's Moon</u>

Originally theories about the Moon stated that it had no magnetic field. However, rocks and samples from the surface display signs of ancient (*remnant*) magnetism, indicating it once had a dipole magnetic field. The Explorer 35 also recorded a magnetic field that is activated by the solar wind.[626] The steady field that has been observed is localized and not global. Some scientists suggest that a large magnetized "body" would be required for this observation, including that it had acquired its field when the global field was much stronger.[237] Yet, according to the Moon's composition this is not possible, and so it has remained unexplained by chemists and physicists, and hence, ignored.[339] Something like a large meteorite below the surface would be required in conventional theory. However, the strongest local magnetic field anomalies are on the other side of the Moon away from the large craters.[626] Auroras, which require a magnetic field to produce, have been observed very close to the Moon's surface.[52,125,345,354,432] The evidence appears to indicate that a time-varying magnetic field exists, but it has long periods of inactivity, is solar linked, and operated more during the Moon's formation than a present.

A number of observations display the typical latitude restrictions of the Field-dynamical Model on the Moon, as well. Mares, large flat surface areas which look like seas, were formed in such a manner that they are presently between the 30° South and 40° North Latitudes. Mare Crisium, Mare Humorium, and others on the dark side were formed at low latitudes close to an older equator than the others.[626]

Short-lived phenomena called Transient Lunar Phenomena or TLP are gaseous, appear reddish, bluish or as brightenings, and are not randomly distributed, but appear in dark craters. They also have strong affinities for Mare edges, which has been interpreted as "implying internal activity."[161] For example, a nebulous white spot was observed in the Linne Crater located at the 27° South Latitude.[531] White spots have appeared suddenly and rapidly, moved like a cloud of steam and issued from the crater called Plato, at 51° North Latitude, or so it appeared (it could be the 15° spread of the Field).[309] In December 1963 there was a lunar eclipse that appeared darker than was observed previously, and could be the result of neutrons.[143] Gas erupted from the crater Alphonsus at 13° South Latitude with the light spectrum typical of a comet, and "could be radioactive heating of organic matter."[433] Such a comment is reflective of hydrogen fusion (and possibly, organic superconductors). Observations by Apollo 15 and 16 disclosed that gamma ray intensity—a possible signature of acceleration processes—was greatest between the 30° latitudes.[626] TLP have been observed from the 6th Century to the 18th Century, as well, and occurred during times of increased solar activity, noted by rich aurora on Earth.[32,128]

Mare Imbrium stretching from 50° to 15° North Latitude has much more gamma rays than elsewhere. This is characteristic of the 15° spread of the Field, which would be at 30° to 35° latitude. Neighboring lava flows of Oceanus Procellarum at 25° North to 10° South Latitudes also emit more gamma rays. Again the spread is similar and this structure was probably the result of a field now displaced from this location, but one that operated during the Moon's formation (see chapters 2, 3 and 10). Gamma-ray levels are ten times more than expected, and cannot be explained geochemically.[339] Meanwhile, particle accelerators routinely produce gamma rays.

Gravity anomalies occur in areas of very large mass concentrations, referred to as Mascons, are noted in five ringed Mares. These Mares are Ibrium (50° to 15° North), Serentiatis (40° to 15° North), Crisium (25° to 10° North), Nectaris (10° to 20° South), Humorium (20° to 30° South) and Orientale (10° to 30° South), and between Sinus Aestuum (8° to 16° North) and Sinus Medii (4° to 20° North). The figures in the parenthesis indicate their latitudes.[537]

Heat or thermal anomalies of the Maria suggest "internal" processes to a number of scientists.[375,668] A scientist boldly comments that the "nonrandom distribution might not be surprising if the craters were formed by internal processes."[668] Alpha particles, a by-product of hydrogen fusion, also known as helium nuclei, were observed in the region of the crater called Aristarchus, at 24° North Latitude. "The result is interpreted as probably indicating internal activity at the site."[319] Orange spots have been seen at Aristarchus that yield the spectrum of hydrogen (H_2) and might be explained by lightning.[437,524,527] The evidence suggests that the hydrogen fusion Field-dynamical Model also operates on the Moon (though more so in its formation).

Other observations confirm this interpretation. Like Mars, there is evidence of water on the Moon. A lake-like area near and on the walls of the crater called Alphonsus, at 13° South Latitude, can be observed. River-like erosion complete with mature meanders from the winding courses of ancient rivers are obvious.[485,723] Like Mars a catastrophic origin of the water "seems possible."[723]

The Moon has almost no atmosphere and no clouds, but lightning was observed six or seven times on 17 June 1931.[302] Airglow, typical of particle flow, was noted shortly after sunset for about two hours during one event.[52,294] One theory claims the glow is due to electrostatic fields that levitate dust above the surface.[52] A red spot or reddish glow was seen at the northern most part, where the shadow was expected to be the darkest (*toward the center of the umbra*). Also, a large red "patch" and at times a blue patch were noted.[49] Temperature (and thermal conductivity) show values much higher than expected, indicating an internal energy source.

The volatile-rich soil contains water, carbon dioxide, methane, hydrogen cyanide, hydrocarbons and other compounds that are produced by ionizing radiation.[301] Flourescent material found on the surface may very well be organic.[432] Its fluorescence may be due to ionizing radiation producing thermoluminescence.

Moonquakes occur at monthly intervals near the closest (*perigee*) and farthest (*apogee*) approach to the Earth, and are also correlated with long term (7-month) lunar variations. One moonquake had an epicenter at 21° South Latitude, and a deeper focus than any known Earthquake at the time (about 800 km or 500 miles).[469] The moonquake cycle of 27 days does not coincide with the anomalistic month, which is between closest approaches (*perigees*), and gravity seems to be ruled out. "The gravitation wave interpretation of the lunar seismic response has the same drawback as the Earth based experiments, namely, the effect is orders of magnitude larger than the theory predicts."[633] A Field-dynamical Model would be expected to produce these observations due to electrostatic levitation and particle cascades within the Earth-Moon system (see Chapter 29 of Volume One, and Chapter 15 and 16 of this volume).

1.12 Triton and Titan

Triton, the largest of Neptune's moons, displays phenomena that suggest that the Field-dynamical Model functions there, too, and suggest how it, like other moons, was formed. Aurora, accompanied by atmospheric disturbances, were recorded by the Voyager. Seasonal ice dominates the polar regions south of 15° South Latitude, and frost forms along the equator. The reddish tint indicates the existence of organic compounds. Together the organic compounds and the seasonal ice generate electrical currents and are likely to be superconducting. Frozen lakes are surrounded by successive terraces that indicate epochs of multiple floods (time-varying).[696]

Like a number of other moons and planets, Triton is more heavily cratered on one side. Against all bombardment theories, it is the side facing Neptune which, because of a synchronous orbit, Triton always shows the same side to Neptune.[696] If Triton were to be bombarded it would have been anything but this side. Like so many other craters this indicates that the cratering was internally produced (see Chapters 2 and 3). Furthermore, Triton has less craters than any other moon whose surface has been observed. The atmosphere is mostly molecular nitrogen with a trace of methane near the surface.[141]

A volcano spews frigid plumes of nitrogen at the South Pole, indicating that much of what is seen is the result of internal processes. According to the way moons are theorized to work, Triton should be "stone-cold dead." However, it takes energy to drive erupting plumes, but Triton should have precious little or no heat to draw on.[419] Some scientists hypothesize that they are like small tornadoes—vortices known as Dust Devils.[382]

Titan, one of Uranus' largest moons, also displays phenomena expected of a Field-dynamical Model, and suggests a recently formed moon at that. For one, Titan displays changes (*aerosol photochemistry and other variations*) that are related to the sunspot cycle.[18] There is aurora-like hydrogen (*Lyman-alpha*) emissions on the night side, and a nitrogen "dayglow." It spins out about 6,000 volts and radio signals, but no significant magnetic field was detected. This means that the radio signal is of a type unrecognized, and not understood in terms of accepted theory. An answer can be found in the fact that radio emissions are generated by particle flow along electromagnetic lines of force. There is a Titan-wide ocean of liquid hydrocarbons, and a smog and rain of hydrocarbons (methane and ethane). Despite the chill of deep space, Titan is a remarkably Earth-like body.[194,280] Like Triton, this scenario requires an internal energy source, and the relatively craterless and unfrozen surface suggests a newly formed object beginning to form craters (the plume).

1.13 Moon-Planet Dynamics

The moons of the planets play an important part in Field dynamics on all of the planets. They interact with each planet's magnetosphere and produce particle cascades along Field lines triggering events in planetary Field dynamics. Moons have, in most cases, resulted from particle flow along Field lines, and thereby, continue to interact at this level (see Chapter 10). For example, this is why the Moon appears to have a fission origin.[108]

Jupiter with its 14 moons is like a miniature solar system. The inner moons are more dense like the Solar System's inner planets are more dense than the outer planets. This alone suggests that there is some dynamical relationship with Jupiter and its moons' formation.

Jupiter's moons have been noted to affect particle flow towards Jupiter. For example, the radio bursts of Jupiter are entirely dependent on the phase of Io.[10,746] A large depletion of ions is associated with Io's plasma (*torus*) region, and this is true of Ganymede, too (*on the inbound wake*).[63,172,705] Io travels through the radiation belt of Jupiter producing (*hydrodynamic*) shocks, causing whistlers (electron precipitation) and Jupiter's magnetic field is "shaken."[478]

Triton is also important in controlling the outer regions of Neptune's magnetosphere. High-energy ions and electrons are dynamically driven by the moon.[444] Triton's inclination is close to 160° and its synchronous rotation allows the same side to face Neptune. Triton's orbit is very close and it revolves around Neptune against the direction of the planet. Five of the six newly discovered moons are within 1º of the equatorial plane and the other is inclined close to 5º, which, like many other celestial objects, form something of a ring. Nereid's orbit is inclined about 30º, which also indicates a relationship with the Field dynamics of the planet in their formation.[696] They continue to affect the dynamics of the planet, as they are all imbedded in the magnetosphere.[444,553,695]

All of the moons observed interact with the magnetosphere of their planets, and affect the dynamics of particle flow. The region of the inner moons of Neptune are associated with a complicated structure of the radiation belts.[553,695] Both the rotating magnetic field, and the position of Io time Jupiter's radio emissions.[746] In fact, NASA'a Infrared Telescope Facility detected a pair of infrared spots traveling across Jupiter's upper atmosphere in tune with the motion of Io, indicating an interaction between Io and the top of the atmosphere. Ion (*cyclotron*) waves are noted in Saturn's magnetosphere near the orbit of Dione (*L shell*).[681] Triton is important in controlling the outer regions of the magnetosphere where there are high-energy ions and electrons. A relative peak in low-energy electrons at Triton's radial distance is noted, and a change in proton spectra with hot protons occur inside its orbit.[444] Triton rotates in the opposite direction of Neptune's rotation, and similar to the dark spot, is inclined 21º to the equator, disclosing furthermore that it is important in Field dynamics. Saturnian and Uranian moons also trigger particle flow and dynamics.[173,174,444] Dione and Thethys, nearly equatorial, pass through the inner magnetosphere and are associated with heavy ions.[262] Pluto and its moon, Charon, share the atmosphere and magnetosphere. It seems that all moons in the Solar System interact with the Field-dynamical Model of the planets, as does the Moon with FEM (further discussion of FEM-lunar linkages can be found in Chapter 29 of Volume One, and Tome Five of this volume).

CHAPTER 2

Planetary Ejections and Cratering *(Mechanisms Responsible for Cratering)*

Ever since the earliest investigations of craters, there has been controversies surrounding their origin. Investigations of numerous craters in the first half of the 20th Century brought about the unwarranted tendency to interpret craters of internal origin (*geoblemes*), as those which resulted from impacts (*astroblemes*).[9,480,495] In fact, internally produced, volcanic-like explosions were originally referred to as cryptoexplosions (below-ground explosions), but now all craters are referred to in this way. Another factor that encouraged the interpretation of craters as impact phenomena is the belief that there was an era of bombardment in the formation of the planets. However, with the new hydrogen fusion model of the planets, there is an energy source to produce cratering as the result of an internal process. Evidence supporting the theory that many craters are internally produced is quite substantial, but almost completely ignored.

2.1 Cratering

Similarities in cratering on different planets and moons are very obvious. Studies of the Earth, Moon, Mars, Mercury and the outer moons suggest that crater characteristics are the same regardless of the target or the size of the proposed "impactor." While sizes range from large complex craters to small multi-ring craters, they display no drastic differences in structure. If craters were the result of different impactors of various compositions, sizes, shapes and speeds, hitting at various angles into objects of different composition, there should be a variety of crater structures. However, there is no great variety.

When there is both a peak and an inner ring in the same complex crater, the peak is smaller than normal, and the peak size decreases with increasing crater size. This is in contrast to experiments and computer models of impacts, which show that the peak should be larger with an inner ring and increasing crater size. Peak and ring structures, from small to large complex craters, disclose that there is an absence of diminished inner-ring diameters for the inner-planet (terrestrial) craters. This is unexplained, because terrestrial craters should be different due to the different compositions, and the differences in atmospheric composition (in some case no atmosphere).[596] These facts alone suggest an internal and uniform process is at work.

Structures displaying rings within (*concentric*) rings are also unexplained. Similar spacing of basin rings inside and outside the main ring indicates that the location of the rings, if not their placement, reflects a common mechanism. The uniformity in groups of ring geometry from planet to planet, moon to moon, or planet to moon supports the idea that some common mechanism controls the spacing of the rings.[596]

The effects of impacts cannot explain this uniformity, but the internal, explosive ejection of material (possibly by a field, such as an electromagnetic vortex), and an energy source involving hydrogen fusion can. Such a mechanism would create no drastic differences, because all would eject relative to the planet's core. An exploding object would create both a peak and a ring or rings with a smaller peak with greater size (resulting from more explosive internal force). Similarities would exist regardless of a planet's composition. Concentric rings with similar spacings would result from the magnetic properties of the material of the planet's composition (*i.e., ferromagnetic constraints due to electrostatic and electromagnetic effects*).

A number of other observations also reveal this internal mechanism is responsible for many craters. Heavy cratering occurs on only one side of the Moon, Titan, Mars and Mercury.[56,519] This may also be true of the Earth for which we presently find more craters in the Northern Hemisphere. However, more land is in that hemisphere and the force of erosion may have obscured the actual situation. Originally, it was thought, according to the "Jovian generator" theory, that Jupiter knocks meteors from the asteroid belt between it and Mars. Yet, the outer planets of Jupiter, Saturn, Uranus and Neptune also have moons with cratering.[519]

Also differences exist between the size distributions of craters for the inner planets compared to the outer planets. The heavily cratered surfaces of Jupiter's moons, Callisto and Ganymede, have less of the larger craters (>40 km. or 25 mi. diameter) and an overabundance of smaller craters (<20 km. or 12.5 mi. diameter). This has been claimed to be due to the different velocities for the proposed impactors in the inner Solar System than in the outer Solar System. Another explanation is that the inner planets were bombarded by objects orbiting around the Sun (heliocentric) mostly in the asteroid belt, while the outer planets' moons were bombarded by objects orbiting the planets (planetocentric).[700]

Size and velocity differences would be better resolved with the understanding that the craters were produced internally. The greater density of the inner planets requires greater velocity for explosive ejection. The size of the object would also regulate crater size, and the inner planets are larger than the outer planets' moons. Finally, the heliocentric and planetocentric origin of impactors cannot resolve the fact that cratering is uniform for the inner planets and the outer moons, but a uniformity of internally generated explosions can.

A good example of this fact is Mercury, because it is more heavily cratered on one side. Mercury's spin is too fast to explain such a distribution as the result of a meteor and asteroid bombardment from the asteroid belt. Therefore, it is essential to find another reason for the one-sided excess.

Mimas, a moon of Neptune, has a huge crater, with a central peak and rings. This crater covers nearly half of the surface of one side! Evidently, had this crater been the result of an impact, the moon would have split into pieces or been pulverized. Meanwhile, an internally generated and electromagnetically controlled ejection could prevent the destruction of the moon.

A number of meteorites have been found on Earth that are from other planets. A golf-ball sized stone found in Antarctica in 1982, and another seven since have disclosed a makeup that is identical with Moon rocks brought back by the Apollo and other lunar missions.[259] Three meteorites have gases that are distinctly Martian and were also uncovered in Antarctica, and may not be explained by an impact on Mars.[123,241,414] It is possible that these meteorites were "ejected" onto the Earth, since impact would require very specific conditions, and there is no evidence of the meteorites being melted twice, once from impact on the Moon or Mars, and a second time from entry into the Earth's atmosphere.

Mars has a basin with a central peak at each pole, where there are Fields. A Mars geophysicist discusses these: "Some comments should be made on the relative similarity between the giant basins situated at each pole. The encircling ramparts sloping down into a basin near the center of which is located a highland area is curiously similar to the morphology of some very much smaller scaled lunar and Martian craters—namely, the ring-walled structures with central mountain peaks."[751]; pp.396-7

He continues by stating that these craters could not possibly be impact sites at each pole. They would require two objects of the same size and speed striking at an angle that would be impossible to achieve (essentially perpendicular to the ecliptic plane). Impacts on exactly opposite ends of Mars is more science fiction than fact. The only answer is that they are of internal origin and controlled by a similar mechanism. Furthermore, what was originally thought to be canals is now known to be aligned craters, again indicating an internal origin for the cratering that is relative to the core.

Another indication of Mars' cratering as internally produced is that they should have worn down long ago. Using spacecraft measurements of wind speeds and patterns, experiments were constructed to duplicate the erosional environment on Mars. The results indicated that the craters, estimated to be hundreds of millions of years old, should have been worn down long ago. This means that the craters are much younger than thought, and that the so-called era of bombardment was not responsible for the cratering.

Venus also displays mysterious craters. A geophysicist studying Venus comments that they are intrinsic (i.e., internal): "Some of the roughly circular forms previously inferred to be impact craters do no show the expected topography or shape, and thus may be of intrinsic origin."[506; p.70]. Venus also has a hundred times more rare (*inert*) gases than the Earth or Mars. These gases are not supposed to be created after a planet's formation. This indicates that Venus was not formed according to widely held theories of an era of violent bombardment of meteors and asteroids (*accretion*).

The Moon also presents some mysteries that do not fit impact theory.[274,626,668] One lunar group comments on the regular distribution of the craters: "On the other hand, this nonrandom distribution might not be surprising if the craters were formed by internal processes."[668] The biggest craters all had formed so that they lied along the ancient equators.[626] Furthermore, small craters have a glassy surface that appears to be due to radiation melting the surface.[50,312] Pressure-induced, shocked rock is richer in rare earth elements and (*alkali*) metals, and are known to have a composition that is similar to meteorites (*carbonaceous-chon-drite-like material*).[534] As discussed previously, the Moon's craters between the 40° latitudes display numerous phenomena that indicate internal energy, such as gamma rays, luminous phenomena, gravity anomalies and more.

The giant fresh lunar craters, those that formed after the formation of mares (large smooth basin like plains), are generally believed to be of meteor-impact origin. However, the distribution of these large, fresh craters is far from random, which is contrary to formation by impact. Some of these fresh craters are on mare borders, such as Aristarchus and Copernicus, which are noted for transient lunar phenomena (TLP); ejected luminous phenomena.

Theories about the surface of both the Moon and Mercury went from volcanic to impact cratering back to volcanic in the opinions of various scientists. Scientists traditionally label such structures as volcanic not realizing that their roots are far deeper than assumed. For example, the great crater of Caloris in the equatorial region of Mercury has flows that cover a very wide area that is not typical of an impact. The supposition is therefore that it may be volcanic. About 20% of Mercury's equatorial region appears to be smooth and may also be volcanic. On the Moon, the Appennine Bench Formation within the Imbrium Basin (15° to 50° North Latitude), which was formed at the ancient equator, may also be volcanic.[416] These few examples demonstrate the underlying conflict of ideas, because there is something that displays the characteristics of both impact cratering and volcanic activity—that is, the only thing that can be said for sure is that force and internal processes are involved (also see Chapters 2 and 3).

2.2 Tektites

Tektites, a word derived from the Greek *tektos*, meaning melted, also display what could be predicted from a hydrogen fusion model of the planets that produces ejection phenomena. Their origin is highly controversial (for more on tektites see sections 28.2 and 28.3 of Volume One). The surface pits, grooves and shapes indicate that they were once molten, and their outer surfaces were aerodynamically molded by flight through the Earth's atmosphere. They possess a magnetism that is the result of ionizing radiation, probably resulting from their formation with lightning-like discharges (i.e., the time-varying electrostatic pulses of the Fields). Tektites also have thermoluminescent properties, which are known to arise from ionizing radiation (i.e., neutrons, gamma rays, etc.).[559-562]

Their composition is similar to sedimentary rock from the Earth, such as sandstone mixed with a little clay, with the exception of one type (*Ivory Coast*). The heavy hydrogen (*deuterium*) to hydrogen ratio observed in their composition is like that of ocean water. The Fields on Earth are situated in ocean water. The lack of air bubbles in the glass is impossible within the short time of an impact, as well. Yet, the two proposed theories of origin are impact on the Earth with the lofted material reentering and melting, or impact or volcanic activity on the Moon spewing material earthward.[559-562]

Meanwhile, both theories are incompatible with the full spectrum of evidence. All of the experts concede that origin by impact on the Earth is inconsistent with the laws of physics that underlie the making of glass, which is a major component of tektites. Furthermore, chemical similarities exist between tektites and Earth rocks, which have not been matched by lunar samples. An expert comments on the force involved, "it appears that such velocities are essentially unattainable without the use of light gases such as helium or hydrogen, or the use of temperatures which would destroy the tektite."[561] When considering the composition is like Earth rocks, isotope (*D/H*) ratios are like ocean water, and the escape speed for the particles to be ejected beyond the Earth's (or Moon's) gravity requires hydrogen or helium, a hydrogen fusion model, FEM, fits all of the necessary conditions, unlike the other theories.

This expert on tektites discusses some of our mistaken ideas about the Moon. He states, "within certain large classes of lunar craters, internal origin predominates over impact."[561] That is, most of the Mare craters with diameters from the 100- to 1,000-meter (330- to 3,300-feet) range are of internal origin.[313,561] He also concludes "that the rate of arrival of meteorite material on the lunar surface has not varied widely [since its origin]; i.e., the notion of an era of violent bombardment is wrong."[561]

Some of the more interesting tektite finds relate to the Australasian tektites. The lunar sample 14425, a speckled sphere eight millimeters in diameter, is similar in composition to the Australasian tektites. Moreover, two tektites were reported from Liberia (west central Africa) and also show a fission track date and chemical composition that coincides with this same tektite event. Another tektite with the same fission track date and uranium content was found in southwest Africa. While the Australasian tektite event is given a date of about 700,000 years ago, there seems to have been an earlier fall in that region about four million years ago.

Another enigma for present theoretical perspectives is the great distances that tektites travel, some travel over distances of hundreds or thousands of kilometers, especially the Australasian and North American tektite strewn fields. Atmospheric resistence is so great for such small-diameter particles that they should fall to Earth within a short distance. In fact, it would be difficult to loft them for hundreds of kilometers, yet they have traveled for thousands of kilometers. The atmosphere would have to be removed to allow for these distances. Such as removal would require tremendous energy (10^{32} *ergs*), typical of nuclear explosions.

If they were produced by an impact it would have left behind a crater of several hundred kilometers in the case of the Australasian tektites. No such crater exists, nor is a crater found for the North American strewn field, both of which are the largest and most recent events.

Again, the Field's release of ionizing radiation would displace the atmosphere, and with FEM's hydrogen fusion there is more than enough potential energy. Interestingly, simulations with atomic blasts are capable of lofting the particles into the stratosphere where they might drift with winds. The finer particles should also fall closest to the proposed impact than the larger tektites, but the observations display the opposite, with the smallest the farthest away. This implies that there is an "atmospheric entrainment scenario;" again, the evidence points to FEM.

2.3 Meteorites and Animal Falls

A similar type of mystery or controversy surrounds meteorites themselves. Organic matter, once thought to only derive from Earth or Earth-like planets, has been discovered in many meteorites.[23,137-139,475,544,722] Carbonaceous chondrites are stony meteorites, which a scientist discusses: "Some of the most volatile substances occur in the Earth's crust in nearly the same proportions as in carbonaceous chondrites. This has led to the suggestion that the inner planets obtained their volatiles from carbonaceous-chondrite-like dust or larger bodies that had formed in colder parts of the nebula."[23] The nebula referred to here is the protoplanetary nebula of the forming Sun that began the Solar System. However, an equally or more plausible explanation is that carbonaceous chondrites originate from the inner planets (and outer moons), which have a core of hydrogen fusion that could allow for ejection. Meteorites of various types and falls of organisms also suggest this theory of ejected materials. Consider the various types of falls consisting of eels, fish, frogs, and other organisms, and Earth-like objects.

A number of historical references describe fish falls. One of the earliest recorded fish rains occurred in the 3rd Century AD, and lasted three days. This was recounted by Anthenaeus in *Deipnosophistae* describing Phoenias writing in *Eresian Magistrates*. Phylarchus, in the Fourth Book, tells us that people had often seen fish rains. Halsted's *History of*

Kent disclosed that a fish rain took place on Easter in 1666. Reid's *Law of Storms* describes a heavy rain that included herrings on 9 March 1830 at the island of Islay, in the town of Argyllshire.[471] These and other types of unusual rains occur in regions near the Fields. It is fitting that these falls should be aquatic organisms, since the Fields are located in ocean waters near coasts.

One source tells us that more than 50 rains of fish have taken place throughout the world. For example, Louisville, Kentucky had a rain of sun-perch in 1835.[471] Many other sources describe similar events.

On 27 June 1901 a heavy rain brought a rain of small fish to South Carolina.[38] Shellfish (*Gemma gemma*) plummeted in a "whirling motion" in a rain storm that hit Chester, Pennsylvania on 6 June 1869.[28] Live snakes rained on Memphis, Tennessee on 15 December 1876.[30] On a calm, foggy day with little wind, fish rained on Marksville, near New Orleans. A day earlier there were either small tornadoes or dust devils.[89] These falls took place in the vicinity of the North Atlantic Field.

In September of 1936 the Island of Guam experienced a brief rain of fish, one of which is tench (*Tinca tinca*) common to western Asia.[347] This location is near the Japanese Field. Falls of black worms and muscles also took place near the Persian Gulf Field.[29,35] As with tektites, light gases, such as hydrogen or helium, or a vacuum created by ionizing radiation could be responsible for these falls. In contrast, physical theories presently employed, and for centuries prior, have been unable to explain these falls.

Other falls of aquatic organisms demonstrate the same relationships. On 24 August 1918, during a heavy thunderstorm at the Henoon suburb of Sunderland, a rain of sand-eels (*Ammodytes tobianus*) fell. These eels are sea creatures common on the sandy coasts near the mouth of the Mediterranean.[512] Near Gibraltar, on 25 May 1915, millions of small frogs fell.[45] The Mediterranean Field's ejection of fusion by-products ejected these eels and frogs into the atmosphere. Hundreds of small frogs rained on Trowbridge, England on 16 June 1939.[129] Seaweed fell on Falkland, Fifeshire, Scotland in a heavy hailstorm accompanied by thunder and lightning on 30 August 1879.[31] These falls, particularly near Gibraltar, which is at the mouth of the Mediterranean, are near the Mediterranean Field.

In Tome One of Volume One it was noted that sulphate-reducing bacteria precipitating iron were found in gelatinous mats in the Field areas. On 11 November 1845 a "meteor" of gelatin 1.2 meters (4 feet) in diameter fell on Loweville, New York.[25] Allegheny, Pennsylvania received a jelly-like mass, on 9 December 1909 referred to as "pwdre ser," that was "almost intolerably offensive in smell."[3] Sulfur, bacteria and decaying gelatin would all be very offensive in smell. Another pwdre ser fell on Ingleborough in 1908, and was reported to be a mass of bacteria.[371] Cambridge, England experienced a gelatinous meteor on 23 June 1978.[196] On 12 March 1867, 6.35 centimeters (2.5 inches) of yellow rain, which was determined to be sulfur, fell on Kentucky.[238] A heavy fall of rain with a red dust, called a "rain of blood," that stained clothes, fell on Britain.[37] This was probably iron oxide (or amorphous selenium). Sulfur rains were noted in Surrey, England on 18 October 1867, Hungary on 17 June 1873, and Mageburg, Germany in June of 1642.[27,307] The Field areas involved are the North Atlantic, Mediterranean and Persian Gulf Fields.

Falls of ice have also taken place. As described, ionizing radiation creates a vacuum, which the upper, much colder atmosphere flows into. This can cause sudden heavy rains, snowfalls, hail or ice, and is the mechanism behind the formation of glacial cover and ice ages (see discussion in sections 15.14 and 15.15).

Consider this ice fall. On 13 December 1973 ice fell on Fort Pierce, Florida, a very unlikely place for ice. It contained silver, the best conductor of all metals; gallium, a semiconductor; germanium, also a semiconductor; mercury, which dissolves in metals and is used in long-lived batteries; and selenium, which reacts with sulfur-like characteristics, reacts with metals, and is photovoltaic and photoconductive. There was also a high salt content reflective of sea water.[595] Here we see the electrical conducting and producing constituents of the North Atlantic Field area in ice that fell on Florida, which is just east of the Field.

On 7 March 1976 three blocks of ice plummeted on Timerville, Virginia. The ice consisted of water, bacteria and yeast-like organisms on a piece of gravel. Long Beach, California, which is near the descending limb of the Fields (i.e., near a plate boundary) had an ice fall on 4 June 1953. It consisted of 50 ice lumps, some as much as 75 kilograms (165 pounds), and totaled about one ton! Again, this is an extremely unlikely place for ice, but it has a relationship with the location where the North Atlantic and North Pacific Fields' descending limbs meet. Likewise, Surrey, England received a 370 square-centimeter (4 square feet) block of ice, on 23 June 1972.[196] As with the other falls their origin has been a mystery.

Other meteorites suggest the same mechanism. A "meteor" fall containing sponges, corals and crinoids, all sea creatures, fell on Knyahinya, Hungary on 9 June 1866.[112] Sedimentary meteorites, one of limestone and another of sandstone, fell on 11 April 1925. The limestone meteorite had a gloss caused by atmospheric friction, and was pure calcium carbonate. It was called the Bleckenstad meteorite. The sandstone specimen was completely ignored after the initial report. Another meteorite of limestone, composed of shells and also having a fusion crust fell in northern New Mexico on 24 March 1933. Two sandstone meteorites fell on the East Coast as well.[203] Others that were water formed (*hydrous minerals; diopside-olivine anchorite*) fell in Egypt in 1911, and near Weston, Connecticut in 1807 (*iddingsite mixture of hydrothermal alteration products of olivine: (Mg, Fe)$_2$ S$_1$O$_4$*).[367] The Cumberland Falls meteorite was composed of fine particles like clay or sand (*enstatite breccia*), very compacted as if it had been under high pressure, such as the ocean bottom, and high in magnesium, similar to ocean water.[517] Meteorites frequently contain petroleum-like hydrocarbons, and organic matter typical of modern organisms.[138] As discussed, ionizing radiation can produce hydrocarbons, and petroleum is found in those areas near the Fields' stems and descending limbs.[513] Meteorites such as these cannot be reconciled with extraterrestrial impactors, since they would have to come from outside the Solar System on a planet with life, water, hydrocarbons, and Earth-like elemental constituents.

Another effect the release of ionizing radiation would have is to precipitate carbon from the atmosphere. On 28 April 1884, a black rain fell on a region of the British Isles between Church Stretton and Much Wenlock. Analysis of the rain disclosed that it consisted of carbon and rainwater. That day an extraordinary darkness made it necessary to use artificial light at 11 am in the morning. A huge black cloud developed a yellowish tinge after which rain, hail and snow fell. Hail and snow in late April, the yellowish tinge, enveloping darkness and carbon can readily be explained by ionizing radiation, while no other theory can resolve the enigma. The storm came from the southwest, which is the direction of the Mediterranean Field.[586]

Other facts about meteorites do not coincide with what is expected either. A number have a short exposure to the supposed effects of cosmic-rays.[22,185,557] One has an unusually long exposure.[277] The unusual composition of some suggest that all meteorites may not form in the Solar System as proposed (*anomalies in C, O, Ne, Xe, Al, Kr, Hg and Mg exist*). However, ionizing radiation processes, such as the hydrogen fusion model of the planets, could explain these anomalies.[246,247,340,508,743]

Numerous observations about variations in the frequency of meteorite and other falls conform to the Field-dynamical Earth Model (FEM). For one, meteor visibility or frequency varies with the solar cycle, with the rates reaching a maximum near solar minimum.[197,365] The flux of micro-meteorites has increased four fold in geologically recent times, while it should be decreasing.[54] Rainfall and meteor rates are also correlated, again suggesting a relationship with the Fields, which control weather.[698] During the Aquarid (early May and mid-July) and Perseid (25 July to 18 August) meteor showers there are fluctuations (*micropulsations*) in the Earth's magnetic field.[163,393] An increase in large "fluffy meteoroids" takes place in the auroral zones (North and South Pole Fields), particularly when the Moon approaches.[268] In New Zealand, near the East Australian Field, an increase of meteor falls is noted during the winter, just following the solstice (June-July). In the Northern Hemisphere an increase also occurs from the solstice to the equinox, in winter (January-March).[365] Likewise, a histogram of 867 recorded falls revealed a peak in early April, around the vernal equinox.[197] Meteors (50 to 50,000 grams) are not randomly distributed in space, but occur in swarms. In one study, two maxima during consecutive months disclosed that swarms took place near New Moon and Full Moon, and were not associated with moonquakes or impacts.[369] Another study disclosed a peak of meteoric influx during the equinox.[88] Meteor showers conform to the peak periods observed in the FEM-lunar and FEM-solar linkages (see Tome Five for discussions on these linkages).

Furthermore, the falls of fish, gelatin and so forth, and others mentioned here with dates, show a maximum near the June solstice, and a secondary peak near the autumnal (September) and vernal (March) equinoxes. The majority occurred within two weeks of these dates, and at Mid-phase, and a secondary peak at Quarter Phase, of the Moon. This is in accord with the discussion on earthquakes, volcanic eruptions and weather phenomena in the following Tome, which have similar peaks that were the result of electrostatic forces and ionizing radiation. This is not to say that all meteor falls are the result of FEM, but all animal, gelatin and ice falls, and meteorites containing life or water formed rocks appear to have no other explanation.

Consider a few unusual events in this regard. On 24 March 1933, three days following the equinox, a passing meteor was accompanied by an electric effect.[251] In Montreal on 7 July 1885, a little more than two weeks following the solstice, and at lunar Mid-phase, a number of people experienced an electric shock as a huge meteor began to fall.[33] It is extremely difficult to imagine how a charge could precede a meteor, if the meteor were creating it. Instead some electrical current on the Earth (geoelectric or electrostatic effect) preceded the meteor. Likewise, on 12 November 1887, at the Mid-phase, a fireball was seen at 46° North Latitude near the North Atlantic Field.[34] A correlation exists between meteors and fluctuations of the geomagnetic field.[171]

Other events show these factors quite clearly. For example, one day with no bad weather there came a sudden squall that blew so hard it knocked a ship's masts over, and lasted for about two hours. Then, in 1845, near the mouth of the Mediterranean (Mediterranean Field), there came a calm accompanied by an "overpowering stench of sulfur and an unbearable heat. At this moment three luminous bodies were seen to issue from the sea."[26] Such events are unexplained by present-day conventional thought, but sulfur and luminous masses of ionized particles are expected of FEM.

When one looks at the cratering record on Earth the latitude restrictions become plain for at least certain craters. Those craters which are of a questionable impact origin were more often formed between the 30° to 40° latitudes when plate motion and polar wander are considered. Furthermore, the dating of the craters indicates that they were formed during geologically active periods, but were not responsible for the plate tectonics and polar wander or reversals that took place.[330] The shifting of the Field system at these times also resulted in internally generated craters, which can reconcile these data, as well as, the Tunguska event.

CHAPTER 3

The Day the Earth Exploded in Siberia (*The Tunguska Event, and Geoblemes*)

On the morning of 30 June 1908, at 7:17 am local time, a blinding light was accompanied by an explosion of tremendous force at Tunguska, in Central Siberia. The light was so intense that it was visible 700 kilometers away (435 miles).[198,199] It had the appearance of a column of fire with a bluish tinge, and left a trail of a bluish streak along its entire path.[453,454,457]

3.1 The Devastation

The blast was so powerful that baromicrographs, sensitive to atmospheric pressure, recorded the event as far away as Western Europe, North America, and other distant countries. The shock wave traveled around the world twice as seismographs, used for recording earthquakes, registered the event around the globe. The blast was so powerful that people heard it 1,000 kilometers away (620 miles).[77,78,79,103,198, 199,272,453,454,564] The series of air waves displayed four maxima within twenty minutes.[205] Four thunder claps and crashes were described by the local people, the Tungus.[453,454,457]

The blast devastated an area of about 2,000 square kilometers (770 square miles) of dense Siberian virgin forest (Yenissi Taiga). Trees were torn up by the roots and in places were piled up in thick layers pointing away from the center. The devastation covered a radius of 30 to 40 kilometers (20 to 25 miles), and the fire from the blast scarred trees for 18 kilometers (11 miles) from the center. Somehow, the central area was devastated with the exception of a ring of upright trees near the center.[205,370,717] Ever since that day, what actually took place has been a matter of deep controversy without any completely satisfactory explanation.

The effects of the event on the people and wildlife confirm its awesome power. A local trader 60 kilometers (37 miles) away from the event describes his experience:

> "I was sitting on my porch facing north when suddenly, to the northwest, there appeared a great flash of light. There was so much heat that I was no longer able to remain where I was. My shirt almost burned off my back. I saw a huge fireball that covered an enormous part of the sky. I only had a moment to note the size of it. Afterward it became dark and at the same time I felt an explosion that threw me several feet from the porch. I lost consciousness for a few moments and when I came to I heard a noise that shook the whole house and nearly moved it off its foundation. The glass and the framing of the house shattered and in the middle of the area where the hut stands a strip of ground split apart."[453]

It should be noted that this description reveals that an explosion rendered the man unconscious. Then when he comes to, he experiences a second explosion. Obviously, this was not an impacting body, because no such time delay would occur if it split into pieces.

Another account says that everyone was in a tent still asleep when "the people were suddenly flung into the air." Hundreds of reindeer, owned by the Tungus, disappeared and others were found as burnt remains. Some people had storehouses, goods and other objects completely destroyed.[703]

One scientist who describes the event soon after the first investigations, nearly twenty years after the event, discusses the mystery: "In looking over this account, one has to admit that many accounts of events in old chronicles that have been laughed at as fabrications are far less miraculous than this one, of which we seem to have undoubted confirmation."[564]

3.2 Enormous Energy

As could be predicted from the new model of the Earth, much about Tunguska is comparable to a nuclear explosion. One scientist states, "the cloud afterward was exactly like an atomic mushroom cloud."[79] Meteors only case minor magnetic disturbances, but the Tunguska event caused a major disturbance, like what occurs after atmospheric nuclear explosions.[370,377,400,758] A local magnetic storm began about six minutes *after* the explosion and lasted more than four hours. These observations resemble those that follow nuclear explosions. Had it been a bolide the magnetic disturbance would have begun as it entered the atmosphere. Nuclear explosions generate nitric oxide, nitrates and disturb climate, and so did the Tunguska event.[474,548,611,717]

Pressure and suction waves were experienced by observers. A pressure wave broke windows inward followed by a suction wave that sucked clods of dirt out of the ground and hurled an iron stove door across a room.[564] Ionizing radiation after a blast would cause a vacuum, producing a suction, but a meteor would not.

The blast was so strong that it was recorded on a seismograph 5,215 kilometers (3,240 miles) away, at Jena. Like a nuclear explosion, the center of the site was 1.5 to 2 times higher in radioactivity than 30 to 40 kilometers away (20 to 25 miles).[694] Mutations at the site were evident, trees doubled and tripled in height, and increased in life expectancy by 2 to 2.5 times. Radioactive elements were found in greater quantities at the center of the site.[782] Natives tell us that the blast had "brought with it a disease for the reindeer, specifically scabs, that had never appeared before the fire came."[703] Likewise, "gray patches" and blisters appeared on cattle after the 1945 New Mexico nuclear test.

Tektites or spherules of iron, magnetite, nickel and silicates were also found.[227,291,717] Similar objects are discovered after nuclear explosions.[548] Meteoritic dust was found far from the epicenter, which some claim is due to its explosion in flight.[210] However, this does not explain the succession of crashes heard. The estimated heat energy indicates a nuclear, not a chemical reaction.[781,783] The Tunguska event has even been compared to a 6,000 megaton nuclear war.[717] An author comments on the energy released: "All of the estimates are independent of one another and show one fact: the radiant energy of the Tunguska explosion comprised several tens of percent of the total energy. But this correlation between the parameters is characteristic only of nuclear explosions."[781] No one in the world had nuclear devices in 1908, particularly in Siberia.

3.3 Not An Impact

Many eyewitness accounts do not give one the impression that Tunguska was hit by a meteor or comet. The event produced a very bright light in the form of what was described as a pipe or cylinder. Such pipe shapes are not typical of meteorites nor meteors. Yet, "the fire-pillar was seen by many people."[445] The event left no smoky trail like most fireballs, but rather scintillating bands that looked like a rainbow or an aurora.

The phenomenon was so bright that those nearby had to cover their eyes, and it was visible at Vitim, 608 kilometers (377 miles), and Bodaibo, 764 kilometers (474 miles), away. These far away sightings are not typical of a meteor in flight.[253] One scientist collecting eyewitness accounts comments: "The explosion was observed from many points in the form of a vertical fountain."[253] Such a description sounds more like the release of energy from the Earth than an impact, because no comet or meteor would fall vertically, and would, in fact, be closer to horizontal, and would not be described as a fountain..

There was a notable geomagnetic disturbance that was more like what accompanied the huge volcanic explosion of Krakatoa.[697] Some described a *subterranean* crash and roar followed five to seven minutes later by a second louder than the first, and finally, a third crash. However, only one luminous phenomena was observed, and if the objection split into three, the pieces would not be delayed by such long intervals, nor described as subterranean. In a quote that still remains timely, a Russian scientist comments: "the nature of the phenomenon that could produce such a special explosion is still obscure."[331]

Throughout Europe and Asia unusual lighted displays were reported in the days that followed the event. They emitted colors described as a "yellowish-green", "orange-yellow", "rosy hue" and "intense crimson." And the colors were not like any comet or meteor. Its broad twilight spectrum was barely equivalent to an aurora, and was commented on at the time as aurora-like.[40,717] It was so bright in the dead of night that, "at 1 AM on July 2nd small print could be read without the aid of artificial light."[205] The luminous phenomena could be seen in faraway cities, such as Copenhagen, Berlin, Vienna, Prague, and numerous others.[135] It was seen from parts of Siberia to Spain and Fenno-Scandinavia to the Black Sea!

If the event had taken place in the United States, in the region of the Great Lakes, the lights could be have been seen as far away as Pittsburgh, Pennsylvania, Nashville, Tennessee, and Kansas City, Missouri. The thunder which had accompanied the event could have been heard in Washington, DC, Atlanta, Georgia, Tulsa, Oklahoma, and western North Dakota.[198,199] The Sun had a halo during the day, and the thunder that had accompanied the event came from a cloudless sky.[453,454,697] All of these observations could be predicted from the release of ionizing radiation, as occurs in nuclear explosions.[548] That is, ionizing radiation caused atmospheric reactions much like an aurora, and the thunder resulted from the vacuum created, much like lightning produces thunder.

Other evidence confirms the presence of ionizing radiation. Nitric oxide was produced in the atmosphere, and climate was affected by its presence.[474] There were reports of a black rain, typical of carbon and hydrocarbon precipitation. Carbon and hydrocarbons are known to be produced by ionizing radiation, such as in a nuclear explosion.[548] No metal, with the possible exception of local native iron, was ever found in the region. This observation, the waves and folds in the region's earth, and other facts are unlike any known meteor impact.[454,457,756,784] Burned and unburned parts of the area, as well as the burned and unburned parts of the same tree indicate a "radiant burn" unlike any meteor or comet fireball.[697,781]

Furthermore, any radiation should not have been measurable after two decades under normal circumstances. The radiation would have been released into the atmosphere with very little reaching the Earth, and it would have dissipated easily and quickly had it been a comet or meteor.[210] Meanwhile, two decades after the event the region was still 1.5 to 2 times higher in radioactivity than the surrounding environs.[697]

Near the center is an area of trees that were protected from the blast, and remained standing.[278,697,782,783] A meteor or comet would have knocked all the trees down. A surge or "tidal" wave on the swampy water (bore) was more like what would be caused by an earthquake.[453] All of these observations can be understood as the release of ionizing radiation constrained by an electromagnetic vortex field, which prevented a circle of trees from being knocked down.

No crater or metal meteorite has ever been found in the area. There are a number of depressions ranging from 10 to 50 meters (33 to 165 feet) in diameter, and 3.5 meters deep (12 feet), but they are not craters nor pits from meteor fragments.[205,455,474] Excavations up to 34 meters (110 feet) deep did not yield any meteoritic material.[198,199] Recently, a high level or spike of iridium was found in the area, but was also noted in the Antarctic ice, on the other end of the Earth, in the same time frame.[63]

The only metal ever noted was described in this way: "Finally, in 1930, native Tungus actually visited the expedition on the site, and reported that, immediately after [the event], they found in the neighborhood of the center pieces of brilliant native iron!"[456] Here is more evidence of the ejection, since meteors or comets are neither brilliant nor native iron.

Another indication of this is that the event was only noted when it was near the ground. Meteors or comets would have been noticed at the time they first entered the atmosphere. Descriptions were unlike these objects, being referred to as a "fountain", "pipe", "pillar" or "tube."[389,456,781,783] Eyewitness accounts and scientific articles say it was a single object. Yet, there were three subterranean crashes, and four plain maximums in air pressure.[205] Meteors are observed at night, but the Tunguska event took place in the early morning, indicating that if it was a meteor or comet, a much greater speed than is typical or even possible for such objects.[142,564,717] Calculations indicate a very low density as well: "It also implies that the Tunguska object was quite unusual in having a very low equivalent density."[717]

Reminding one of the cratering and water flows on Mars, the Moon and elsewhere, the Tunguska event produced water. "The indubitable traces of an inundation are in agreement likewise with the testimony of witnesses, who relate that, at the moment of the [event], the ground was struck through, and that for several days afterwards water shot up out of the Earth."[455]

As commented on in Volume One, the Earth has maintained ocean salinity at a constant level in spite of salt production, and the photodissociation of water in the upper atmosphere. Fresh water was also observed in the region of the North Atlantic Field. Hydrogen fusion, as it exists in the new model, could produce water. The solar-terrestrial linkage is revealed in the fact that a disturbed Sun was noted at the time, with a solar prominence reported by the British Astronomical Society, and the event took place nine days after the June solstice.[697]

Due to the estimates of its speed and density, a comet or comet fragment has been proposed.[59,60,103,717] However, the density does not match that of either a comet nucleus or tail.[59,717] No comets with the appropriate orbit are known either.[656,717] Furthermore, comet matter is too fragile to survive long enough to reach so close to ground level, and the other data on comets do not fit many of the facts about the event.[59,656] The skyglows were not characteristic of a comet, nor even a single object, since one object could not distribute so evenly.[60]

Other theories have failed to explain all of the evidence. Another hypothesis is that the event was due to a miniature black hole.[389] However, this would have shot straight through the Earth, but no such event was noted in the other hemisphere.[98,370] The fact that this theory even surfaced displays how inadequate the effects of either a comet or meteor are for explaining the event. Likewise, the theory of antimatter was proposed for the same reason.[198,199] Moreover, the antimatter theory fails to explain how antimatter could penetrate so deeply without exploding on first entry into the atmosphere.[370] Another theory claims that the explosion was due to trapped gases of heavy hydrogen (*deuterium*) in a detached shock front.[210] This should have left meteoritic dust in the immediate area, but the dust was found far from the epicenter. It also does not explain the subterranean crashes, the sighting of the object near ground level, nor the radioactive levels two decades later. The failure of these theories can be understood in the fact that most articles published in recent times claim that the event was due to a meteor or comet, but these theories fail to explain all of the evidence.

An author whose theory involves a nuclear explosion makes the problem evident: "Substituting for years of painstaking investigation, an unsubstantiated allegation is made that the Tunguska event results from impact of a huge meteor. A meteor impact simply does not fit all know facts sufficiently well to render the allegation credible. Repetitious publication tends to cause unqualified explanations to become accepted without challenge."[104]; pp.95-96

And so, the most prevalent theory, that the Tunguska event was the result of a meteor impact, is repeatedly claimed in newly published manuscripts. Yet, the facts indicate the Tunguska event resulted from a nuclear-like explosion from the Earth, but present models of the Earth prevent scientists from coming to that conclusion.

Another somewhat similar event took place in the upper reaches of the Brazilian Amazon, near Brazil's border with Peru, on 13 August 1930. Like Tunguska, meteors are observed at night, but this event took place in the early morning (about 8 am), indicating a much greater speed than is typical or even possible for such objects. It occurred in the middle of a forest knocking trees down everywhere. The event was accompanied by a triple shock similar to the rumble of thunder and lightning. The three distinct explosions, each stronger than other, were heard hundreds of kilometers away. Two particularly puzzling features were the fall of dust *before* the fireballs were observed, and the lack of any mention of a blast wave.

Another strange explosion took place near the Russian town of Sosova, about 350 kilometers (217 miles) southeast of Moscow, on 12 April 1991. Several ideas about its true nature were proposed, but all were abandoned with the exception that it was endogenic (i.e., internal) in origin. For years prior to the explosion there were signs of increased tectonic activity, fireballs, and slow ground deformation. Animals became anxious, people fell ill, railroad communications failed, and about a minute before the event, noise was heard on radio receivers, which eventually jammed all radio stations. Up to several hundred kilometers away from the epicenter, people felt a heat wave and suffocation.

Like Tunguska, the damage was unusual and selective, affecting the town and a village 20 kilometers (12.5 miles) away with what looked like tornado damage, but a tree 10 meters (33 feet) from the epicenter was undamaged. There was no damage at ground level at a distance of about one kilometer from the crater, but at large distances there was damage. At the time of the explosion some people were carried by an unknown force. Electric light bulbs and hollow plastic toys exploded, and inner windows were smashed while outer ones were undamaged. Soil was thrown about, and there were light phenomena.

CHAPTER 4

Blazing Balls of Fire
(Comets: Their Structural Dynamics and Occurrence)

Observations of comets have revealed that they too are structured like other objects in the Universe. Magnetic fields, aurora, time-varying characteristics, and latitude restricted phenomena have been observed. A solar linkage with the Interplanetary Magnetic Field (IMF) is observed in many forms. Another comet-sun linkage involves what are called Sungrazers, which are comets that smash into the Sun. This may mean that the Sun is less finite than has been assumed. Comets, such as Halley's Comet, also reveal the Field-dynamical Model's structure and dynamics.

4.1 Comet Structure

Comets have a magnetic field and a rotating nucleus. The nucleus rotates on its axis in about 14.614 days according to the initial results of observations. Yet, its brightness varies in 7.3 days, and observations of the curvature of its jets suggest a rotation of about two days.[273,395, 678] The differences in these periods are due to time-varying components, as well as rotation.

Comets also have a magnetosphere that interacts with the solar wind and the Interplanetary Magnetic Field (IMF).[556] Due to the shape of the nucleus of Comet Halley (*ellipsoidal*) its rotation should have been reoriented, but it has not done so, indicating a dynamic system.[412] Its pole is not located as expected, but instead it is in the general neighborhood of the pole of the orbital plane (rotation and revolution in the same sense).[757] Observations indicate that comets, in particular Halley's Comet (it has been observed more), have properties that are typical of the Field-dynamical Model, including a solar linkage.

The nucleus wobbles, generating electric and magnetic forces (*due to electrodynamic and magnetohydrodynamic effects*).[657] The inner cometary head or coma is filled with whistler radiation, like the high-frequency waves observed on those planets with lightning.[658] At times there are short-lived, star-like outbursts near the nucleus of bright comets. Outbursts are followed by an elongation in a tailward direction that sometimes gives the appearance of a multiple nucleus.[468,757] Usually, after a few hours, a concentric halo forms around the nucleus.[468] These facts and the complex motions indicate a time-varying component, and spherical electromagnetic and dipole magnetic fields that produce this dynamic structure, along which particles flow.

Other observations support the Field-dynamical Model. Electrostatic forces on the surface levitate dust off the surface even at large distances from the Sun.[358] The motion of the nucleus is not chaotic, indicative of stabilizing dynamics.[273] Highly specific conditions, including non-gravitational forces, seem essential to the formation of Halley's, as well as other comets.[412] The surface (*mantle*) of comets contains organic molecules that are electrostatically charged, and undoubtedly superconducting, particularly in the far reaches of outer space.[17,296,425] Together these facts suggest that comet surfaces are superconducting, and their structure and observed behavior are more stable than purely conventional theory would have us suspect.

Also indicative of a more dynamic comet structure is the activity that takes place far from solar influences. Brightenings, tails, outbursts and other cometary activity has taken place at distances beyond where solar heating and/or solar-wind action could have any significant influence. This challenges the dirty snowball theory of comets, as comet activity is supposed to be triggered by solar heating and the solar wind. Likewise, cometary comas have been known to contract as the Sun is approached, when the coma should expand under solar influences, according to the dirty snowball theory. Comets may well have an intrinsic nuclear energy source, as well as more dynamic structure than present theory allows.[197]

Hydrogen, carbon and oxygen particles are in much lower levels outside the inner head or coma, revealing a structural arrangement in its composition.[178] Likewise, the material of comets has a lower heat conductivity and capacity, and some activity on the night side suggests more structural stability than was originally assumed.[412] Yanaka 1988r, a comet that passed close to Earth in early 1989, displayed an even greater deficiency of carbon, putting the leading theories in doubt. Also, Halley has relatively stable, slightly increasing, non-gravitational effects in orbital motion, and other comets have shown wild effects often associated with outbursts.[324]

As occurs with other objects in the Universe, there are latitudes that display more activity. For one, comets have auroras, and most of the activity takes place in one main active spot on the nucleus.[273,516] This feature is larger and brighter than other sources on the comet, suggesting an unsuspectedly structured arrangement. A broad central plane exists near the pole that is often pointed toward the Sun as it emits dust. An exception to this scenario occurs during its closest approach to the Sun when neither pole receives input and dust emissions take place elsewhere on the comet.[616]

Jets form mostly on the sunward side of the nucleus.[657,757] Such discreet jets consist of gas and dust that come from "vents" (i.e., Fields) that cover about 20% of the surface. Gas jets extend more than 60,000 kilometers (37,000 miles) from the nucleus in spite of pressure from the solar wind and the speed of the comet. The jet mechanism is unknown, but does involve organic components.[467]

In spite of solar wind pressure, gas production is strongly oriented toward the Sun. Yet, the surface (*crust or mantle*) prevents free disintegration (*sublimation*).[397,412] Likewise, a spike-like extension spreads within 20° of the sunward direction for about 700,000 kilometers (435,000 miles). A streamer was observed nearly 180° from the spike, which suggests a mutual structural relationship with the spike-like extension (i.e., a cone-shaped Field would displace the emission from exactly 180°).[658] Mass ejection rate and brightness increase suddenly and outbursts of hydrogen (*Lyman alpha*) have a period of about 2.2 days (i.e., time-varying acceleration).[742]

A funnel-type flow, typical of the Field, was observed for the dust jets that issued from what has been called a small "vent." The jets have discrete features, such as spirals, semi-circular halos, pinwheels, envelopes, arcs, and straight anti-sunward jets. Molecules (*radicals*) were emitted in spiral patterns, though they were not observed to be emitted directly from the nucleus, and are thought to have emerged from an unknown source.[468,657] The most frequent coma pattern was spiral arcs, and the stronger spirals evolved two to three days later into envelopes surrounding the nucleus.[468] Gas jets were accelerated in an anticlockwise direction, and at least two distinct sources of dust emanated from the comet.[467] The ions that formed drifted perpendicular to the magnetic field, producing a ring.[398] Jets were an unexpected discovery for the dirty snowball theory of comets.[467] Here we see polar and mid-latitude phenomena, including a ring, spirals, and time-varying phenomena (including a 3- to 4-day period) that are typical of the Field-dynamical Model.

On closest approach to the Sun, water increased and displayed a 7.4-day (*sidereal*) period, as did certain molecules (*radicals*) and dust. All of these phenomena take place together in phase. Afterward, a structured component of about eight days was noted that consisted of a broad 3- to 4-day peak, followed by a narrow one-day peak.[692] The 7.4-day rotation period is unexplained, and shows up in the changing fanned structure of the jets. Yet, the sharp curvature of the jets appears to require a rotation of about 2 days. When considering these observations a conflict occurs in the same data if everything is to be explained by the effects of rotation and gravity.[678] Meanwhile, the answer to this apparent dilemma lies in a rotation of one duration and jets with their own time-varying phenomena, which can be expected of the model.

This is also why time-varying phenomena are plentiful, and often appear to be contradictory. The crater-like feature consists of a dark basin surrounded by a brighter rim whose brightness varies with a period of about 7.3 to 7.4 days.[549,653,678] These brightness variations do not vary in a way that could be predicted by the comet's distance from the Sun, displaying periods as short as one to two days, and even one to two hours.[324] Hydrogen (*Lyman-alpha*) shows outbursts in a cycle of about 2.2 days.[746] Water, hydrocarbons and other compounds (NH_2, *CN*, *OH and HCN*), characteristic of the model, display various periods that are often in phase.[504,692] The structure in the head of Comet Halley in 1910 can be explained if the rotation of about 52.5 hours existed, disclosing a slower rotation in the present.[273] Doubtlessly, this change in rotation is a result of a solar-comet linkage, including the greater solar activity of the present.

4.2 Solar-Comet Linkages

Comets exhibit definite relationships with the Sun. Halley's Comet became more active at about the distance of Jupiter's orbit (*6 AU*).[757] The outer surface or mantle is a transient phenomena, and none exists at distances between the Sun and the Earth (*less than 1 AU*).[657] Outburst episodes, in the 1910 and 1986 encounters, produced halos and anti-sunward jets that took place at about the same distance from the Sun, near the Earth's orbit (*0.8 to 1.1 AU*), both before and after its closest approach to the Sun (*perihelion*).[289] The tail is driven by solar radiation pressure, though an anti-tail or sunward tail occurs after closest approach to the Sun. After closest approach organic grains or molecules were also emitted.[657] Brightness flares occur at large distances from the Sun when high-speed, solar-wind streams pass the comet.[324] Hydrogen production peaks both before and after the closest approach to the Sun.[201] Aside from these phenomena there are other phenomena triggered by the Sun and the solar-comet linkage.

Disconnection Events, when the comet's tail becomes disconnected for a brief period, are one result of its interaction with the Sun. The possible causes for Disconnection Events include solar flares, other solar phenomena (*coronal mass ejections, coronal holes, high-speed plasma streams, etc.*), and particularly, interaction with the Interplanetary Magnetic Field (IMF).[489,618] The tail becomes disconnected as the IMF changes polarity, resulting in a reversal of the comet's magnetic barrier.[610] Many Disconnection Events take place around times of brightness variations.[324] The old plasma/magnetic structure disconnects from the head in a Disconnection Event. Then the comet forms a new plasma/magnetic structure with a new polarity.[133] As occurs with other objects in the Solar System, there is a dynamic interaction between the comet and the Sun.

Other observations confirm this dynamic interaction. The solar wind flows across the IMF, causing the newly born cometary ions to rotate around the comet's local magnetic field. Unexpectedly, a strong (*magnetohydrodynamic*) flow of cometary ions is ejected into the solar wind.[289] Ions of high energy (*500 KeV*) require an acceleration process in addition to what could be predicted by present models.[133] The IMF drapes around the nucleus several times as its polarity reverses, creating a symmetrical arrangement of magnetized sheets.[610] A magnetically-channeled outflow of ions from the ionosphere moves down the channel formed by the tailward closing of the IMF's draped magnetic field.

Outflow results from the IMF (*flux-ropes*) interacting with the comet, producing outflow on both sides of the comet.[767] The whole process reminds one of the IMF-FEM linkage producing outflow in the mid-latitude Fields on Earth and other planets.

Comets that disappeared or fragmented near the Sun are called Sungrazers (see Plate 4.5). Eight Sungrazers and an additional three or four possible ones were discovered from ground-based telescopes. Another six appeared in United States Air Force satellite photographs taken during solar studies. NASA's Solar Maximum Mission satellite had uncovered two more, and a possible third.[240,243,244] Likewise, recent bursts of gamma radiation from three points in space have been theorized to be due to comets or their split fragments falling onto other stars.[569] Together these facts indicate that comets are dynamically inter-related with the Sun (and other stars), and that they return some of the mass lost in the solar wind.

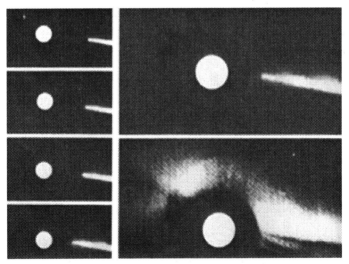

Plate 4.5. Sungrazer. Here are a series of photos showing a comet crashing into the Sun.[NASA]

This is also suggested by the fact that comets appear with greater frequency during times of increases in solar activity, including the equinoxes and solstices. A catalog of comets and the dates on which they were sighted, from 161 BC to AD 1699, shows that more were discovered in February-March, particularly from the 16 to 25 March or four days before and after the vernal equinox. A secondary peak is November-December and another for August-September.[344] These peaks include the month prior to and the month of both equinoxes, and the December solstice. The June solstice may also be represented if Southern Hemisphere sources were available, but this study only included Northern Hemisphere sightings. These periods are times of increased solar activity, particularly the vernal equinox (see Chapter 13).

Likewise, an analysis utilizing the years of sightings reveals that more comets are observed around the times of recognized solar maximums.[344,648] These periods were characterized with social disorder and more severe weather, making it likely that more comets existed than those that were observed. Thereby, this observation, at the least, establishes the reality of solar maximums and solar eruptions (flares, etc.) at these times, as well as a greater number of comets. Tome Five, and *In Defense of Nature—The History Nobody Told You About* discuss the historical record, and show the histograms displaying the correlations.

Other observations confirm this solar-comet linkage. The Morehouse Comet in September 1908 came at a time of sunspot activity, and the greatest geomagnetic storm occurred on 11 September 1908. Geomagnetic storms occurred in the autumn of 1835 as Halley's Comet exhibited a striking change. The great geomagnetic storms of 3 December 1846 took place just a few days before Biela's Comet split in two.[46] Geomagnetic storms increase at times of greater solar activity.

The activity of comets also increase at the equinoxes and solstices, and Halley's Comet is a good example. There is strong evidence on 7-8 March 1986 for a reversal of the comet's magnetic barrier.[556] After close approach to the Sun, in late March around the equinox, a complex set of jets issued from the more active hemisphere, producing strong counter-clockwise spiral patterns every other day.[468] Short-lived variability also took place on 23-25 March 1986, just days after the equinox.[164] From late April to early June a spike-like extension in the general direction of the Sun projected 700,000 kilometers (434,000 miles). This observation was completely unexpected, because the outflow was towards the Sun in spite of solar wind pressure.[658] Jets of a hydrocarbon gas (*cyano*) were most prominent in late April, following the vernal equinox. In an earlier encounter, a halo developed following the solstice in January 1836, and expanded for a few weeks to 460,000 kilometers (285,000 miles).[657] One of the most significant outbursts for which many Disconnection Events are associated occurred on 12-15 December 1985, days before the solstice.[324,634] On the first day of this period a strong outburst with a jet took place near the nucleus.[746] The biggest surprise was when Comet Halley was returning to deep space, and a bright emission burst occurred during solar maximum, around the June solstice. All of these events are within days to weeks of the solstices and equinoxes, and confirm a discussion on remote influences on solar activity (see Chapter 13).

Comets are associated with the meteor showers of April 461, September 647, and April through June 1539.[440] These dates are within the historical cycles of AD 250-450, AD 550-750 and AD 1400-1600 that are discussed in *In Defense of Nature—The History Nobody Told You About*, and encompass the equinoxes and June solstice. Around the time of the present maximum, numerous comets were observed by the Solar Maximum Mission satellite and other observatories. For example, a new comet, Comet Austin, appeared following a peak in solar maximum in April. According to orbital periods, one comet remained unobserved for 10 trips and then was "rediscovered" during solar maximum in 1990. As with other objects in the Solar System, comets have a dynamic linkage with the Sun.

4.3 Solar System Dynamics

Comets formed with the Sun, planets and asteroids from interstellar grains imbedded in the gaseous pre-planet cloud surrounding the Sun, known as the protoplanetary nebula.[757] However, it is questionable whether interstellar grains could survive early Solar System processing.[397] The irregular shape of comet Halley's nucleus suggests a break up of a larger body, possibly by collision during its formation.[412] The mixed composition of comets (*huge heterogeneity of the icy component*) and dust size implies that they were formed in various outer parts of the Solar System.[757] This observation contrasts the theory that all comets formed about halfway between our Sun and the next star (i.e., Oort's cloud).[222]

The composition of comets brought about a few surprises. The crust consists of radiation processed ices and organic molecules.[397] For example, amino acids (*pyrimidines and purines*) make up the organic component of Halley.[425] Organic materials in comets Wilson and Halley show the most pristine version of organic grains that condensed in the sheltered environments of star-forming regions.[17,296] The spectra of the organic grains in Halley and Wilson closely resemble each other, suggesting a common composition for all comets.[296] A superconducting surface aids in the comet's dynamic interrelationship with the Interplanetary Magnetic Field. The amino acids observed could not produce life (*Spermiae*), as some have suggested.[362,425,563,760] While methane and ammonia were noted in the coma, methane (CH_4) was unexpectedly low, which may be due to irradiation and evaporation (*and subliming*).[19,178,231] The extremely small grains of carbon, hydrogen, oxygen and nitrogen are strong evidence for organic grains and radiation levels that support an "interstellar-connection" theory (i.e., formation and interaction between star systems).[657]

The lack of uniformity in composition (*homogeneity*) in the nucleus suggests that it formed in the neighborhood of the planets, as does its battered appearance. Yet, its low density favors formation in gravitationally less active regions. The formation from and on interstellar grains brings the comet nucleus into perspective as a "primordial rubble pile."[757] The evidence, however, seems to encourage a slightly different label: a primordial refuse collector and recycler. After all, comets have been noted to collide with the Sun and have been referred to as Sungrazers.[240,243,244,523] Other observations suggest that comets typically collide with stars, because gamma-ray bursts have been tied to comets hitting stars.[569] As discussed, comets appear with more frequency at times of increased solar activity.[46,344,648] All of these facts indicate the probability that comets are more or less scavengers that maintain a steady state balance in the Solar System by retrieving the mass lost in the solar wind, making it a less finite system.

Another indication of this scavenger effect are the recent observations that some asteroids have turned into comets during solar maximum. Chiron with its unusually distant orbit for an asteroid, between Saturn and Uranus, recently began acting like a comet with increases in brightness. Comet Schwassmann-Wachmann 1 was initially identified as an asteroid, but then brightened by a factor of 100. The asteroids Olijato, Adonis and Phaethon are associated with meteoroid streams and resemble comet nuclei. Others (1984KB, 1982TA, 5025P-L and Olijato) are closely associated with the Taurid meteoroid stream.[417] Most of these observations have been noted within the last solar maximum, as were the Sungrazers. The material generated by these events are gravitationally driven back toward the Sun, as well as accelerated by the sunward polarity of the IMF.

Even comet orbits display the restrictions of the common structure of the Field-dynamical Model. There are 152 known original orbits of comets with a period longer than 10,000 years. Their orbits (*aphelia*) avoid three zones of the celestial sphere: the two galactic poles and a strip along the galactic equator. This arrangement is not random, and characteristic of conventional theory it is claimed to be due to a "vertical galactic tide" with a maximum at 45°.[219] Meanwhile, orbital observations of comets reveal forces that restrict cometary phenomena in relation to the poles, 40° to 45° latitudes, and an equatorial ring.

Similar restrictions can be noted in the Solar System itself. Thirteen well-documented cases indicate that the locations where long-period comets have split always takes place close to the ecliptic plane (the Solar System's ring plane).[343] Sixty well-observed, very-long-period comets shows that most intersect at nearly the same point near the Sun (*on the heliocentric sphere*). That point is along the asteroid belt, which is itself a ring around the Sun.[730] Some Sungrazers circle the Sun in orbits tilted about 40° to the ecliptic plane, and with the orbits' major axes aligned in the same direction. Again, a ring and 40° phenomena exist in the Solar System, and demonstrate the dynamic solar-comet linkage.

CHAPTER 5

Galactic Visions
(Active Galactic Nuclei and Galaxy Structural Dynamics)

Galaxies display a remarkable uniformity of structure, though at first glance some might appear quite different. Jets, lobes, hot spots, and extended structures form relationships and angles with other components of the galaxy. In the centers of galaxies are active nuclei with associated radio, gamma-ray and X-ray emissions typical of acceleration processes. These emissions are revealed by the spectrum of light emitted. Super massive black holes are often believed to be a part of active galactic nuclei, but many theoretical and observational problems force us to question the assumption of super massiveness. Other components include quasars, Blazars and disks. The variety of phenomena galaxies produce include ejections, collisions, explosions, plumes, starbursts, cannibalism (or coalescence), and multiple nuclei. As with other celestial objects, a Field-dynamical Model is conspicuous.

Active galactic nuclei (AGN) can be considered the core of galaxies. Often star-forming, thermal and nonthermal activity are found together, which indicates highly structured dynamics. Periods of star formation take place in all spiral galaxies, as well as other time-varying phenomena.[682] There are thermal relativistic electron plasmas, (*magnetohydrodynamic*) plasma waves, and shock waves that could be reckoned with the aurora of planets, though on a much larger scale.[479,645] Relativistic electrons are moving at speeds close to the speed of light. At times components appear to be traveling faster than the speed of light, which are referred to as superluminals.[579] All of these phenomena will be discussed in this Chapter.

This quote reflects the state of present theoretical constructs and observational limitations: "Active galactic nuclei (AGN) continue to be poorly understood objects, not least because our current technology usually permits measurement of only the *integrated* nuclear properties."[765; pp.(2)27]. The nature of the energy sources of active galactic nuclei and quasars have not been made clear.[149] No theory of what mechanism or "central engine" causes the activity in active galactic nuclei (AGN) has been generally accepted nor supported by all of the observations and members of the scientific community.

5.1 Jets

A jet is a long thin feature of bright emission extending from the nucleus of a galaxy. Jets have much stronger constraints and confinement than had been theorized.[593] Radio jets appear to consist of a hot gas or plasma-clouds of energetic electrons traveling in magnetic fields at speeds close to the speed of light. These relativistic jets are not understood, and according to the observational evidence, are more stable than theories predict. It is difficult—if not impossible in the conventional framework—to understand how relativistic jets, well structured (*collimated*) over great distances (*hundreds of kiloparsecs*), could remain stable when encountering various instabilities in the environment. Scientists concede that jet structure is more on the order of manmade linear accelerators used to accelerate particles.[149] The Fields of the Field-dynamical Model are time-varying linear accelerators.

All AGN emit jets of some sort. One-sided jets, known as asymmetrical jets, occur in powerful quasars (*e.g., 1007+417, 1150+497 and 0800+608*), and weaker nearby radio galaxies (*e.g., NGC 6251, M87, 1759+211 and NGC 3310*).[639] Quasars can be described as extragalactic objects that look like a point of light, somewhat similar to a star, but emit more energy than 100 supergiant galaxies. Jets are observed as long thin traces of radio emission that connect lobes and hot spots with the central source. As extended regions of relatively low brightness and steep spectra, radio lobes often exist symmetrically displaced on either side of the galaxy. Hot spots are regions of enhanced brightness within the lobes. A (*flat spectra*) radio source radiates at the core of the galaxy.[603] Jets are the stem of the Field, and the lobes and hotspots are the top of the funnel or cone shape of the Field.

Extended radio-source patterns usually display one or more extended components that have offset structures or asymmetries that are matched by asymmetries in the core. The extended components of quasars display lines that join lobes to the central core at angles between 9° and 13°. The half-angle of the cone of superluminals must be 20° to 30° (*to make the linear size problem go away*). The position angle defined by the source's motion means that it can be up to 20° to 30° away from the true axis of the cone.[145] Therefore, the true axis of the cone can be at an angle of 30° to 40°. This cone is why radio jets and counterjets sometimes appear to be slightly misaligned. The jet is a time-varying linear accelerator that is in a dynamic relationship with the core or AGN, much like the Fields are in a dynamic relationship with the core of a planet at an angle of 30° to 40° (i.e., Field latitudes).

Like time-varying linear accelerators, jets are known to accelerate particle bundles or aggregates in what is referred to as pressure gradients and shocks. Internal shocks in jets disperse through the medium along with a pressure gradient.[265] Some sources (*i.e., 3C 273, etc.*) have a jet that is perpendicular to the main body and is compressed by shock waves.

Polarization is the degree to which an electric and/or magnetic line of force (vector) changes in a regular fashion. Thereby, polarization reveals the extent of structural arrangement and time-varying characteristics. Compression of magnetic field lines and varying visual effects (*opacity*) are part of the rapid polarization changes that reveal the acceleration of particles.[721] The optical and infrared polarization of optical jets shows strong evidence for on-site acceleration of electrons.[498] Particle bundles are periodically accelerated by time-varying linear accelerators, as occurs with the Fields of the Field-dynamical Model of the planets.

One-sided radio jets, always on the same side of the nucleus, display the same dynamic mechanism. Accepted theory (*relativistic beaming*), which predicts more random behavior, fails to explain this uniformity.[399] Apparent changes in jet direction of about 180º in some sources (*e.g., 3C 395, etc.*) uncovers a dynamic mechanism that present theory cannot explain.[671] The observed X-rays from jets and lobes, and the correspondence of X-ray and radio structures (*e.g., M87, Cen A, etc.*) "require incredibly efficient particle acceleration."[102]

Enormous energy output, short timescale variability, and high and variable polarization have suggested to some that a massive central core exists. Observations of objects frequently reveal the existence of beams of gas "squirting" from the core. Two narrow funnels free of material around the rotational axis are essential to the formation of most jets. Radio-quiet quasars may be supported without dominant magnetic fields, while radio-loud quasars, Blazars and superluminal sources are associated with electromagnetically powered jets.[160] However, a massive core is an assumption based only on circumstantial evidence, as will be discussed.

Superluminal, or "faster-than-light," components that are observed (*with Very Long Base Interferometry; VLBI*) could be shocks dispersing down a pre-existing relativistic jet. When shocks form in the jet, the magnetic field is compressed as part of the acceleration process. In the axis, the compression of the magnetic field is aligned perpendicular to the jet axis as the shocks flow down pre-existing jet lines.[20] These magnetic pulses are similar to the time-varying accelerators noted for all of the planets. The apparent "faster-than-light" speeds are an illusion created by the source itself traveling at the speed of light with one or more populations of particles accelerated to or near the speed of light in the same or opposite direction; it is a matter of relativity (and non-cosmological redshifts, as will be discussed).

Radio jets twist through space with the strength of about 100 million stars. When there are two pairs, each pair twists and tangles together accelerated, high-energy particles, and magnetic fields explode from bright spots that appear to be turbulent, pan-shaped regions.[170] The Field pans out in a funnel shape, and using electromagnetic bursts accelerates different populations of particles in a spiral or helix (as were noted in planetary auroras).

Radio jets emit enormous amounts of energy generated in the nuclei of giant elliptical galaxies and quasars. Jets often reach well beyond the boundaries of the galaxy and end in the radio lobes. Certain (*flat-spectrum*) quasars (about 50%) display an extended structure composed of a jet with one hot spot. Some luminous quasars display a curvature of the jet, while some have wiggles with symmetry between the jet and counterjet. Others have wiggles or bends over 90° or are perpendicular. Present theories, involving the gravitational effects of super massive black holes, have not been able to explain these observations of highly structured phenomena.

Before many jets terminate into radio lobes there is a large amount of curvature. The magnetic field is aligned parallel to the jet axis, and follows the wiggles and curvature. The lower luminosity field is parallel to the jet axis near the galaxy, then flips to perpendicular, and can be explained by a spiral or helical field. Many luminous, core-dominated radio sources display a jet linked to a hot spot, and some show superluminal phenomena. Long, one-sided jets have prominent lobes and hot spots on both sides of the radio core.[279] This stability and structure calls for an electromagnetic vortex—the Field—whose wiggles, curvature, bends, lobes and hotspots result from the structural arrangement of the Fields, and are partially the result of interaction with other fields (primarily, intergalactic fields and plasma layers).

Other observations support this interpretation. Jets (e.g., NGC 6251) have been known to decrease in brightness much more slowly with increasing distance than if the jet had heat loss or gain (*adiabatically*), as would be expected (when not considering the existence of the Field). The expansion conserves the flow of relativistic electrons and the magnetic field. Relativistic particles have to be re-accelerated and/or the magnetic field must be amplified for the jet to extend such great distances from the core (*many tens of Kpc*). A helical field fits most of the data and calculated density.

The observed temperature and total mass of the hot gas is similar to what one would observe if the gas were undergoing confinement. Likewise, the fields involve a large-scale magnetosphere and field structure outside the jet.[136,155] The Hubble telescope disclosed that galactic centers have about 400 times more stars near their centers (as in NGC 7457) than was previously suspected, which suggests that an outer core structure exists. All of these observations could be expected of the Field-dynamical Model.

Galaxies are capable of emitting or ejecting objects, it seems even quasars. Spiral, also referred to as Seyfert, galaxies at times (*i.e., NGC 1097*) eject as many as three narrow, straight jets at points away from the galaxy's nucleus. These jets are often equatorial, while many radio galaxies eject jets along the axis or "poles." The ejected material is only gas (*hydrogen-alpha*), and does not interact with the dust or the stars, which indicates a magnetic field that structures the gas.[64-72,150,152] It has been suggested that powerful galactic jets may eject even the "central engine" from the nucleus of its galaxy.[639]

No general agreement on the mechanism that produces radio jets, nor the physics of spherically-symmetric ejection of matter has emerged.[151] It could be that the nuclear restoring force dominates the jet thrust, and likewise, that the basic super massive black-hole, accretion-disk model of the jet is incorrect. The reason for the failure to explain the observations is that conventional theories do not consider the Field-dynamical Model, and are focused on gravitational concepts too strongly.

5.2 Acceleration Processes

Lobes, hot spots, jets and the nuclei of galaxies emit radio, gamma-ray and X-ray spectra, all of which are the signature of acceleration processes.[269,603] A time-varying mechanism is revealed by the compact radio sources at the centers of galaxies, which are active at times and inactive at others.[494] In many cases it is commonplace to find variability on timescales of days, months and years, which are seen in optical wavelengths, and high and low radio frequencies.[149]

One starburst galaxy, M82 (also known as NGC 3034), displays a radio emission at the center that is associated with a molecular ring and star formation.[719] M82 has a central radio source that is time-varying on a scale of months to years.[448] This scenario is fairly typical of starburst galaxies.

Radio emissions are produced by linear accelerators, and radio emission is common in galaxies. For example, M82 has 40 discrete radio sources that vary from a few months to a few years.[448] Elliptical galaxies have radio cores, and the optically brighter galaxies tend to have higher radio strengths in the core.[306] All strong radio sources have or will probably show apparent superluminal characteristics.[145]

In the case of Blazars (also known as BL Lac objects) the bulk acceleration takes place towards the observer, and varies on a timescale of hours to weeks. Strong correlations exist between the X-ray, optical and radio emissions, which appear to be commonplace. The radio source is in conflict with some proposed mechanisms (*i.e., the Compton limit on temperature brightness*), and some have apparent superluminal motion.[62,304] As the result of similar acceleration mechanisms, all galaxies have radio and X-ray emission.[363]

More than 90% of quasars are radio quiet, and this has been claimed to be due to what has been called "thermal bubbles" radiating from the central objects.[261] Another possible way of explaining this observation is that quasars display our best view of confinement within a galactic core. As will be discussed, quasars are the beginning stages of galaxies.

Radio emission is typical of particles accelerated along a magnetic and/or electric field. Linear features of radio emission are present in spiral galaxies, and in the disks of all galaxies.[434] In some objects (*such as SS433*) there is very rapid radio variability.[126,132] In other objects (*M87, Cen A, etc.*) the observed correspondence between the X-ray and radio structure requires extremely efficient particle acceleration.[102]

Some spiral (*Sbc*) galaxies display radio emission from their spiral arms and nuclei. Radio emissions are the result of relativistic electrons spiraling in from and out into the spiral-arm magnetic fields. It is not clear where the electrons originate in conventional theory, but the production of electrons is commonplace with thermonuclear fusion. Meanwhile, normal galaxies spend several percent of their lifetimes in the Seyfert or spiral phase, and therefore, we are seeing a mechanism that is common to galaxies. Central radio sources exist in double galaxies, which are four times stronger than in isolated galaxies.[216] Seyferts have ionized gases emerging from, and high pressure components in, their nuclei, which is the result of acceleration by Field dynamics.[583]

If the electric field in the plasma is very strong, then electrons will gain more energy than they lose. Some electrons may be accelerated to very high velocities as a result (*runaway electrons*). An electric current often produces electrostatic double layers that are associated with an almost discontinuous jump in voltage, which would be ideal for acceleration.[11] The source of the electrons is the thermonuclear fusion produced by confinement in the Field-dynamical Model, which also guides the electrostatic layers.

The signatures of particle acceleration indicates that all AGN are powered by the same central engine (*coexistence of monoenergetic power law and Maxwell-Boltzmann electron energy spectra*).[645] Particle acceleration also discloses the existence of astrophysical magnetic fields, and a differential rotation of plasma associated with a hierarchy of vortices. The current requires that electrons acquire very, very high velocities (*i.e., ultrarelativistic*) in a common direction.[146] The evidence indicates a mechanism that is highly structured, and functions as the Field does in other celestial objects.

A peculiar elliptical galaxy (*NGC 1275*) has two nuclear components, one of which is a compact radio source that produces a point-like X-ray source, and rapidly varying flares.[85] Quasars and Blazars of one (*S5*) survey disclosed that of thirteen sources, at least twelve show bulk relativistic motion. Therefore, a common mechanism exists that produces this phenomenon in core-dominated radio sources.[766] Such uniformity of structure seems impossible to arise solely from the accumulation of matter in the massive black hole scenario.

Polarization is the degree to which the direction of the electric or magnetic line (or vector) changes in a regular fashion. This indicates the degree of structural dynamics in a given source. Waves in which the electric field lines are entirely horizontal with respect to the direction of motion are called linearly polarized. Synchrotron radiation is strongly polarized.

Synchrotron radiation is electromagnetic radiation from very, high-energy electrons moving in a magnetic field. It is nonthermal, and explains the radio emissions observed. Synchrotron radiation moves in a helical path, which might also be described as pairs of spirals or vortices curving in opposite directions, something like the separate proton and electron streams that make up the aurora.

A superluminal source (*3C 345*) shows that high linear and low circular polarization occurs in extragalactic radio sources. The dominant emission mechanism is (*incoherent*) synchrotron radiation from relativistic electrons in a magnetic field. This electron-magnetic field interaction also produces X-rays. A cone-shaped plasma beam emanates from the central core, revealing that the funnel-shaped Field is involved in accelerating plasma from the core. The core has a magnetic field that is uniform, and relativistic electron density varies smoothly with distance along the jet.[111] An electron density gradient would allow for efficient plasma acceleration and is characteristic of linear accelerators, such as the Field.

Luminous X-ray sources radiate from quasars and other AGN. X-ray emissions disclose evidence of what appears to be relativistic beaming—the helical emission of synchrotron radiation—in more than 30 sources. However, relativistic beaming alone fails to explain the uniformity and stability observed. Furthermore, at least one source (*3C 273*) has an unexplained excess of X-ray emission.

When it comes to X-ray verses optical, and X-ray verses radio emissions, the correlations indicate radio-loud quasars and superluminal radio sources are similar.[771] A common structure that accelerates particles in the form of a helix exists in what might otherwise appear to be completely different objects. Likewise, a helical beam of electrons and protons produce the auroras of the planets.

Compton emission is a theoretical concept used to describe the interaction between a light-particle, or photon, and a charged particle, such as an electron, in which some of the photon's energy is transferred to the particle. Synchrotron-Compton emission is an undefined and somewhat random (*incoherent, random-pitch-angle*) synchrotron radiation in a magnetic field. The evidence simultaneously supports and rejects the Self-Compton hypothesis, leaving other interpretations open.

A theoretical problem surfaces when discrepant timescales of the variability in the (*mm to far*) infrared and X-ray are considered. The

Self-Compton hypothesis is a "highly unlikely mechanism for the X-ray emission mechanism."[497]; p.292. Accepted theory fails to explain the observations of what is readily explained by the Field's acceleration processes.

X-rays are observed in most sources, and the bulk relativistic motion due to acceleration is common in core-dominated radio sources.[766] Luminosity results mostly from X-rays or soft gamma-rays. Seyferts, more commonly known as spiral galaxies, emit hard X-rays. QSOs (quasi-stellar-objects), also known as quasars, discharge faint X-rays. Blazars are soft X-ray sources. The nearest radio galaxy (*Cen-A*) displays variable X-ray and gamma-ray emission. Our galaxy radiates a weak X-ray source. Starburst galaxies are strong X-ray sources.[264] No clear picture of the mechanism has emerged, but because the gamma-ray and X-ray variability is on short timescales the mechanism must be highly efficient.[93] This predominance of X-ray emission means that a common acceleration mechanism is operating in these various objects.

Other observations also uncover the characteristics of the Fields' time-varying acceleration mechanism. Theoretically, any field that emits X-rays can also absorb them. For example, quasars absorb X-rays, as well as emit them. Low-luminosity, active galaxies have large X-ray absorbing columns, while high luminosity ones do not. The size of the X-ray emitting region, which increases along with X-ray brightness, is smaller in high-luminosity galaxies.[777] All objects seem to emit and absorb X-rays, which is typical of the Field with its time-varying acceleration.

Variability rarely reaches less than a day, while variability is one day in four quasars, and six to eighteen months in nine of twelve objects.[777] Seyfert flares show a rapid component of tens of days, a second component of several years, and a third component of decades. This is also true of quasars and Blazars in Seyfert nuclei.[490] The Fields are time-varying, linear accelerators, which produce X-rays, and are obviously variable.

Magnetic fields along the jet axis are parallel and then become perpendicular in the outlying parts of the jets. A helical structure keeps the jet together as it bends.[155] Quasars have components that show motion along a twisting, helical or straight path.[748] Jets (*such as NGC 6251*) display more stability than expected. The particles must be re-accelerated and/or have magnetic field amplification over great distances from the core (*many tens of Kpc*). A helical field fits the data and density, while the temperature and total mass of the hot gas expose the fact that it had undergone confinement.[136]

AGN, which are crudely spherical, like the cores of the planets, have emission lines that extend far and wide. Violent, dramatic motions are evident. Warm, ionized plasma fills the central region. Strong emissions (*Fe II emission relative to all other broad lines*) reveal that an additional heating source must be present (*other than photoionization*). The line-emitting material is organized in filaments or clouds, which could be preserved and/or created by confinement in the core (AGN). High temperatures originate from non-radiative processes, such as compression shock waves or an earlier passage through a zone of strong radiative heating.[446,447] AGN are the core of the Field-dynamical Model whose fields accelerate plasma and charged particles toward and/or away from the core, where strong radiative heating results from confinement and thermonuclear fusion.

The strong shocks in the central region of AGN are due to particle acceleration and confinement. Nonthermal halos around AGN provide a strong argument for very efficient proton acceleration and proton-photon injection (*of ultrarelativistic pairs*). A substantial fraction of the radiation from AGN is due to an acceleration process that does not only result from heated plasma (*i.e., nonthermal ultrarelativistic electrons, and relativistic or ultrarelativistic protons*). A significant flux of relativistic neutrons, typical of thermonuclear fusion, can stream out of the central region of AGN, as well.[666,670]

Elliptical galaxies are no different in overall dynamics. They show that points of equal light intensity twist from a given source (*isophote*). Also they have slow rotation, shells, and are not perfectly elliptical, but show edge-on disk components. Strong changes in character take place from the inner to the outer regions with a shallow velocity gradient in the outer regions, and a much steeper gradient in the core (as in NGC 4494). Or a very small rotation is noted in the outer region, but a strong rotation in the interior (as in NGC 3608).[390] This arrangement is ideal for the acceleration towards or away from the core, and confinement in the core or nucleus.

The reversal of rotation direction in isolated galaxies is difficult to explain conventionally (as in NGC 3608). One galaxy (NGC 4494) shows strong evidence that the central engine is not a massive object, such as a massive black hole. Furthermore, no evidence exists (*photometric or otherwise*) of the aggregation of smaller bodies, which is referred to as an accretion event.[390] Accretion is the gravitational collapse of matter onto the proposed massive black hole.

Accelerated electrons in jets and the magnetic field that issues radially from Blazars is not really all that different either. For this structure the magnetic field has to be "remarkably" uniform.[682] Bundles or "clouds" are re-accelerated (time-varying) in AGN, including quasars.[168] Again, parallels exist in linear accelerators that accelerate bundles of particles.

Acceleration processes (on site synchrotron) are seen in the AGN polarization of Blazars, quasars, jets, Seyferts, and radio galaxies.[498] X-rays and gamma-rays emanate from Seyferts, quasars, and Blazars.[264] The presence of gamma-rays on the same timescale as X-rays and associated with AGN indicates a highly efficient, "unknown" acceleration mechanism.[93]

Quasars' field dynamics are revealed by the fact that they have a funnel shape, often referred to as a "nozzle formation," or "cone." Reinforcing this interpretation are the relativistic jets, and the nonthermal emission. Quasars emit infrared to X-ray, even thought some are generally radio quiet (at least from our observational viewpoint).[261] The vortex structure of quasars requires a differential rotation of plasma that is accelerated along the same line, much like the magnetic field reconnection required to produce solar flares. Scientists concede that solar flares and auroral processes are remarkably similar, and are also like galactic jets.[146,593]

The Broad Absorption Line of quasars could be an ionized gas that is ejected in a cone oriented close to our line of sight. This can also mean that the variability in quasars can be due to beamed ionizing radiation; an idea which has not been investigated theoretically.[150] However, ionized gases in a cone-shaped field, and beamed ionizing radiation are expected from the Field-dynamical Model.

Much of what is causing the misinterpretation is the mind-set involved when observing the phenomena produced by something like a magnetosphere. These phenomena are claimed to be evidence of a massive black hole. As the problems encountered with the other magnetosphere models, too much is trying to be explained by a dense rotating body (the massive black hole) producing a dynamo-generated dipole magnetic field. This theory has continued in spite of very little success in explaining all of the features observed.[487]

Magnetospheres include a rotating conductor, a magnetic field, and a current carrying plasma. A vacuum region would allow the acceleration of particles by electrostatic fields. Scientists admit that the observations have counterparts in the Earth's, Jupiter's and other planets' magnetospheres, the solar wind, and winds from (Ap) stars.[593] As we have seen with the Earth and other planets, and in later sections on the Sun (Chapter 13) and stars (Chapter 10), the answer lies in a Field-dynamical Model.

5.3 Blazars or BL Lac Objects

Blazars or BL Lac objects are extremely energetic, violently variable extragalactic objects that resemble quasars, but lack emission and absorption spectra. All are radio sources. Light and radio waves are strongly polarized, indicating intense magnetic fields with rapid variations in both strength and direction (time-varying acceleration). The lack of spectral features may mean that the chemistry of the material is so extreme that it cannot form spectral lines, or a field transcends our theoretical bias, instrumentation, and the presently accepted laws of physics.

Blazars show a bulk, relativistic acceleration towards the observer, and are time-varying radio sources (periods are on the order of hours to weeks). Some have superluminal characteristics.[62] The very-compact (*infrared-to-millimeter*) spectrum can be explained in terms of synchrotron radiation from a constant-temperature (*isothermal*) relativistic jet aligned close to our line of sight. Isothermal means that radiative and expansion losses are replaced by a *re*-acceleration mechanism, indicating time-varying acceleration.[295]

Emission lines come from a region about 10,000 times larger than the variable core, which results from the funnel-shape of the Field. The continuum of Blazars' properties indicate that there is an extremely compact, active, synchrotron and radio component. At least 70% of the known superluminals are identified with Blazars, which discloses that our line of sight has much to do with the observed properties.[379]

One Blazar (*PKS 2155-304*) is variable at all frequencies, with the variability increasing with increasing energy. The differences in the timescale of the variability strongly suggest that the dimensions of the source(s) responsible for the various emissions are different. The X-rays emanate from a substantially more compact source than the radio or optical emissions. This rules out the typical explanation that is embodied in the massive black hole scenario (*relativistic beaming*), but not a field producing a helical emission.[714]

The detection of Blazars is intimately connected with compact radio sources. "It is possible that *every* compact radio source will eventually be found to have both apparent superluminal motion and

Blazar properties."[379]; p.234. Again, the observations reveal a uniformity of structure for what might otherwise be considered dissimilar objects.

Blazars are the best example of the AGN power source with continuous ultraviolet through radio spectra. Most Blazars have beamed emission, and the most luminous show the highest polarization and the most variable polarization position angles. Of all AGN, Blazars are most likely to reveal the true nature of the central engine. Fluctuations of 100% per day are not unusual for Blazars, which indicates a very compact emission region (*i.e., no more than a light day*).[132] The observations indicate that Blazars are objects which are viewed down a Field towards the core (AGN) in the Field-dynamical Model.

5.4 Spirals and Disks

A flow of material moves into and out of the nuclear regions of barred spiral galaxies. An interaction exists between the nucleus and the associated galactic disk, which can be considered a ring. Particularly frequent is the presence of faint (*annular*) structures that form a ring.[67,587] Like planetary rings, galaxies have ring-like structures.

Spiral galaxies have linear radio emissions with some features present in their disks.[434] Predictable for the Field-dynamical Model, one Spiral Galaxy (MR 2251-178) has the disk structure inclined about 40° (features extend from 30° to 50°). When not equatorial the structural dynamics would produce this angle (similar to Neptune's rings). Spiral and Seyfert galaxies' material flows into the active nuclei.[587] Similar to other spiral galaxies, a radial outflow of hydrogen takes place from parts of the Milky Way, along with star formation and supernovas.[682]

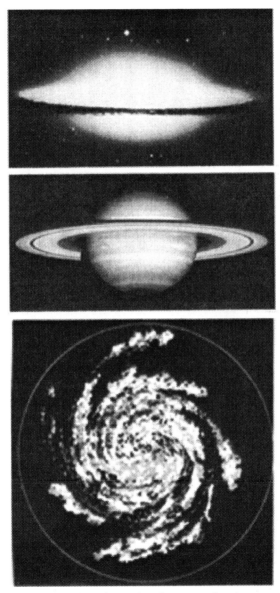

Plate 4.6. Similarities of Celestial Objects. The image of the galaxy NGC 4594 (top) [ESO] is quite similar to the image of Saturn (middle)

[NASA]. This spiral pattern (bottom) is a computer simulation of the stars in a spiral galaxy [by permission of Philip Seiden, IBM Watson Center]. Note how this spiral pattern is also noted in Plate 4.2, Mars polar plates, and in Figure 5.1, the Topknot of the Sun. This pattern is also noted in the polar region of Halley's Comet, sunspots, and much more.

The existence of spiral forms in a large fraction of all galaxies has been known for a long time. How such structure is generated and maintained over many periods of rotation is one of the great unsolved problems facing scientists. The standard explanation is that it is generated by spiral density waves, which is a gravitational theory. However, as with planetary rings this purely gravitational theory does not explain all of the features observed. In fact, the theory requires that a great deal of matter exists that cannot be seen (see section 5.8). The satellite galaxies at the ends of spiral arms (*as in M51*) also remain an unexplained feature, unless we consider ejections from galaxies (see Chapter Six). Another is the existence of magnetic fields in the spiral arms.[687] The excess of S-shaped spirals over anti-S-shaped (Z-shaped) spirals has no explanation in the gravitational perspective, either.[151]

Rotation curves for spirals generally rise very steeply within a certain distance from the center (*a few kilparsecs*), then flatten, and remain at an almost constant velocity. This is rather startling in the gravitational framework, because the luminosity in galaxies is falling rapidly at large distances. If the light and mass were distributed similarly, then the rotational velocity should fall off (*as defined by Kepler's laws*). Only a few galaxies display the expected falloff, and therefore, as with planetary rings, some confining mechanism has to be present.

Luminosities of both spiral and elliptical galaxies are correlated with their internal motions, particularly rotational velocity. If the spiral pattern was merely tied to matter distribution, differential rotation would destroy the pattern in only a few rotational periods, but this does not happen.[363] These observations encourage the idea that the pattern results from ejection and electromagnetic confinement, not gravity.

A careful analysis of the generally accepted spiral density wave theory indicates that it cannot be correct. In order to fill the void left by these theories it is necessary to find new theories, and plasma theories offer an answer. Electric fields exist parallel to magnetic fields in space and produce double or more complex layers of plasma (as is probably the case in superluminals). One of the first indications of electric double layers in space was the discovery of an unexpected electron distribution, indicating an electrostatic acceleration in space. From that point on, much in the way of evidence has accumulated that electric fields parallel to magnetic fields exist in space.[13] The disks and spirals of galaxies are quite similar to the rings around planets, and show that the Field-dynamical Model is also active in galaxy dynamics, formation and evolution (see Plate 4.6).

5.5 <u>Black Holes</u>?

A black hole is theorized to be an astrophysical body whose surface gravity is so strong that nothing can escape from within its grasp. On another level, according to the special theory of relativity, nothing can travel faster than the speed of light. In black holes, light is not capable of escape, and therefore, all other material must be trapped, as well. The surface of a black hole is like a membrane where material may fall into it, but no information or energy can come out. The mass of our Sun packed together to about the size of an atom would be similar to a massive black hole, though on a much smaller scale. In theory, a black hole could continue to collapse into what is called a singularity, a body of zero radius and infinite density.

The equations of relativity also allow for the existence of white holes in which matter flows into another universe or time and/or place (space-time) through what is referred to as a wormhole.[547] Some astrophysicists have suggested that quasars are white holes in which matter from another universe or space-time flows into our own. Another possibility is that what appears to be a black hole is a field system which transcends our present understanding of physics (i.e., the Unified Field).

Blazars are the best example of the active galactic nucleus' (AGN) power source. It is not unusual for continuous emissions, from ultraviolet through radio, to undergo changes of 100% per day, which only *suggests* a massive black hole.[132] Notwithstanding, massive black holes offer many unresolved theoretical problems, as an astrophysicist comments: "The preferred scenario at present is a massive black hole whose origin is never really worked out, and an accretion disk."[149; p.370]. This is the state of this theory after three decades of attempts at affirming it.

It was about thirty years ago that the understanding had been reached that the energy released in the nuclear regions of galaxies and quasars is either due to gravitational energy, or energy released from some creation process.[149,360] The latter proposal—a creation process—has been largely ignored on the grounds that "new physics" should not be considered until all the possibilities associated with conventional (gravitational) physics have been exhausted.[149] Yet, after about three decades we find this statement: "In summary, it seems to me that the idea that a black hole or an accretion disk actually gives rise to what we see is fraught with serious physical problems which have not been solved, but have simply been ignored."[149; p.371]

Massive black holes may be capable of producing thick disks and radio jets without evoking new physics, so theories have focused on this scenario, but without any real success. Likewise, it is claimed that the acceleration in jets may be due to radiation pressure or electrodynamic processes, resulting from a massive black hole.[169,487] The mass flowing into a black hole, if it exists, *may* have a rotational energy that ejects a radio-emitting flow of plasma in the form of a helix.[625] This is the present theoretical framework, which fails to explain the observations fully.

An event in a Seyfert or spiral galaxy (NGC 5548; *Type Sa*) displayed large amounts of helium that have been assumed to be a star being engulfed in a gravitational (accretion) event.[589] However, this galaxy is known to vary dramatically on a short timescale (by a factor of 2 or 3 in a few weeks). Highly ionized gas responds quickly to the surges of the central source, while less ionized gas does not. Such an observation suggests that the clouds have a complicated structure with the highly ionized gas in the center, and the less ionized gas in a ring or disk.[182,356,588] At best the evidence is only circumstantial, and other possibilities have been largely ignored.

Space-time curvature is no longer needed to explain the force of gravity, and therefore, massive black holes need not exist, even at a fundamental level. Rather a field energy exists, possibly relayed by particles (*e.g., gravitons*), that bring about the force of gravity by the exchange of momentum due to a time-varying linear accelerator. Furthermore, a massive black hole seems to be an impossibility, because in the stage where matter is attracted to a dense body it will lose its mass (i.e., its gravitational pull) as it races in at the speed of light. As a result, the central body may not be able to accumulate the extra mass to produce a super massive black hole. The observed scenario, however, has close similarities with the effects of an electromagnetic field. Furthermore, jets and spiral arms of galaxies are ejection phenomena, thereby making the accumulation of mass and the proposed super massiveness all that much more unrealistic.[66,67] A massive black hole is not required to produce all of these observations, while an electromagnetic vortex or helix, the Field in a Field-dynamical Model, could readily produce what is observed.

If a black hole grows beyond a certain size its surrounding disk of matter becomes super heated and radiates energy so intensely that it blows away any additional incoming matter. This outflow, in the form of jets, prevents the black hole from gaining more mass. A shell of luminous gas was observed circulating around two galaxies at speeds and patterns that suggest a black hole about 100 times more massive than this limit. The observation was made while studying the whirling disks of gas emanating from one galaxy (*NGC 6240*).[119] This observation goes against all theoretical constraints on the size limit of the super massive black hole, and requires a totally different model that allows for these motions without such massiveness.

A rotating black hole can be produced by a strong magnetic field. Therefore, the existence of a large-scale external magnetic field can be related to plasma acceleration. Most astrophysicists assume that the magnetic field is the product of a massive black hole generating a magnetic field through dynamo processes. Meanwhile, magnetic vortex tubes can accelerate charged particles.[146] Furthermore, dynamos only reinforce existing fields, they do not create the initial fields.[571]

The supposed massive black hole at the Milky Way's center is time-varying. It was observed in 1977 in gamma rays for two years, and then switched off. A month later it turned on again. A vast river of gas was observed funneling toward the galaxy's core. Also, about 0.7° off the center were lower energy gamma-rays and X-rays.[357] How does a massive black hole turn its effects on and off? How does a massive object produce energy that is off center? The gravitational physics of massive black holes find no truly convincing explanation for these and other questions.

It is difficult to see how a luminous quasar with the presumed high mass and accretion rate could evolve into a low-luminosity object, such as a Seyfert galaxy, yet this is being suggested. Meanwhile, "the existence of rapid gas motions does not necessarily indicate the presence of a massive black hole."[683]; p.18. The existence of dust and gas, including shocked molecular hydrogen, around the outside of a cavity (*of 2pc radius centered on IRS16*) may indicate a powerful outflow. This would then indicate that gas motions are non-gravitational.

Many estimates vary for the size of the central black hole in a given AGN. The massive black hole is not seen, but inferred from speculation and circumstantial evidence. Yet, understanding the data is where artistry enters astronomy, especially because all samples are arbitrary and approximate to some degree.[683] It seems that our artistry is too bound up by the misperception that strong gravity has to be responsible for what we observe.

Estimates of massive black holes are getting smaller and smaller in some cases. There appears to be strong evidence for massive black holes in M31 and M32, and less so, M87. The crucial evidence comes in the form of rapid changes in luminosity of approximately one day. Thermonuclear processes could supply the energy output, but some scientists suggest that about 10,000,000,000 solar masses are required for quasars. Meanwhile, M31, M32 and M87 are the best studied cases, and there are still conflicting theories involving the existence of a massive black hole.[233] In contrast to standard theories, these objects display a great deal of evidence for ejection phenomena instead.[66-68,72]

M87 is a giant elliptical galaxy located near the core of the Virgo Cluster, and it has a nucleus that is a radio and X-ray source with optical jets emanating from it. The light in the central spike is stellar, rather than nonthermal, and a star cluster appears to inhabit the core. Yet, the findings tells us that it is an educated guess: "*A central black hole is not required by the available data.*"[233]; p.225. A rising velocity distribution in the core levels out with no significant rotation, and a central spike is dominated by nonthermal light.[233] Again, the existence of a massive black hole does not explain the data.

Astrophysicists have not been able to construct (mass to light ratio) models that are consistent for either M31 or M32. In M32 there is a rapid rotation, a central light spike composed of stars, and a rapid rise in velocity distribution. Alternative theories to the super massive black hole in M31 is a stellar bar (*dimension about 1pc*) that points along our line of sight, or the center of M31 could be disk-like. These ideas, however, are considered "very exotic," mainly because they are not massive black hole theories. Both M31 and M32 have little or no activity in the nucleus, and therefore, must have an efficient cleaning mechanism to prevent the proposed black hole from capturing the gas lost from evolving stars. The alternative that is often scoffed at is that a mechanism exists other than a massive black hole.[233] This non-massive black hole option is suggested by studies of the morphology of ionized gas in M31.[176]

There are processes with enormous energy in the centers of galaxies. The nuclei of galaxies look brighter than their outer regions in all ranges of the electromagnetic spectrum. This suggests to some astrophysicists that a massive black hole is present and compels stars to move around it, tearing stars apart and swallowing them whole. An alternate explanation, that the nuclei are confinement chambers, would also explain this brightness.

One galaxy (NGC 6251), for example, has a jet of matter emerging from the nucleus. In addition, a jet axis lies along the same line as the axis of the two radio-emitting lobes outside the galaxy's main body. This structural arrangement suggests that they are produced by one and the same phenomenon that "pumps" matter along that line (i.e., time-varying acceleration along a linear path; i.e., a time-varying linear accelerator).[612]

The evidence for massive black holes comes in the form of X-ray variability in a minority of quasars and Seyferts, central light cusps, stability in radio axes, and superluminal expansion. Regardless, these observations are only indirect evidence for "spinning, relativistically deep potential wells."[116] This scenario does not demand that a massive black hole exists. Observations could be due to electromagnetic and electrostatic fields accelerating plasma, gas, and stars.

Comments by astrophysicists reveal the failure of present theory, which has been worked on for about three decades: "How AGN form and evolve is still, in many respects, mysterious."[613] "The notion that active galaxies are powered by accretion onto a massive black hole is almost as old as the discovery of quasars. As is well known, the observational evidence in favor of this view is circumstantial and not even by the laxest standards of scientific proof can we claim to have *demonstrated* the existence of a black hole within the nucleus of any galaxy (or indeed anywhere else)."[116]; p.282

Furthermore, strong gravity is an untested assumption, "Thus, this argument for black holes is unsatisfyingly circular, because their existence depends on the untested validity of *strong-field* general relativity and, at the same time, the acid test of general relativity is the existence of black holes."[502]; p.(2)191

The proof of general relativity does not necessarily require the existence of *massive* black holes, an acceleration mechanism, known as a white hole or wormhole, will satisfy the equations. However, the above quote shows the unwise perspective of present circular reasoning on the subject in order to leave current, gravitational physics intact. Peculiar (*temporal or broadband*) spectral phenomena have been considered, which could result from an acceleration process, but these phenomena are obscured by the (*magnetohydrodynamic*) effects of gas motion.[502]

Many features of quasars, the beginning stages of galaxy formation, remain enigmatic, and it is hard to fit them into a pattern. Many say it is not known why most quasars peak at a certain redshift ($z = 2$), while others are at another redshift ($z = 3$). Yet, why not at other redshifts? Furthermore, detailed modeling is very controversial. Another possibility is that it is thermonuclear energy, though few subscribe to that theory. Yet, the Field-dynamical Model produces thermonuclear energy in the core.

The orbits of stars pass near the centers of nearby galaxies causing an enhanced concentration of stars around the center. The giant elliptical, M87, has a similar distribution which is believed to be non-stellar. M87 is a radio source and has a well-defined jet with efficient acceleration that varies with distance from the nucleus and cannot be solely accounted for by synchrotron emission, as is theorized.[316]

For M31, velocities rise toward the center and the flattened stellar system is rotating. The inner most region is misaligned with the overall major axis and the light distribution is not symmetrical around the dynamical center. These observations are not what could be expected of a massive black hole. Notwithstanding, M31 is a good candidate for a massive black hole, and a number of articles even claim that such exists.[613]

Other examples indicate the need for new theories. NGC 4494 shows strong evidence that the central engine or black hole is not a massive object, such as a neutron star. Furthermore, there is no evidence (*photometric or otherwise*) of an accretion event.[390] Likewise, prototypical quasars (such as 3C 48) display nebulosity around the nucleus that is measured at about the same redshift as the nucleus. Therefore, no gravitational fields can be responsible, because the gradients in these fields would give different redshifts for different parts of the galaxy if this were the case.[66]

A gigantic, exceptionally bright star, which scientists thought could become a black hole, is shedding mass at such an staggering rate that it will eventually disappear. If such massive stars are losing so much mass, they will not form black holes. This discovery casts doubt on both stellar and black hole theories.

Alternatives cannot be excluded, though massive black holes are a natural inference from standard gravitational theory. If the stars where not closely packed near the center of the galaxy, there would be no evidence of a massive black hole. Meanwhile, field structure could just as easily account for these closely packed stars near the center. No observations have been unequivocally made of the sudden release of gas as a star is tidally disrupted by the proposed massive black hole. For example, the predicted luminosity is not observed in M31, though it is a prime candidate for a massive black hole, and some claim that a massive black hole is present.[613]

Recently, an eruption of energy equal to about a million years of sunlight erupted from the center of a distant galaxy. The energy was in the form of X-rays that came from a "common" quasar, known as PKS

0558-504. The Ginga satellite had been observing this quasar for months when a sudden and powerful eruption occurred. This eruption was interpreted as a large blob of matter that was sucked toward a massive black hole, but this explanation is merely an attempt to support predetermined theory.[617] X-rays are routinely produced by linear accelerators, and by high-energy charged particles colliding with other charged particles. This observation only reveals the acceleration of matter away form and/or into the core, and any other explanation is a conventional, gravitational interpretation of *circumstantial* evidence.

Observations exist that would cast considerable doubt that AGN are powered by massive black holes.[116] One is a well-defined rotational period or regular time-varying characteristic. This may be obscured by interaction with a system of interstellar or intergalactic fields, and/or our line of sight. However, NGC 6814 displays regular time-varying characteristics that include the regular recurrence of flares (*12,000 seconds*) observed in medium X-rays, a dip in medium X-ray (*by a factor of 2 in about 300 seconds*), and variable low-energy absorption in the X-ray spectrum on a timescale of months. This object is considered "unusual," but this Seyfert is almost face-on from our viewpoint, and all may display similar variability if viewed at this angle.[134]

The Seyfert, NGC 4151, has time-varying phenomena (*with a period of about 131 seconds and a few thousand seconds*) with a spectrum that becomes softer at the same time.[276] No explanation exists for this uniformity of two components in the massive black hole scenario. This object and NGC 6814 are only considered unusual because they provide evidence against the preconceived notion that massive black holes produce what is observed.

Another blow to the massive black hole scenario is proof that a large class of quasars are genuinely quasi-stellar, and not surrounded by galaxies or protogalaxies. This has been demonstrated particularly by one astrophysicist in what has been termed the "redshift controversy," where some redshifts are non-cosmological (this will be discussed in

Chapter 6). Furthermore, superluminals may be the result of two integrated components that are simultaneously expanding and contracting, or the object is moving in one direction and accelerating material along that line at the same time or in the opposite direction.

Determining that extended radio sources are contracting, while it has been assumed that extended radio sources are expanding, is another blow to the scenario. Spiral galaxies appear to have matter flowing both in and out of their AGN. There may well be two components, one contracting and the other expanding. Furthermore, if some redshifts are not indicators of distance, for which there is sufficient evidence, then some sources are not expanding when they have been claimed to expand (see Chapter 6).

Less central mass than indicated by theory (about 10^6 solar masses) is another observation that would cast doubt. However, estimates of the central mass are derived from circumstantial evidence. The underlying problem with finding new interpretations is apparent in this statement: "Does a central body with high mass-to-light ratio have to be a black hole? This is not self-evident."[613] Many objects do not have the predicted central mass or far exceed the proposed limit.

Furthermore, protons must be accelerated with high efficiency, and a significant flux of relativistic neutrons stream out of the central regions of AGN.[670] This scenario means that mass loss or creation is taking place, and in a way not explainable by the conventional massive black hole and accretion scenario (*relativistic beaming or radiation pressure*). There is also sufficient evidence of ejection phenomena in galaxies.[64-72,152,388,421,546,581,587,606,702]

Galaxies display a remarkable uniformity of dynamics and structure. Furthermore, the different types may very well be similar objects, but at different stages of galaxy evolution, or simply viewed from different lines of sight. Can we really continue to foster a theory that has, for more than three decades, failed to produce any verifiable proof of its existence that is not merely circumstantial? How can one object be considered the result of a massive black hole while another object with a similar structure be proven *not* to be the product of such? Clearly, there is a need for new ideas, if we are open to them.

5.6 Unified Mechanisms

Massive black holes are the most promising conventional theory to explain AGN, including quasars, and other related objects. Yet, the theory is unable to explain all of the observations. However, the presence of an electromagnetic field is more realistic in terms of supporting the evidence.[258] Optical and infrared polarization of AGN could be due to a structured component that accelerates electrons, other charged particles, or organic dust grains.

Blazars have the strongest polarization. About 3% of quasars have high polarization. Two of thirteen broad-line radio galaxies have high polarization. In Seyferts, such as NGC 1275, there is high polarization.[498] This preponderance of high polarization suggests that a similar mechanism is being seen at different viewing angles or stages of galaxy evolution.

A quasar, such as 3C 273, is the archetype of superluminals. Observations of 3C 273 could be explained by relativistic expansion of a double structure, and equally well by intensity variations (time-varying acceleration) of stationary components in a more complex structure.[780] This scenario requires a field that accelerates particles *very* efficiently.

Galactic nuclear star formation complexes, as in NGC 5253, NGC 7469 and M82, indicate that a remarkably uniform magnetic field is required for the helical (synchrotron) emission.[682] Present (*synchrotron relativistic beaming*) models are known to be insufficient and other mechanisms are required.[92] Likewise, one-sided radio jets are always on the same side of the nucleus, suggesting the same mechanism, while this observation departs from massive black hole models.[399] Meanwhile, as could be anticipated from the new model's system of Fields, X-ray halos around early-type galaxies "seem to require an external confining medium."[318]

Viewing angles can obscure common structure. For example, quasars probably do not expose their active nucleus when seen edge on, and are less symmetrical and bent than radio galaxies. Apparent bends of 90° produce large variations in brightness. Quasars could be radio galaxies seen end-on, while most or all radio galaxies are quasars that are not seen pole-on.[644]

What appears to be different types of Seyfert or spiral galaxies may only be different viewing angles. Type 1 Seyfert galaxies could be viewed along the axis. Type 2 Seyfert galaxies could be viewed along the equatorial plane.[447] This is evidently true for all of the various galaxies.

The parameters of active sources strongly suggests a common mechanism. Outbursts are due to time-varying particle and plasma acceleration. Polarization is characteristic, and large jumps in position angle between 70° and 90° is typical of the Field-dynamical Model. An astrophysicist states the inevitable: "There is some unifying, regulating process."[623]

Blazars disclosed that all event components were ejected at a very similar position angle of just over 180°, which implies that there is a very narrow "nozzle" (field) attached to the central engine. An extended structure with very weak emission was observed. The ejecta are driven by a hot "piston"—time-varying bursts—that compresses what appears to be an initially random (*ambient*) magnetic field. Blazars are the only example of deceleration components that are possibly due to collision with an intergalactic magnetic wall.[539] The large-scale structure of the Universe can also obscure the structural uniformity of the various types of galaxies.

A lower luminosity field is parallel to the axis of the galaxy and then flips to perpendicular. As exists in the Field-dynamical Model, this observation can be easily explained by a helical field, but other interpretations are lacking.[102,155,279] The magnetic field is always parallel to the flow with very complicated electromagnetic processes involved in the expansion.[585] Evidently, this also involves an electric field parallel to the magnetic field, and the acceleration of plasma.[13]

Radio galaxies generate enormous amounts of energy that orthodox theory is at a loss to explain, when considering a mechanism. It is unknown how the energy reaches from the central parent galaxy to the lobes hundreds of thousands of light-years away. Again, this must involve parallel magnetic and electric fields, which have long-range force, interacting with plasma.

In addition to gravity there must have been hydrodynamic and electromagnetic effects, or there would not be such order.[11] One mechanism can account for the findings and is anticipated: "It can be hoped that, perhaps by a process of elimination, one general mechanism will be shown to operate."[117]; p.326. The mechanism is the Field-dynamical Model.

Highly polarized BL Lac objects and somewhat similar QVV quasars, known collectively as Blazars, may be due to a relativistic jet seen end on. The difference between radio-noisy and radio-quiet objects, which appear remarkably similar in other respects, could be simply due to the environment.[570] Common structural (field) dynamics exist for all objects.

Seyfert or spiral galaxies display time-varying nuclei in all emission lines from radio to X-ray. Mass accelerating toward the core has been observed.[739] Seyfert galaxies are spirals, but broad-line radio galaxies, which are ellipticals, have optical properties much like Seyfert 1's. Narrow-line radio galaxies are very much like Seyfert 2's (*Liners*). Most groups of galaxies show definite evidence for nonthermal phenomena, either as radio emission, hard X-rays and gamma-rays, or optical (*a quasi-nonthermal phenomenon of a power law continuum*).[570] "The different types of activity are closely related, presumably to some sort of perturbation that sends down material to feed the monster lying in wait in the nucleus."[570]

Astrophysical effects of strong magnetic fields, whether created by gravitational collapse or not, can explain a wide variety of observations. Strong magnetic fields (10^8 to 10^{14} *gauss or more*) at the core of AGN, including quasars, in combination with rotational energy, determines the general appearance (morphology). The ratio of magnetic field energy to rotational energy is highest for quasars and Blazars. The ratio then decreases as one goes down to radio galaxies, Seyfert spirals, Markarians (most are spirals), and ordinary spirals, until it is near zero in ellipticals.[327] This observation suggests that an evolutionary sequence exists that begins with quasars and/or Blazars, evolves to spirals, and ends with ellipticals.

The excellent agreement between the oxygen (*O III*) emission and radio axes suggest the beaming of ionizing radiation along the disk's rotational axis. The hydrogen (*-alpha*) emission and radio axes are also strongly correlated, but not as well. A hydrogen (*-beta*) velocity field appears to be dominated by rotational motions. For Seyfert galaxies with double, triple or jet-like radio sources, there is a close correspondence between the radio ejecta and the high-excitation, high-velocity gas (*of the Narrow-Line Region; NLR*). Circum-nuclear star formation and nuclear activity are often found together in active galaxies. However, it is unclear why both should exist together when considering the massive black hole, which stimulates the search for other scenarios.[765] In contrast, all of these observations could be envisioned from an understanding of the Field-dynamical Model.

5.7 Field-dynamical Model and Structure

A Field-dynamical Model of AGN should reveal a number of observable phenomena. There would be heating and nonthermal activity present that are unexplained by massive black hole theory. Evidence of confinement should be exhibited with an excess of hydrogen and helium beyond what would be due to stars alone. Very efficient acceleration and X-ray, radio and gamma-ray emissions would reveal a mechanism beyond what is expected from the proposed massive black hole. One observation supporting this more efficient mechanism would be the existence of gradients in the core when viewed from a particular line of sight. Phenomena in dynamic relation with the core should produce angles of approximately 0°, 30°-40°, 60°, 90° and 180°, as well as within 15° of these angles due to plasma layers and the funnel shape of the Field (or the "cone's half-angle"). These should include mid-latitude, polar and ring phenomena. All of these observations would be analogous to the planets and plasma processes in the auroras.

Common structural dynamics exist in various types of AGN. Star formation complexes in AGN have a remarkably uniform magnetic field for the proposed synchrotron emission.[682] Just as the Fields are in dynamical relationship with the core, observations indicate that AGN or galactic cores are at the base of jets.[497] Common characteristics among the various types of AGN suggest that the same structure is operating, but we are seeing it from different angles or stages of evolution.[117]

Seyfert 2 galaxies appear to be Seyfert 1 galaxies that have expelled the dust from in and around one region (*i.e., the Broad Line Region; BLR*). In order to see this region one needs to look down a narrow cone, and the opening angle of the cone increases with time as the dust content decreases. As a result, dust content (*i.e., opacity*) is one of the physical characteristics that produce the observed variety of AGN.[120] Organic dust or bacteria covering the nuclear region is defused exposing the AGN, which results in luminosity shifts (*from infrared to ultraviolet and optical*).[683] Furthermore, organic dust or bacteria have superconducting properties, and would align in magnetic fields.[359] These processes could account for the decrease of dust through time and explain the variety of AGN as the result of a common mechanism at different evolutionary stages, or different viewing angles.

AGN emission lines extend great distances, and their large widths bespeak of violent, dramatic motions. AGN are crudely spherical, and warm ionized plasma fills the central region. Confinement leads to photoionization, which is responsible for a major portion of the line emission. Moreover, the strength of one (*the Fe II*) emission, relative to all other broad emission lines, indicates that an additional heat source is involved. The additional heat can be explained as the result of confinement and thermonuclear fusion (other than stars). The line-emitting material is organized in filaments or clouds that could be preserved or created by confinement in a field system. Characteristic of thermonuclear fusion, the high temperatures could be due to an earlier passage through a zone of strong radiative heating.[447] All of these observations clearly could be anticipated from the Field-dynamical Model.

Observations of luminous (*IRAS*) galaxies reveal that all are extremely rich in molecular hydrogen, and most are recent galactic mergers or splits, analogous to a biological cell.[120,421] Approximately half of the interstellar matter is contained in the galactic center (*i.e, the central Kpc*). The gas concentration has resulted in the formation of a massive central star cluster.[421, 654] The central star cluster need not be the result of massive black hole gravitation, but instead can be due to confinement. Helium abundance is in uniformly high levels in galaxies so that it is difficult to suppose it merely came from ordinary stars.

X-ray observations of some isolated massive early type galaxies are of the type that are likely to generate powerful radio sources. Hot gaseous halos surround these objects, and an even hotter, low-density, intergalactic mechanism exists.[762] X-ray halos around early-type galaxies are evidence that an external confining medium is present.[318] X-ray halos are the result of a magnetic field in conjunction with parallel electric fields and plasma layers that produce supernovas and heat the halos (this will be discussed).

Elliptical galaxies display twists, slow rotation, shells and are not perfectly elliptical (*isophotes*), but show edge-on disk components. A shallow velocity gradient in the outer region is associated with a much steeper gradient in the core (*as in NGC 4494*). In other galaxies, a very small rotation in the outer core is associated with a strong rotation in the inner core (*as in NGC 3608*). The much steeper gradient in the core is strong evidence that a massive black hole cannot be responsible, and this is why no evidence of an accretion event exists.[390]

The reversal of direction of rotation in the interior of an isolated galaxy (*e.g., NGC 3608*) is difficult, if not impossible, to explain by standard massive black hole theory. It suggests, in the gravitational sense, that either there was an interaction with or ingestion of a smaller companion with the opposite rotation.[390] In cases of high rotation or counter-rotation, and dispersion in the nucleus, it is not always aligned with the major axis of the "host galaxy."[233] Meanwhile, these observations can be more readily explained by a field accelerating two populations of oppositely charged particles, plasma and/or stars, analogous to the opposite spirals of electrons and protons in the aurora.

Strong shocks from the central region could be due to particle interactions resulting from acceleration. For example, the nonthermal halos around AGN provide a strong argument for proton acceleration.[666] Furthermore, the protons must be accelerated with very high efficiency. A significant flux of relativistic neutrons can also stream out of the central regions of AGN.[670] Proton acceleration was noted in planetary aurora, and protons are important for thermonuclear fusion, and neutrons were observed in planetary magnetospheres and are by-products of thermonuclear fusion.

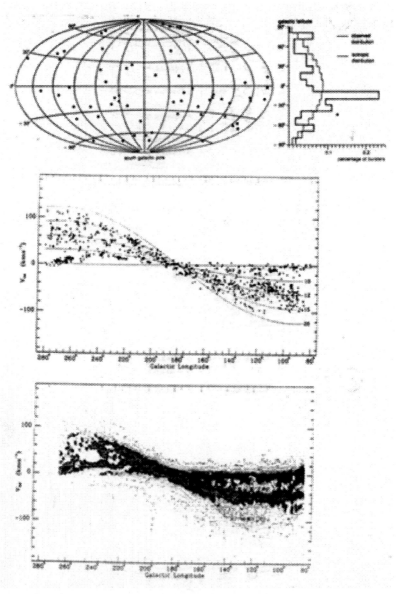

Figure 4.4. Galaxy Features. A feature of GRBs is a significant excess along the 20° to 30° South Latitudes, as shown at the top.[from

Audouze, J., Israel, G. (1988) Cambridge Atlas of Astronomy. NY, Cambridge Univ Press]. Galaxies also display a warp in a sine curve (something like a "S" on its side) [from reference 772]. FEM also displays a sine curve with the arrangement of its Fields, as discussed in Volume One, and we will see that this is also true of the Universe itself.

Angles typical of the Field-dynamical Model of the planets are evident for galaxies and their components. For example, in one source (*3C 345*) a component (*C4*) changed position angle by about 45° (from 135° to 87°) relative to the core as it moved some distance (*0.3 to 0.8 mas*) from the core, which could result from a cone- or funnel-shaped field. Other components show angles of 86° (*C3*), 74° (*C2*), 64° (*C1*) and 31° (*3" Jet*), which are similar to other sources.[111] The peak polarization of Blazars is about 0° and a second peak is about 90°.[379] Observations demonstrate the similarity between objects: (1) there are nonthermal ionized gases at 90° in relation to the radio jet; and (2) a polarization of 90° in relation to the rotational axis (as for example, in the Milky Way).[682] Around some quasars there are large halos or disks (rings) of ionized material with an inclination angle of around 40° (features extend from 30° to 50°).

Also observed are what could be referred to as poles and rings. A lower luminosity field is parallel to the axis near the galaxy, and then flips to perpendicular, which requires a helical field.[279] Some galaxy's jets (*e.g.*, *NGC 1097*) are equatorial, while radio galaxies eject jets along the axis or poles.[57,65] Also interactions exist between the nucleus and the associated galactic disk, which can be considered a ring. Especially frequent are the presence of faint (*annular*) structures which form a ring.[587]

5.8 Missing Mass?

The outer parts of spiral galaxies display rotation curves that do not fit conventional gravitational theory (*i.e.*, *Kepler's Laws*), as well. The rotation curves suggest that there is a large amount of mass, but it emits no light.[109] Yet, such an idea is completely untested, even untestable, which is as far from a scientific perspective as one can get.

Can we really claim to know the Universe when more than 99% of it does not speak to us? Yet, this reasoning is what is being suggested in order to preserve orthodox, gravitational theory! Meanwhile, a magnetic field similar to the Sun's Interplanetary Magnetic Field (IMF), on a much larger scale, would eliminate a need for the missing mass. Once again a gravitational perspective does not explain what is foreseeable of the Field-dynamical Model.

Alternatives to this dark matter hypothesis, abbreviated DMH, have been offered, because the theory has proven to be a big disappointment. A physicist discusses the problem:

"The one thing the DMH can boost of is that it leaves current physics intact. Various routes are open once we question conventional physics. Any link in the chain of arguments leading to the mass discrepancy may be questioned: the velocity-wavelength relation, the exclusive role of gravity in the dynamics of galactic systems (allowing for new interactions), or the validity of Newtonian dynamics (the laws of gravity and inertia)."[526]

Genuinely scientific theories offer predictions that are unavoidable when tested (i.e., they must be verifiable). However, DMH offers no prediction, not even the dark matter will eventually be found. Such an approach is distinctly unscientific and is, in fact, more like a religion, because it requires only faith in the theory, not observable and testable phenomena. Furthermore, a modification of the distance dependence of gravity is inadequate for explaining what appears to be a mass discrepancy.

Meanwhile, a long-range force couples to angular momentum. An external-field-effect (EFE) consists of a micro-dynamic system placed in an external field of dominant acceleration that is affected above the tidal effects. To put that in simpler language, plasma—the micro-dynamic system—is accelerated by the external field—the ring plane of the Field-dynamical Model—which overcomes gravitational effects. The EFE can explain open star clusters, the outskirts of globular clusters, and the workings and fate of dwarf-spheroidal satellites of the galaxies.[526] The Field's curvature of space-time and the associated electrostatic effects of plasma layers produce the EFE, no missing mass is required.

Plasma theory is necessary to understand the structure of spiral galaxies. Confinement of plasma takes place in the nucleus. The emission of arms from a spiral galaxy resembles the streams of solar wind in the Interplanetary Magnetic Field, as well as planetary rings, planets around stars, and moons around planets. Ionized matter and plasma flow along the magnetic field of the spiral arms.[285]

Dark matter in Sb and Sc Spirals does not explain their flat disk rotation. Also the rotational velocities of ellipticals are very difficult to explain even with modifications to Newton's Laws.[724] A physicist who received a Nobel Prize for his theories on plasma dynamics comments: "Furthermore, we need not have some exotic mechanism (like black holes) to produce the ejection. A system of electric currents and the production of torsional hydromagnetic waves may be enough. There is no 'missing mass.'"[13]; p.216

Torsional hydromagnetic waves are hydrogen plasma waves in a twisting or helical motion, something like a flattened electromagnetic tornado. This arrangement is quite similar to the aurora of the planets and is typical of the Field-dynamical Model's equatorial ring. There is no factual support for the claim that galaxies contain the missing mass that adherents to conventional theory are always looking for, but never find.[13] An astrophysicist considering relativistic plasmas "on the safe ground of theory" notes that the high-energy break in the spectra is a major outstanding problem for the theory, and that, "this could be related to the unknown mechanism of electron acceleration."[479]; p.(2)553. Electrons are accelerated along the Field lines of the Field-dynamical Model.

No theory has accounted for the extremely energetic (10^{62} *ergs or more*) violent outbursts from some galaxies and quasars. One suggestion: "Perhaps the cosmic gusher of a white hole is simply the 'other end' of a black hole, recycling material which has been swallowed up somewhere else in the Universe (or in a different universe altogether)."[329]

Spiral galaxies are becoming more widely known as sites of ejection. "Thus it is not inconceivable that spiral arm condensates pre-exist as dense plasma which undergoes gradually weakening confinement. By locating dark matter in molecular clouds therefore, we seem to be led rather directly to the old idea that spiral arms evolve from short-lived jets."[184]; p.345

Therefore, temporal or time-varying confinement in AGN can account for what is thought to be missing mass.[184] A flow of material into the nuclear regions of barred spiral galaxies suggests that this is, in fact, taking place.[587] Seyfert galaxies could be white holes and/or have ejection phenomena that produces the spiral arms and jets.[69,328,329,639]

Some scientists are going as far as to say it could be that "our present theoretical description of gravity—general relativity—is incomplete."[637]; p.282. Present interpretations of physical laws do not explain why spiral galaxies are flat even into the inner regions. The observations require modifications to our interpretation of the theory of general relativity for gravity, which would include the addition of Unified Field theory (i.e., the Field-dynamical Model's curvature of space-time, which includes wormholes).

No evidence for significant mass discrepancies in spiral galaxies exists. While there appear to be no local contradictions to general relativity there may be at large distances from mass concentrations or very weak fields. Those modifications should include a long-range force that is associated with a (*scalar*) field and a cosmological constant.[637] "It is fair that we still don't have an acceptable relativistic theory of stronger gravity on extragalactic scales, but any such theory will probably involve a field in addition to the usual tensor field of general relativity where the effects of this additional field are somehow locally suppressed."[637]; p.290

"Locally suppressed" may not be the best choice of words, while locally obscured would be more to the point. The basic problem with conventional interpretations involves the perspective that gravity is curved space-time when there is no convincing explanation for how this is accomplished. Curved space-time is the result of the Field with its polarization of the vacuum, which leads to mass concentrations and gravity, not the other way around, which is what often dominates the mind-set.

5.9 Superluminals

The *apparent* velocity of superluminals is at least two times the speed of light. This, however, is in conflict with the special theory of relativity, which states that nothing can travel faster than the speed of light. Obviously there is some sort of illusion involved. The unbeamed counterparts of superluminals (*and all core-dominated, flat-spectrum radio sources*) are normal extended double radio galaxies and quasars.[566,579,594,604] Nearly all superluminals' components move outward towards the brightest outer features.[660] All things considered, superluminals are probably those objects that have their axis along our line of sight, while the object is moving at the speed of light, and also accelerating particles or plasma at the speed of light.[352]

These observations are evidently due to parallel electric and magnetic fields moving at relativistic velocities. At the same time these fields accelerate double or more complex plasma layers to relativistic velocities in the same direction. In the case of superluminal expansion the fields and plasma layers may move in opposite directions. The relative motions create the illusion of superluminal motion.

To put this in the form of an analogy, consider this scenario. A car is moving at 60 kilometers per hour on a transparent sphere that is moving in the same direction, also at 60 kilometers per hour. If we were to view this car and sphere from a remote position, it would appear to us that the car was moving at 120 kilometers per hour. However, the driver of the car would see himself traveling at 60 kilometers per hour, and his speedometer would also read that speed. The whole situation involves our relative viewing positions, as do superluminals.

Typical superluminals have a one-sided core-jet morphology, and components that expand and fade as they move away from the core. The ejecta is driven by a "hot piston" that compresses an initially random magnetic field.[539] This is typical of a time-varying linear accelerator that accelerates particle groups or "bundles" with surges of electromagnetic energy (the "hot piston"). The Fields of the Field-dynamical Model are time-varying linear accelerators.

Superluminal motion in the cores of some extragalactic objects may be a consistent, steady rapid motion between components. Some have extended structure on both sides of the core.[602,603] The two or more components produce the illusion of an object moving at speeds faster than the speed of light.

3C 273 can be considered the archetype of superluminal sources. It is a radio source with a bright quasar. Changes in the nucleus (3C 273B) can be explained by relativistic expansion of a simple double structure, or variations in a more complex structure. Observations indicate the existence of at least three dominant components. Changes could be explained by the motion of a single knot away from the nucleus in the direction of an optical and radio jet (3C 273A). The jet is continuous from the core to beyond the limit of the optical jet. The jet curves about 20° (*within the first 10Mas*) from the nucleus and then becomes aligned with another (*arcsecond*) jet. Four superluminal components have been traced in the nucleus (3C 273B). As could be anticipated from the Field-dynamical Model, motions occur along fixed tracks, and are dominated by pressure, or more accurately, acceleration gradients.[619,780] The jet is very narrow, and perpendicular to its width the jet material is compressed by shock waves.[721] The observations clearly indicate the acceleration and ejection of matter.[66] Again observations show a helical emission that is stopped by an external confining medium: "The components may be whirling along a helix or sliding along a straight path until they hit a 'brick wall.' Twisted motion has been convincingly demonstrated in 3C 273 and in 3C 345."[748]; p.529

Observations demonstrate time-varying acceleration is involved in superluminal source characteristics. The quasar, CTA 102, is the fastest superluminal known, and varies at low frequencies with a timescale of a few months. Superluminal time-varying acceleration explains the observations.[748] The rapid increase in size of another source, 3C 454.3, is followed by a period when some features appear nearly stationary. This behavior appears to be more complex than the typical superluminal.[575]

All strong radio sources probably have superluminal motion. Likewise, every known superluminal source has an extended radio emission, which is indicative of a field(s) and the acceleration of particles and plasma. Meanwhile, no obvious radio emission features distinguish superluminals from other general radio sources. This resemblance could mean that superluminals are viewed from a specific angle, and all sources could produce similar observations if viewed from a different angle or during certain points in time (evolution).[145] For example, some originally identified non-superluminals (e.g., 4C 39.25) have shown recent superluminal events.[497,643,660] Radio emission is typical of an acceleration mechanism, and all objects may eventually display superluminal characteristics.

The variety of superluminals are many. The extended triple sources are steep-spectra, extended objects in which the main emission regions straddle the optical galaxy or quasar. Most of these objects have a very weak central emission. Compact steep-spectra sources are characterized by low radio variability and low polarization. Superluminals also have a "core-jet" structure where the core is relatively weak, and most of the emission comes from the steep-spectra jet. These asymmetric, core-jet sources are mostly objects dominated by compact, flat-spectra radio components that coincide with the optical object. They generally have high levels of polarization and are highly variable (*in flux density*). Unresolved and barely resolved sources appear to be more compact versions of the asymmetric core-jet sources, including similar levels of polarization and variability. Compact double sources appear to consist of two well-separated emission regions of comparable brightness and have steep-spectra at high frequencies. They have low polarization and little variability, and are identified with faint galaxies.[580] These observations indicate highly structured time-varying mechanisms viewed at different angles.

3C 395 is unique with a superluminal moving towards two stationary components. It has a highly variable core, and an apparent change in jet direction of about 180°.[671] A change in jet direction such as this is impossible to explain when using the standard models (especially, as a superluminal).

No superluminal compact doubles have been observed. 3C 216 has a large-scale triple structure with a jet that is roughly perpendicular. There is also interaction between the radio source and the galactic environment.[580] However, compact doubles are superluminals, it is only our orientation and the environment that cause what appear to be different radio properties.[352]

Blazars have a very narrow nozzle attached to the core, and an extended structure.[539] At least 70% of superluminal sources are identified with Blazars (BL Lac objects). "It is possible that *every* compact radio source will eventually be found to have both apparent superluminal motion and Blazar properties."[379]; p.234. Observational results are consistent with different viewing angles of a thin disk confining gas motion near an active core. Peak polarization is about 0° and a second is 90° typical of the Field-dynamical Model. The properties of Blazars indicate that there is the dominance of an extremely compact, active, helical component.[379] Superluminal expansion is very common among the brighter sources, as well as compact radio sources.[115,118]

The half-angle of the cone of superluminals must be 20° to 30° (*to eliminate the linear size problem*). This means the position angle defined by the observed motion can be up to 20° to 30° away from the true axis of the cone (i.e., the funnel-shaped Field). The extended structures of quasars display lines that join lobes to a central component at angles between 9° and 13°.[145] This observation means the source could be 20° to 30° away, at about 30° to 40°. The Fields of the planets have a similar half-angle, as phenomena were noted 15° away from the true axis of the Field due to its funnel shape. As shown with latitude phenomena on the planets, there are fields at the 30° to 40° "latitudes" of superluminals

CHAPTER 6

Discordant Redshifts
(Anomalous Redshifts and Ejection in Galaxies)

The Doppler Effect, described in terms of redshifts, derives from an experiment performed in the 1840s. Christian Doppler staged an experiment with a band of musicians on a moving train, while others with perfect pitch stood by and listened. The notes were higher when the train approached and lower as it departed. Also the faster the train traveled the greater the pitch shifted. This understanding has been applied to the light emitted by objects in the Universe. If the object is moving towards us the spectral lines will be blueshifted, while those moving away are redshifted.

All quasars are redshifted. According to accepted interpretations, quasars are believed to be the fastest and farthest objects. However, a number of observations indicate that this velocity and distance interpretation is flawed. In fact, redshifts have never been fully proven (i.e., quantified) to represent what they are believed to portray.[545]

The Doppler Effect involves both wave and particle viewpoints, and therefore, magnetic fields or electrical discharges also play a role. However, when the magnetic field is considered the effects of boundary conditions (i.e., plasma layers, etc.) are often neglected.[11] Magnetic fields themselves are also usually not considered in order to simplify a hypothesis. In short, the Doppler Effect as it is usually interpreted may be more on the level of science fiction in some cases.

These incomplete perspectives could account for the apparent superluminal motions, and what are called discordant or non-cosmological redshifts in some objects. Helical or "handed" fields or materials (*chirowaveguides*) are known to create two different redshifts, even when the object and observer are at rest relative to each other.[150] The Field-dynamical Model accelerates and ejects helical plasma layers, and therefore, could easily produce what are called discordant redshifts (see **Conclusions** for further discussion and references).

6.1 Quasars and Controversies

Quasars have been a mystery since their discovery. Immense energy is packed into a relatively small object that may actually be the energy cores of recently born galaxies. Approximately double the size of our solar system, they emit hundreds of times more energy than an entire galaxy. Quasars appear to move at enormous speeds when the standard interpretation of redshifts is considered. According to relativity theory nothing should travel faster than the speed of light, and therefore, the apparent superluminal—faster than light—speeds of some objects alone brings into question the standard interpretation of redshifts.

Meanwhile, quasars appear to be the result of interactions between galaxies. The apparent distance of quasars may be illusionary, and they could be nearby. In fact, a good deal of evidence demonstrates that redshifts cannot be trusted as indicators of distance when it comes to quasars.[64-72,149,587] In Grand Unified Theory most or all radio galaxies are quasars that are not seen pole-on.[644] Therefore, quasars, and at least some galaxies, are revealing a more complete understanding of AGN and their activity, which includes non-cosmological redshifts, and interstellar and intergalactic fields.

Compact radio sources are located within quasars, which suggests a highly structured acceleration mechanism.[74] Most quasars are known to have "nozzle" formations (i.e., funnel-shaped fields), relativistic jets, and nonthermal and infrared to X-ray emissions.[261] These observations alone suggest that quasars are young objects just beginning confinement and acceleration processes that lead to new galaxies.

Redshifts could be due to the high-velocity helical motion of matter, and therefore, provide evidence that standard interpretations of redshifts are flawed. Theories that galaxies eject luminous material, which causes redshifts of a different character, are more than three decades old, but have never been widely accepted. One astronomer who originated the idea was even ridiculed, and inhibited in his research and publications, and eventually left the United States for Germany in order to continue his work. This was simply because such theories are not conventional. An electromagnetic-field wave "pumping" matter could be responsible for ejection phenomena.[66] This time-varying acceleration effect could produce different redshifts in objects that are near each other, or even within the same object.

Quasars peak near very high redshifts ($z = 2 \text{ to } 3$) in both optical and radio wavelengths. An orthodox hypothesis designed to interpret this fact is that quasars with massive black holes have spewed energy throughout most of the history of the Universe, but are now dying out. Observations suggest, in the conventional sense, that massive black holes are found at low redshifts in quasars and Seyferts, but not normal galaxies. However, the recent discovery of a very young quasar (Cygnus A; 3C 405) discredits this viewpoint totally.

Meanwhile, most galaxies could harbor currently inactive quasars that have been short lived or flare up, and do not have massive black holes. Observations of quasars and infrared galaxies suggest that the majority of quasars are formed through galaxy collisions and/or ejections, and all quasars begin at the ultraluminous-infrared-galaxy phase. In fact, some evidence suggests quasars evolve into the nuclei of Seyfert and radio galaxies.[68,683]

A major problem of present theory is that there is the ad hoc assumption that gas motions are gravitationally driven, as can be seen in this statement. "The existence of rapid gas motions does not necessarily indicate the presence of a massive black hole."[683]; p.18. As in other cases, a gravitational focus when interpreting the observations can lead to an erroneous conclusion.

Pure luminosity evolution models for AGN luminosity have gained popularity. Luminosity evolution states that the rate of fading is independent of the source's brightness, which is measured in magnitude. However, the luminosity function appears to maintain its shape only over a certain higher redshift range ($0.6 < z < 2.2$), but not the lower redshifts.[590] Density evolution indicates the probability that the source that is extinguished is luminosity independent. In other words, it will fade at its own individual rate according to its composition and dynamics. Pure density evolution over predicts the local density of weaker sources, because it is inconsistent with low and high redshift data.[117,118] These observations of luminosity and density evolution alone bring into question the standard interpretation of redshifts.

Quasar formation or evolution could depend on the environment (e.g., a field system) and/or a long-range force associated with other objects. Quasar-quasar clustering exists on smaller scales (*less than 10 Mpc*), and for low redshift ($z < 1.5$). At higher redshifts the clustering seems to be absent on any scale (*from 4 Mpc to 1 Gpc*), inferring strong clustering evolution.[426] These observations indicate quasars form in epochs of ejection (i.e., time-varying acceleration), and also reveal the fact that redshifts are not good indicators of distance, at least in the case of quasars.

According to two recent, independent surveys quasars mostly have redshifts that are between 2 and 3. In the theoretical framework of an expanding universe, redshifts are proportional to the distance from the observer, recessional velocity and age. Therefore, it seems that quasars occurred in an epoch between 1.9 and 3.0 billion years, considering an age for the universe of about 15 billion years. This has left cosmologists with an enigma of why quasars were "born" and flourished in such a small time slot. However, there is no enigma if the redshifts are not what they are purported to be.

Likewise, redshifts come in bunches and are related to each other, which implies a link between galaxies and quasars. The rational conclusion of a number of astronomers seems to be that quasars, peculiar galaxies, and probably even normal galaxies, are related and form some sort of evolutionary chain.[64-72,150,151] One such example is NGC 1275 (also known as 3C 84.0), which is a system that contains objects that appear to have different radial velocities; the different redshifts within this galaxy contradict conventional interpretations.[606]

Quasars are closer to bright galaxies, particularly spiral galaxies, on the average than would occur randomly. Of nearly 500 quasar-galaxy pairs most are lying very close to each other (*within 10'*). Consider a few examples. The density excess of quasars near one spiral galaxy, NGC 1097, with an inner and outer ring, appears to coincide with the axis as defined by extensive optical jets. One very bright quasar is linked to the spiral galaxy, NGC 4319, by a luminous filament. An optical bridge exists between Mk 205 ($z = 0.07$) and NGC 4319 ($z = 0.0057$), though both objects have very different redshifts. The ultraviolet knot and gas show an outward radial velocity relative to the nucleus of NGC 4319. The entire central disk exhibits hydrogen (*-alpha*) emission lines that suggest shock excitation, and/or explosive depletion of disk gas (i.e., time-varying acceleration).[68,72,702] Here is powerful testimony that redshifts are not good indicators of distance, but result from ejection, and involve interactions between objects due to intergalactic fields.

Another source (NGC 2237 + 0305) has one or more quasars still in the center of its "parent" galaxy. Two of three quasars near an elliptical galaxy (NGC 3842) are optically identified with a pair of X-ray sources almost aligned with the nucleus of the galaxy. Quasars with large redshifts are physically associated with both faint and bright galaxies that have lesser redshifts.[68,72,702]

Some spiral (*Sc and later-type*) galaxies are closer than their redshifts indicate. One (*Sc I*) type has the largest components of non-velocity redshifts. A particular source (*NGC 262 or Mark 348*) has a large neutral hydrogen (*H I*) envelope, and is the largest galaxy known when standard redshift interpretations are considered. However, without the standard redshift interpretation it would be classified as a low surface-brightness, hydrogen-rich dwarf with an active nucleus.

Sb galaxies have the lowest, and Sc galaxies the highest, redshifts. Such an observation reveals an evolutionary shift that also changes redshifts. Both normal and barred spiral galaxies evolve in form from the "early" type, designated Sa, through Sb to, "late" type, Sc. Furthermore, there is a luminosity dependence between supernovas and the redshift of a given galaxy, which also demonstrates the fact that changes in redshift take place along with evolutionary transitions.[68]

High-redshift quasars are more luminous than low-redshift quasars. At high redshift there is a preponderance of flat spectrum radio sources. In conventional theory, the predicted angular size/redshift relationship disagrees with radio observations of distant sources.[740] Again, this indicates redshifts are evolutionary phenomena, or different Field-controlled components, not indicators of distance.

A survey of redshift effects in systems of different scales and levels of hierarchy even reveals a higher strength of redshift within systems than between them. The Local Supercluster is one physical system consisting of a huge flattened cloud of galaxies and clusters of galaxies, including the Milky Way, centered on or near the Virgo Cluster. The strength of redshift is higher within the Local Supercluster than for the Universe as a whole.

In clusters, groups and pairs of galaxies, redshift depends on the type, compactness and status. High redshift is usually connected with features pointing to the youth of a galaxy. Redshift is also a function of the positions in the systems, which indicates strong intergalactic fields that influence redshifts. Likewise, individual galaxies display redshift gradients from their inner to far limits.

A redshift field is also found in the plane of the Milky Way with what conventional theory would call an "expansion" (*Hubble constant*) that is ten times higher than the Universe as a whole. Such a notion seems ridiculous, because constants are supposed to be just that, constant. However, because this effect is stronger within systems than between them, it demonstrates interaction is taking place, not expansion.[388,581] This is against one of the tenets of the massive black hole and Big Bang scenarios, which require expansion, while the results of reinterpreting redshifts would resolve the missing mass problem.

Bright quasars are six times greater than the average density of the Universe, as a whole, in the direction of the Local Supercluster. Radio quasars have a redshift ($0.7 < z < 1.1$) in the same direction, and show an ordered magnetic field system associated with the Virgo region, within the Local Supercluster. The density of quasars are distributed throughout space in association with the brightest, apparent magnitude galaxies. These observations also indicate interaction is producing the different redshifts.[69]

Variability poses many severe difficulties for energy generation in quasars or AGN, and therefore, alternate theories are needed (*due to the Compton paradox*). Either large bulk relativistic motions are present, or the objects showing these effects, nearly all of which are quasars, are much closer than suggested by their redshifts. A number of quasars are associated with nearby bright galaxies, and are so close to the nucleus (*within 2'*) that it cannot be due to chance. Therefore, many of the quasars are nearby, and much is not understood concerning violent ejections in galaxies (especially superluminals). Most astronomers tend not to look at objects independent of redshifts in order to maintain conventional theory, and so, the evidence has been relatively ignored. The conclusion is inevitable: "New physics or even a new approach to cosmology is indicated."[149]

A variability problem can be seen in one Blazar (*AO 0235 + 164*) that is considered "extraordinary." This conclusion is reached because of its extreme variability in radio and optical emissions, its complex and variable (*21-cm*) absorption, and its very unusual optical spectrum. The radio source is very compact. Soon after this Blazar was considered faint it underwent a dramatic increase in brightness by more than a factor of 100 in a week or less. There was also a similar radio outburst (*at GHz frequencies*). While the object was bright, in 1975, two different redshifts (*0.524 and 0.851*) were seen, but no emission lines. Data between 1975 and 1985 disclosed various emission lines (*Mg II, O II, O III and Ne V*) all at a yet another redshift ($z = 0.940$). Other emission (*O II and O III*) and absorption (*Na I D-line*) lines were at quite a different redshift ($z = 0.524$). Certainly, all of these very different redshifts in the same object cannot be due to the different distances and velocities proposed by present interpretations.

This Blazar (AO 0235 + 164) is an object whose spectra shows two emission redshifts and two absorption redshifts. The two emission redshifts make it a uniquely "peculiar" object, but only because standard methods fail to explain the entire data. However, the unconventional hypothesis of non-cosmological redshifts due to two interconnected objects does explain the data (*companion connected with a OVV object*). Most astronomers would reject this interpretation because it is unconventional, but fail to offer a conventional explanation for all of the data.[151]

The nature of this Blazar itself shows unique properties. It varies violently and periodically by at least a factor of 100. The object has a very compact radio source, a steep optical continuum, and is an X-ray source. Its radio polarization changes on a timescale of months in a way that can be interpreted as the rotation or periodic variation of some structure within the source, which is likely to be accelerated plasma. Certain (*VLBI*) maps show that the central source has an extension about 180° from the direction in which the companion object lies; it is nearly in a straight line. The components which have the multiple (*21-cm*) absorption at one redshift ($z = 0.524$) vary in relative intensity in periods of 0.25 to 0.5 years. This object's properties might be due to viewing the source along or near the axis of a jet, and include the acceleration of plasma that emits radiation in the form of a helix (i.e., chirality produces the different redshifts).[150]

6.2 Aligned Objects

All of these observations and much more have led to the discordant redshift hypothesis, which has three essential elements. Galaxies are the sites of ejection events involving compact objects. These compact ejections are newly born galaxies, known as protogalaxies. Redshifts are an intrinsic and evolving property of these ejected protogalaxies. The scenario includes some first generation galaxies that eject compact objects of much higher redshift, which are observed principally as quasars. The quasars gradually evolve into increasingly less compact forms. Within this evolution, star formation begins transforming the quasars into galaxies. As this takes place, the ejected objects' redshifts decrease to their appropriate Doppler values, while other physical changes occur. An observational sequence of forms suggests a correlation between compactness and an excess of a non-Doppler redshift component.[65,66]

Numerous surveys have shown excess redshifts for many companion galaxies. These include M31, M81, and interacting companions, nearby groups, Southern Hemisphere companions, hydrogen (*H I*) companions, (*Karachentsev*) spiral companions of ellipsoidal galaxies, and many others. For example, in M31 and M81 all companions are systematically redshifted with respect to the dominant galaxy, and this also contrasts the conventional proposal that these objects contain massive black holes.[66]

"Anomalous" arms are in the plane of NGC 4258. The eastern arm magnetic field lines are directed toward the center, and the western arms are directed away from the center. On both sides there are magnetic field lines that follow the radio components.[373] This arrangement is very similar to the Interplanetary Magnetic Field issuing off the Sun. That is, there is both inflowing and outflowing magnetic field lines in what is referred to as a bisymmetric magnetic field, which is typical of all spiral galaxies.[285] Usually, bisymmetric field structure coincides with the optical arms (as in M33, M81, etc.). This scenario suggests that there was the compression of a preexisting, possibly primordial, bisymmetric magnetic field that helped form the galaxy (i.e., a system of fields existed prior to the galaxy's formation).[439]

Some of the most surprising manifestations of galactic turmoil are the galaxies that eject large amounts of matter. Conventional physics even leads to the possibility that the massive black hole itself could be ejected from the nucleus.[639] However, it is less massive ejections that take place, forming new galaxies.

Observations of gaseous or filamentary structures also suggest ejection. The line of quasars west-southwest of M33 is rotated about 20° with respect to a line of hydrogen from M33. Both the line of hydrogen and the quasars appear to have been ejected. In NGC 300 the line of hydrogen is rotated about 25° from the line of quasars. Both M33 and NGC 300 are large spirals. The evidence indicates that a track in the early Universe created a line of objects, and ejection from the formed objects has taken place.[66,67] Typical of the Field-dynamical Model: "The observations imply instead that material is ejected outward from the nucleus in a fairly well-collimated jet or cone of fairly narrow opening angle."[66]; p.122

Evidence of time-varying acceleration ejecting material is plentiful. For example, occasionally, quasars flare sharply as did 3C 279 in April 1937, when it peaked with a greater value than any known object.[520] The scene involves a system of fields in the Universe along which an object forms, then it ejects other objects, which do the same and so on, leading to the still evolving Universe.

Recently, an eruption of energy equal to about a million years of sunlight erupted from the core of a distant galaxy. The energy was in the form of X-rays that came from a "common" quasar (PKS 0558-504). This was interpreted as a large blob of matter that was sucked toward a massive black hole, but this interpretation is merely in support of a predetermined theory.[617] X-rays are routinely produced by linear accelerators, and by high-energy charged particles colliding with other charged particles. This observation could then be interpreted as the acceleration of matter away from and/or into the core (AGN).

The redshifts of all companion galaxies are positive, and the three hydrogen clouds are negative, with respect to M31. The companion galaxies are satellites in a plane orbiting around M31, and we see this plane edge-on. The line is closely oriented along the minor axis of M31, which is more suited for ejection, as this galaxy has rotational symmetry.[66]

Numerous observations indicate ejections do take place. Galaxies with exploding or ejecting appearance have the strongest association with radio sources—the signature of acceleration processes—many of which are quasars. Brighter galaxies are associated with bright quasars, while faint galaxies are not (see Plate 4.7).

Two of the brightest quasars were found astonishingly close to massive elliptical galaxies in the core of the Virgo Cluster. Elliptical galaxies have a distribution of predominately older stars that form a smooth, ellipsoidal arrangement. They are mostly all bulge that is somewhat flattened, and appear to be the final stages of galaxy evolution. The observations suggest that quasars have been ejected by the most "mature" galaxies, which are giant spirals and elliptical galaxies.

Galaxies evolve into more and more complex forms until they reach the giant spirals and elliptical stage, and then they eject quasars that form into galaxies; an analogy can be drawn with a mature organism giving birth.

High-luminosity quasars are associated with high-mass galaxies in the Virgo Cluster, which indicates a physical interrelationship. Spirals are less massive than elliptical galaxies, and in a bulk sense tend to be companions. Companions and spirals tend to be younger and more active in producing radio ejections and explosions. Characteristics of youth and activity can be empirically linked with the excess redshifts.[66] Again, the evidence suggests that young objects are ejected, and subsequently develop into galaxies who then eject quasars that also form into galaxies.

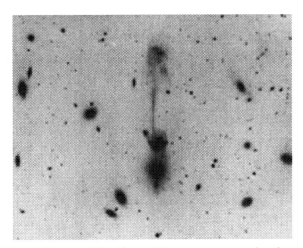

Plate 4.7. Galaxy-Quasar Ejections. Here we see an ejection from galaxy NGC 3561. One of the ejection filaments appears to puncture a spiral galaxy. Nearby is a quasar with a very different redshift. [Figure 9.10 from reference 66].

If we interpret the redshifts conventionally then an obvious problem arises. "The point is that the redshifts of the chain members differ sufficiently that if they are interpreted as velocity, the chains will fly apart in a time much less than the presumed age of the galaxies."[66]; p.147. Many examples of chains of galaxies are seen throughout intergalactic space, and the chains are aligned with radio ejections.[66,72] In fact, some groups, such as M96, display large intergalactic hydrogen (*H I*) clouds that indicate large-scale intergalactic fields.[646]

Emission filaments and radio arms of spiral galaxies are caused by clouds ejected from the nucleus in two opposite directions in the equatorial plane (as in M87 and NGC 5128). Spiral arms are characterized by young, hot, bright stars, which demonstrates relatively recent star formation has taken place. The magnetic field along the spiral arms appears to turn ejection velocity into rotational velocity.[*]

The arms may also be the tracks of the same ejections that mark the birth of quasars. For example, all three quasars in NGC 1073 lie along the general line of major spiral arms in galaxy ejection activity, which is characteristically accompanied by X-rays.[66,72] This scenario is very similar to the discussion on planetary rings and the formation of moons, and as will be discussed, the rings around protostars and the formation of planets (see discussion on protoplanetary and planetary nebulas in Chapter 10).

The lines of gaseous hydrogen that extend outward from galaxies (such as M33 and NGC 300) could arise as the result of ejections. One galaxy (NGC 5128) has a jet in its center that is strong and narrow (*about 90 parsecs*) with both radio and X-ray emissions. The jet points outward from the nucleus closely in the direction of the narrow emission filaments. Observations of the gaseous emission filaments indicate that they have been excited by physical collision (shock excited) due to

---

[*]   Afterward, the conventional hypothesis of density wave compression and random walk (*stochastic*) star formation may take over to some degree.

time-varying acceleration. Gas temperatures are greater than what could be caused by hot-star radiation alone. The gas would have to have been accelerated outward from the center of the galaxy.[66]

Another example is the disturbed neutral hydrogen in the galaxy NGC 3067 pointing to the quasar 3C 232 with a different redshift.[166] The Field is evident in that the filaments are extremely straight and narrow—collimated—with particular characteristics that are not easily explained otherwise. An astronomer comments on this in a way that describes the Field-dynamical Model:

> "There is the extremely difficult and unsolved problem of inventing a mechanism which can so strongly collimate a beam emerging from the nucleus. There is the extremely difficult problem of pumping the particles in the beam with enough energy to get them traveling so close to the speed of light. Finally, there is the simple but appalling question of what happens when the beam arrives at the distance from the galaxy at which it should have an 'extended lobe.' The basic observation is that the energy has to go from small (cross section) to large, but rather suddenly, not gradually. The 'hot spots' in the extended radio lobes are supposed to be the impact points of the beams on an external medium. But why just at this particular point?"[66]; p.140

It is obvious from this statement that there is more structure than either conventional or even unconventional hypotheses would dictate. These comments can be explained by a cone- or funnel-shaped field that widens suddenly at a certain distance from the core, as do the Fields of the Field-dynamical Model, producing the extended lobe. Other galaxies also have narrow jets or counterjets that can be explained in this way (*M87: 60 & 120 pc jets; and NGC 1097: 200 pc jet*). It is along these jets that we find aligned objects and phenomena that suggest ejection.

Consider some of those aligned objects. All the elliptical galaxies in the vicinity of M87 clearly define a line that is almost exactly the line of M87's jet. Material can be seen on one side, and the effects of passage on the other side of the jet and its counterjet. Likewise, the strongest X-ray sources found in the center of the Virgo Cluster lie along a line of elliptical galaxies.

Another example is how all the bright galaxies around NGC 5128 form a line. It has a very compact high-energy jet aligned in the direction of a large radio source, and young stars are found along the direction of the major axis. The line of galaxies are also along the major axis close to the direction of the ejection of radio material and the narrow emission filaments from NGC 5128. Its slight misalignment is probably due to rotation or an offset ejection due to the Field's funnel shape. There are also some spirals and non-elliptical galaxies in the NGC 5128 chain. In addition, many of the active galaxies along the lines in both M87 and NGC 5128 tend to be double galaxies.[66,152]

The jet of quasar 3C 273 points in the same direction of the major axis of a cloud with a different redshift. 3C 273 is aligned with M87 (also known as 3C 274) across a galaxy (NGC 4472) in the center of the Virgo Cluster. The jet of M87 points directly at the radio galaxy 3C 272.1. This indicates that these objects were ejected from M87, and the cloud is an ejected beam from 3C 273, which has two radio sources on each side.[70,316]

The more recently ejected companions show a greater tendency to be multiple interacting objects that are not in equilibrium. This could mean that fissioning and/or multiple compact protogalaxies are characteristic of the earlier stages of galaxy formation. Typical of the Field-dynamical Model, ejections may be intermittent, and epochs of galaxy formation widely separated in time (i.e., time-varying acceleration). The elliptical galaxies in a line near M87 are old and have, like other elliptical galaxies, ejection phases producing other younger galaxies. This is why the spirals composed of generally younger stars in the vicinity of M87 form an approximately oval distribution around the line of elliptical galaxies. Furthermore, the center of the oval is more or less empty of spirals. It would be extremely difficult to explain this distribution on any other grounds than that of ejection. The Virgo Cluster is undoubtedly similar to other clusters.[66,71]

6.3 Redshifts and Ejection

The redshifts of the elliptical galaxies in the line of M87 are different, and they are believed to be among the oldest galaxies known. These galaxies should have been separated from their alignment if the redshifts indicated distance and velocity. The fact that they have not become scattered is an indication that the redshifts are not due to velocity and distance.

If galaxies with a certain redshift (*between $3100 < cz_0 < 5100$ km s^{-1}*) are plotted they form one huge filament stretching over more than 40° across the sky. The filament is centered on the bright, relatively nearby Sb spiral, M81. All the brightest, apparent magnitude galaxies in the northern sky show that 13 out of 14 in uncrowded regions have similar lines of high redshift galaxies.[66] This scenario indicates redshifts are not the result of distance, but are due to ejection.

No direct evidence exists to indicate what it is that is ejected to give rise to the radio lobes, though the emission is helical (synchrotron). The radio lobes are probably due to accelerated electrons and/or discrete self-gravitating objects ejected from the nucleus. A property of the powerful radio sources is the large amount of optical emission lying along the radio axes, which indicates the ejection of mass (*most common is H-alpha [O III] and Lyman-alpha [in objects with large enough redshifts]*).[152]

Inevitably, and once again, a description of the Field-dynamical Model surfaces: "It appears that the extended emission requires ejection from the center, interaction with an extended gaseous medium, and an ionizing source at the center. There is also evidence in some cases for discrete optical synchrotron sources not associated with the nucleus and for young stars in some of these extended regions. Thus we have evidence that the central machine in the radio galaxy is able to eject far more than just beams of relativistic particles."[152]

High-redshift quasars require an intergalactic medium that would have to have been highly ionized at the early epoch, and could not have been shock-heated, as in the Big Bang scenario.[530] In many cases a quasar close to a galaxy has the same redshift as the galaxy. However, there are many cases when the quasar has a redshift much larger than the adjacent galaxy. The only reasonable explanation for such high kinetic energies is that the body has emitted plasma and/or radiation in one direction (which is necessary for there to be no blueshift). Therefore, quasars are accelerated by an internal release of energy through a field that produces the apparent, discordant redshift.[11]

It has long been known that discordant or anomalous types of redshifts are explainable by the Unified Field Theory. Furthermore, the handedness of the components of a helical emission, known as chirality (*i.e., chirowaveguides*), can produce different redshifts even when the object and observer are at rest relative to each other.[150] The Field-dynamical Model is a system of Unified Fields that produce helical emissions.

6.4 Time-Varying Phenomena

A systematic relationship exists in the redshifts of radio galaxies and spirals. Correlations are observed between redshift and radio emission in galaxies. A survey of radio galaxies in the Coma, Virgo and Hercules Clusters makes it evident that radio galaxies in all these clusters have significantly higher redshifts on the average. This observation provides excellent evidence for the existence of non-cosmological redshifts.

When the central galaxy is reasonably radio-quiet, possibly due to no acceleration processes, then it is the lowest redshift member of the group, and the companions are higher in redshift (as in M31). However, if the largest galaxy is a radio source, then it tends to be the highest redshift in the group (as in M87). For pairs, groups, or clusters, the spirals have systematically higher redshifts whenever galaxies are demonstrably at the same distance (as in the Virgo Cluster).[66] These data suggest that the radio sources, especially spiral and elliptical galaxies, are actively ejecting recently formed galaxies, and radio quiet sources are inactive. A time-varying acceleration process in galaxies ejects quasars that form new galaxies.

A magnetic "bridge" between two galactic clusters has been discovered. Magnetic fields spanning entire clusters also exist, such as in the Coma Cluster of galaxies.[449,450] Speculations have been arising that a large-scale field system might exist, but current techniques and instrumentation cannot detect it. However, the observations are clear: "Such alignments clearly suggest ejection along an axis which has a memory."[72]

Other facts indicate a similar interpretation. The difference between the (*Faraday*) rotation of our galaxy, and the rotation of distant extragalactic radio sources is known as the residual rotation measure (RRM). This RRM has uncovered the existence of large-scale magnetic fields out to high redshift objects (*z of 2 or more*). Strong optical absorption of ionized hydrogen correlates with an "excess" of RRM among a sample of 37 quasars. In those with weak or no absorption, the larger RRMs are less numerous. Some have fairly weak (*a few to a few tens of microgauss*) magnetic fields, while others have extremely strong fields (*approaching the milligauss levels*). The strong RRM absorption correlates with radio extended and compact sources, suggesting large dimensions in terms of conventional redshift interpretations. However, they are more consistently explained as extended ejection along a field system (*at least as large as the extended radio emission scaled from z_e to z_a*).[448]

A sample of 116 quasars discloses that the higher-redshift quasars have a larger RRM than those at low redshift. The result of these studies correlates the RRM with redshift, so that the intervening systems may be pre-evolved precursor systems of large galaxies at a significantly earlier time (i.e., a primordial system of fields). In addition, the apparent regularity of the RRM change suggests a large-scale component to the magnetic field. These results indicate that there are pre-evolved intervening galactic systems associated with high-redshift quasars (*up to z of about 2.5*).[448] A system of fields exists along which ejections from astrophysical objects take place in a time-varying process.

Redshifts correlate with the morphology of the galaxy. Companions in groups show an excess of redshift components in the most luminous, and usually, spiral galaxies. Those with the most peculiar morphology show the largest redshift, and may be recently formed galaxies that are evolving to become more structurally arranged.[702]

Late-forming galaxies, which appear to be abundant in the Universe, also suggest this model is at work. Oxygen absorption lines are about ten times more abundant in quasar spectra than expected from counts of visible galaxies. This excess means that we are looking through extended regions where star formation has begun, but which have not yet formed into galaxies.[773,774] This observation departs from present theory and leaves open the question: why do some galaxies form later than others? The answer lies in the time-varying characteristics of the Field-dynamical Model, producing epochs of galaxy formation.

The optical jet in NGC 1808 stretches over great distances (*4 Kpc*) on opposite sides of the nucleus and is perfectly aligned, not only in form (morphologically), but also in respect to radial velocity. The jet is perfectly aligned on both sides of the core, making it a clear indication that a dynamic relationship exists with the nucleus. Furthermore, the direction of the jet agrees very well with the major structure in the nuclear, hot-spot region. This confirms that there are ejections from the nucleus. The surrounding faint galaxies seem to be related to NGC 1808, and show indications of non-cosmological redshifts due to helical acceleration processes.[647]

Some studies show radio sources outside the optical disk, and sources aligned with the nucleus. Other astronomers come to the same conclusion: "We believe that the results summarized here indicate that ejection activity occurs in spiral galaxies."[702] Ejection could be an ongoing process producing new galaxies, because an excess of smaller companion galaxies are near bright spiral companions, and their axes and spiral arms.

Late morphological type galaxies violate every observational test of their redshift distances. The larger the redshifts, the larger the discrepancies. These tests include the linear redshift-apparent magnitude relation (*Hubble relation*), distances derived from rotational velocities (*Tully-Fisher*), the association of other objects of "known" distance, and the absolute magnitude and frequency of Type II supernovas.[68]

Redshifts appear to be due to the variable mass hypothesis of gravitation, not distance and/or velocity. This hypothesis states that there is a long range of uninterrupted series of graduated steps (*i.e., scalar*) of interaction between any two particles in the Universe that generates a change in velocity for each (inertia).[72,546] A time-varying linear accelerator involves a series of graduated steps that accelerates matter (also, the Field polarized the vacuum, which has effects on gravity, inertia, redshift, and grand unification).

In addition, almost all cosmic plasmas that have been studied in detail seem to be penetrated by magnetic fields. This implies that considerable electric currents also exist, which have long-range force.[167] Ejections from galaxies demonstrate that "an unknown force" overcomes gravity.[11,66]

The different redshifts for interacting objects must arise because of different times in the creation of matter. A way of approaching the creation of matter in recent epochs is to consider that AGN have a time-varying source of energy. The reverse of a black hole is possible within the framework of physical theory. This "white hole" or something like it, could eject matter (i.e., a core entailing electromagnetic confinement, producing thermonuclear energy; a Field-dynamical Model).[66,72]

Likewise, quasars are not passive objects, but stir up their environment very fiercely, and eject heavy elements.[570] Observations of one type of quasar (*broad-line absorption quasars or BALQSOs*) indicate that differences may merely arise because the gas is ejected in a cone that is along our line of sight. With a strong magnetic field controlling charged particle motions, violent activity and energy release from the center, it would lead to the complex absorption profiles noted. Hydrogen (*broad-Balmer*) lines in Seyfert galaxy nuclei vary on a timescale of the order of one year, and in some cases less than a month. Such variability remains controversial, because it does not support the massive black hole scenario. Meanwhile, the consequences of beamed ionizing radiation have not been investigated theoretically, but would explain this variability.[150,151] All of these observations are what could be expected from the Field-dynamical Model, including the time-varying beamed ionizing radiation down a cone- or funnel-shaped field.

Additional redshifts are noted in those galaxies behind a cluster of galaxies. The distribution of quasars is marked by several peaks, indicating successive waves of formation (i.e., formation is time-varying). An astronomer comments: "Unless the quasar formation has appeared in successive waves, it is very hard to explain this effect in a natural way."[581; p.306]

The best evidence for non-cosmological redshifts involves the brightness and uniformity (*isotropy*) of the X-ray background of quasars, and microwave background of the Universe. The planes of quasars' halos correlate with the supercluster's equatorial plane, showing a ring exists for superclusters, as well.[769] Projected onto a two-dimensional sphere of the Universe, quasars display definite features. The equatorial region is void of quasars, and the 30° to 40° latitudes show a greater concentration.[581] Typical of the Field-dynamical Model a ring effect and time-varying ejection at the 30° to 40° latitudes arise.

CHAPTER 7

Action At A Distance
(Long-range Forces in Cosmology)

In relativity, distance has no real meaning, while location in quantum physics is also meaningless. Gravity can be described as warped space-time. Inversely, time-warps can be produced by gravity. Time runs measurably faster in space, away from matter and gravity, than on Earth or any object. A "stretching" of time goes along with reduced distance. Superluminal speeds could turn space-time inside out, twisting time into space and space into time. Time-space warps are a tenet of Unified Field Theory.

In other words, the simplistic classical relationship between the whole and its parts is totally inaccurate. The quantum factor and relativity force us to perceive particles of matter only in relation to the whole. A network of interrelationships reaches out to other objects and transcends our observational bias.

A network of fields connecting objects can explain most of the present-day enigmas about the cosmos. For one, empty space can become electrically polarized in the presence of an electric and/or magnetic field. Polarization of the vacuum is now known to have had, and continues to have, an influence on gravity. It can even bring about the existence of matter and result in the fine-tuned physical constants.

Gravity is warped space-time, or a distorted emptiness. Yet, the other three forces can be represented by fields of force extending through space-time. What this all means is that what takes place here on Earth or any other point in space-time effects the entire Universe at some point(s) in time and space.

7.1 Plasma Dynamics

Almost all cosmic plasmas that have been studied in detail are integrated with magnetic fields. This implies that considerable electric currents also exist, which have long-range force. Such currents have a strong tendency to flow in sheets, and narrow structures, known as filaments. Filaments are subject to the pinching action of electric currents, generating mass.

A fundamental type of (*Bennett*) pinch consists of a cylindrical and fully ionized plasma that carries a current directed along its axis. This current, together with the resulting "doughnut-shaped" (*toroidal*) magnetic field, gives rise to a pinching force directed radially inward towards the axis of the plasma cylinder. In cosmic plasmas, other and more complex types of plasma (*cylindrical and plane-parallel*) pinches may also exist.[167] Plasma pinches cause a plasma to collapse upon itself, leading to filaments that induce nebulous masses, which eventually give rise to celestial objects along with their Field-dynamical Model characteristics.

One type of phenomena that can indicate the presence of electric currents is filamentary structures. Filamentary structures are fairly common in our galaxy, and in galaxy/galaxy and galaxy/quasar associations. Many are connected with the exploding stars that leave behind supernova remnants. Others are the molecular clouds or dust clouds out of which stars form. A problem that is connected with these clouds entails how they could be collected into filaments before the gravitational forces became strong enough to contract them. Here, the pinching effect, due to electric currents, offers a mechanism for generating aggregates of mass.[167]

Current carrying plasma filaments have considerable self-organizing properties (*Bennet, cylindrical and plane-parallel pinches*). This allows considerable electric currents to flow along magnetic field lines in the ionosphere and magnetosphere of the Earth, or any other planet. These currents often exist in sheets in the auroral zone (*Birkeland currents*). The solar atmosphere consists of a highly conducting plasma in which considerable electrical currents flow, as well. The low mass of filaments means gravitational, Coriolis and centrifugal forces are of negligible importance in the early evolution of objects.[167] Plasma contraction can also bring about the formation of interstellar clouds by the action of the electromagnetic forces present.[11] These interstellar clouds have been observed colliding, and eventually producing stars.

A current system in a rotating and magnetized galaxy serves as a unipolar generator. Electric currents flow in the equatorial plane of the galaxy, follow along the axis of rotation, and close at large distances from the galaxy. This resembles the current system of the Sun's Interplanetary Magnetic Field coupled with the solar wind.[167] Plasma clouds flow along this field, and produce stars and other objects (e.g., planetary systems, binaries, clusters, etc.).

All of these objects constitute a whole. Their interaction and level of influence depend only upon the strength of these forces and their direction. A network of interrelationships exists between *all* of these objects, and a system of fields appears to be necessary for generating the plasma that form these objects.

7.2 Long-Range Forces

The properties of a cosmic cloud depends on the whole circuit in which the current flows, and can be a long-range force. If an electric current flows in the cloud, then the properties of the cloud may be determined by processes taking place far away. If the current through the cloud is part of a general galactic current system, the cloud may be energized by an electromagnetic field located in another part of the galaxy, and its evolution is determined by conditions far away. Action at a distance can occur between galaxies, and beyond or within.

The circuit may keep the cloud magnetized until the whole circuit is dissipated there. If gravitational forces begin to dominate the contraction of the cloud, then the gravitational energy will convert to circuit energy, and produce a magnetic field. Electromagnetic forces are believed to counteract the contraction, but could also aid in the contraction.[11,84] A plasma physicist asserts: "The cloud may contain small active plasma regions in which most of the energy exchange takes place. These regions may convert gravitational and kinetic energy into electromagnetic energy, so that they are self-supporting." [11]; p.113. Those regions are a field system which curves space-time along which currents flow, and produce everything from planets and galaxies to superclusters and the large-scale structure of the Universe. In fact, this is why the angular momentum verses mass relationship shows a saturation of the electromagnetic force.

Surface electric currents parallel to magnetic fields, such as those produced by double layers, have long-range effects. They give space a general cellular structure, which is relevant to theories on galaxy formation. The chemical composition of matter may be different on both sides of the current layer in interstellar and intergalactic space.[13] As a result, long-range forces are active throughout the Universe.

The idea of an action-at-a-distance creation process was suggested long ago, but it was not widely accepted.[360] Now we know that quasars emerge through interactions between galaxies in a creation process.[64-72,388,587,662,713] Some double nuclei galaxies (e.g., Mkn 266, 463, 673 and 739) contain at least one Seyfert nucleus each. About 75% of the Seyferts have companions, and the more active companions are closer. Starburst activity in companions offers strong evidence of interaction with Seyferts.[283,429] An action at a distance creation process has not been widely accepted, because it suggests a purposeful cosmos, which many scientists are immediately opposed to (see Chapter 12).

Yet, more evidence supports a system of fields, and creation at a distance. There appears to be an excess of close pairs of quasars with a radio-loud component. Interrelationships between galaxies evidently exist, because there are 400 quartets of galaxies in the sky. Galaxy/galaxy associations include dense groups (such as H82), isolated triplets, and isolated pairs. Surveys indicate discordant redshifts in 25% or more of these galaxy/galaxy associations. Likewise, an excess of satellite radio sources near spiral galaxies, which have radio arms, indicates ejection from the nucleus has taken place (as in NGC 4258).[702]

7.3 Gamma-Ray Bursters

Gamma-ray bursters (GRBs) have largely remained a mystery, because of the absence of low-energy counterparts. Undetermined are such crucial properties as source distances (cosmological redshifts), magnetic field strengths, energy sources, transportation mechanisms, and whether binaries are responsible for the bursts. Once gamma-rays are produced, only a small fraction of the total energy becomes hot as the result of interaction with nearby matter. The lack of interaction with nearby matter could signify that a field system exists. Likewise, strong magnetic fields are suggested by the observed low-energy features. In fact, a highly magnetized GRB was observed, and the term magnetar was coined.

Conventional interpretations of redshifts (*z of about 1 or 2*), like those of quasars, are not fully supported by the observations.[569] The often suggested super massive black holes, otherwise referred to as neutron stars, fail to explain the data, as well.[325,515,569] Instead, the observations support an interpretation that the bursts occur along a field system with long-range force. Scientists suggest this view: "It is possible that the GRB sources are indeed distant objects, and that the optical flashes are either unrelated to the GRBs themselves or that they arise by some other mechanism."[515] "It may well be that the origin of gamma-ray bursters is not related to neutron stars at all."[569]

Likewise, a team of astronomers studied Hubble telescope images taken on 26 March and 7 April 1997 that show a gamma-ray burst that is offset from the center of a "fuzzy" object that looks something like a galaxy. The object and the GRB are aligned. A team member states, "Hubble's unmatched resolution was crucial in pinpointing the fact that the gamma-ray burst is away from the center. This would rule out massive black holes, thought to dwell in the cores of most galaxies, as the source of these incredible explosions."[**]

The data reveal the most characteristic properties of GRBs, which do not fit proposed theories. The burst duration varies (*by tens of milliseconds to hundreds of seconds*), and though the bursts are extremely diverse one can identify several of the most characteristic time structures. Burst profiles do not exhibit a clear periodicity with the exception of a few objects. However, those few objects indicate that a neutron star cannot be responsible. A clear periodicity would not occur if it were indeed a neutron star. A "peculiar" trend in the time structures is that when the process of burst development takes place faster, then the details of its time profile become depressed accordingly. Again, a neutron star cannot resolve such a time feature, which requires a dynamic mechanism beyond that of gravitation. Forerunners or precursors up to a few tens of seconds prior to the main burst-phase appear typical. Time profiles are one of the most important considerations for determining the source.[465,500,569] These data support the idea that GRBs result from a system of fields, giving rise to the present theories that GRBs are remote and/or local.

Gamma-ray bursts appear like a nuclear explosion.[521] Protons are accelerated along with gamma-ray bursts, where the proton acceleration and electron production are responsible for the gamma-ray bursts.[411] Optical emissions can be explained by thermonuclear explosions, or

---

[**] From http://oposite.stsci.edu/pubinfo/pr/97/20/PR.html, accessed October 1998

plasma in a magnetosphere, much like an aurora. These observations support the Field-dynamical Model on a large scale, and that GRBs result from a system of fields.

Energy spectra exhibit a fast evolution, and an highly uniform pattern. The existence of broad emission features suggests that the bursts of radiation consists of a hard and a soft component (*possibly cyclotron and annihilation lines*). The shape of the hard component could originate from plasma moving one-dimensionally along magnetic field lines (*i.e., annihilation emission of non-equilibrium electron-positron plasma*). The emission mechanism is not known, but it is thermal in nature (*bremsstrahlung or synchrotron*); that is, thermonuclear. Total energy yield in bursts is not constant, grow with increasing event duration, and may be accompanied by optical flashes.[500]

GRBs at first were thought to be distributed in the galactic plane but instead are distributed uniformly over the celestial sphere. The sources could be local and associated with the Solar System (i.e., the Oort cloud of comets), and are at non-cosmological distances, as in the case of the redshift controversy surrounding quasars. Or, they may be at cosmological distances or both.[500,569] The evidence appears to support that they are both remote (galactic) and local (Solar System or Oort cloud) sources, again indicating the existence of a system of fields.

The true nature of GRBs is still a matter of debate. Some possibilities are that they originate from nuclear explosions of matter ejected from the interior of the object, and the switching on and off of a strong magnetic field. Likewise, the amount of (*photospheric*) expansion increases with increasing gamma-ray levels. The most plausible models are thermonuclear, not the often suggested neutron star or massive black hole.[500,515] That is, typical of the Field-dynamical Model, there is time-varying acceleration causing ejection, and thermonuclear energy. Furthermore, an efficient mechanism is required: "Since the radiation cooling time is incomparably shorter for any emission mechanism than the burst duration, an extremely efficient mechanism of energy or matter replenishment should operate in the source."[500] GRBs are not associated with any known astrophysical objects. Therefore, theories that claim they originate from strongly magnetized neutron stars are observationally unsubstantiated.

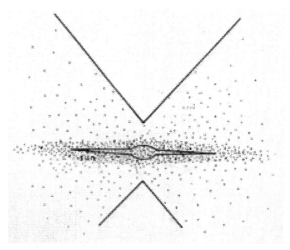

Figure 4.5. Gamma-Ray Bursters (GRBs). A large-scale distribution of gamma-ray bursters (GRBs) discloses a funnel devoid of GRBs along the pole or axis of the galaxy.[from reference 325]. Another feature of GRBs is a significant excess along the 20° to 30° South Latitudes, both of which are typical of the Field-dynamical Model.

The Gamma-ray Observatory revealed that the distribution of GRBs is fairly uniform. However, other studies indicate the possibility that GRBs are crowded toward the galactic disk and center, and also display a concentration between the 30° Galactic Latitudes. Furthermore, a funnel-shaped region where no GRBs exist is observed at each pole of the galaxy (see Figure 4.5). GRBs appear to have originated in or near the galactic plane and core, and then spread (were ejected) throughout the galaxy.[315,325,500] A scientist comments on this phenomena: "Some mechanism may prevent all (or most) in the galactic disc to emit gamma-ray bursts parallel to the galactic plane (possible nonuniform emissions)."[325] Notwithstanding, we do not observe GRBs in our neighboring galaxies, suggesting that the GRBs are even closer, possibly arranged in a halo around our solar system in the vicinity of the Oort Cloud of comets.

CHAPTER 8

Our Galactic Home: The Milky Way (*Structural Dynamics of the Milky Way*)

If the Field-dynamical Model is really involved in galactic structure and dynamics then some distinctive characteristics should be observable in the Milky Way, our own galaxy. Gas or plasma should be moving in and out of the core. Polar, 30° latitude, and ring (disk) features would be observable. Magnetic fields should play a role in the structure and formation of the galaxy. In association with this magnetic structure, plasma dynamics should be observed. A roughly spherical magnetic field or shell producing a halo, a magnetic field similar to the Interplanetary Magnetic Field (IMF), and ejection phenomena should also be apparent. Evidence of confinement, ionizing radiation, and ionized gases would exist. Other characteristics should include radio, X-ray, ultraviolet and infrared emissions, as by-products of acceleration. Aside from what has been discussed previously, these features are quite evident everywhere in one form or another.

8.1 Core and Polar Phenomena

Evidence for a massive black hole in the Milky Way comes primarily from observations of gas moving towards the center, and is, therefore, circumstantial. A unique radio source lies at the dynamical center. However, the source does not lie exactly at the center of the peculiar arm-like gas features near the core (*2 parsecs*).[613] The linear polarization at the center is indicative only of a magnetic field, not necessarily a massive black hole.[614] Clusters in the Milky Way consist of two components, the halo and disk systems.[667]

Large, highly organized radio structures can be seen near the galactic center. The magnetic structures reveal a substantial "polar" or "axial" (*poloidal*) component to the magnetic field at the central region of the galaxy. The reason for the observed ionization is unknown according to massive black hole theory. Conventional theory also has no explanation for the relationship between the dust, which appears to be organic, and the ionized gas. The arched filaments suggest thermal processes, while the vertical filaments indicate nonthermal processes. A helical convex curvature exists within the galactic plane. Radio structure, not due to star formation nor supernovas, can be noted on the filaments when they cross the galactic plane. Moreover, the radio arc is a physically unified phenomenon analogous to solar prominences.[775] The ionization, filaments, radio arc, the gas-dust association, thermal, and nonthermal properties can easily result from the system of fields of the Field-dynamical Model.

Unexpectedly, filaments were discovered by chance during a study of star formation in the galactic core. Radio astronomy revealed long filaments of ionized gas about 150 light years long curving up out of the galactic disk about 30,000 light years from the Earth. Emitting radio energy just like planetary radiation belts, these filaments are apparently kept in place by a strong polar, galactic magnetic field.

Radio emissions from atomic hydrogen near the galaxy's edge indicate that some of the gas is heading toward the galactic center. Some clouds are also moving outward.[200] This gas motion is like the polar (outflowing) and solar (inflowing) winds that produce aurora on the planets, and the inflowing and outflowing winds as a star forms. Radio emission, filaments and ionization are expected of the Field-dynamical Model's time-varying acceleration and thermonuclear fusion. Furthermore, a characteristic funnel-like region forms around the rotational axis of the galaxy where no gamma-ray bursts are observed (as shown in Figure 4.5).[325]

At the center of the Milky Way are radio, X-ray, ultraviolet and infrared emissions, as well as ionized gases (*also He II and Lyman-alpha*). Nonthermal ionized gases can be noted 90° from the radio jet. Polarization also takes place 90° from the rotational axis. Radial outflow in parts of the Milky Way coexists with star formation and supernova remnants.[682] Throughout the full range of objects, from planets to stars and galaxies, there is bipolar outflow, indicating a similar structure.

The Milky Way's core reveals an array of huge plasma blobs close to compact radio waves. The orientation and location of these blobs of hot, ionized gas suggest that they were ejected in opposite directions from the core. The blobs were ejected out along with strong particle and radiation winds coming from the center. Revealing the path of ejection is a trail of hot, ionized gas that points directly away from the central radio source.[776] Ionized gas is often seen in the nucleus, and is the result of photoionization by a central energy source, which includes star formation (*as in H II regions*).[363]

Material cascades down towards the center in a series of stages, typical of linear acceleration. A ring of neutral hydrogen continues to feed ionized gases into the center along a complex series of spirals. A gas cloud and supernovas display complex relationships that reveal a large-scale structure sending a stream of gas toward the center.[736]

The center of the Milky Way displays time-varying activity. Gamma rays were observed in 1977 and continued for two years, shut off, and then switched on again a month later. Vast rivers of ionized gas were observed funneling toward and away from the galaxy's core.[233,357] Ionized neon in rapid motion has been noted in the center of the galaxy, and it is surprisingly cool. The region's nucleus appears to be crowded with stars from the viewpoint of our Solar System, thousands of light years away.[554]

8.2 Latitude Phenomena

The Milky Way consists of a disk-shaped spiral with a bulge in the middle. This is a structural phenomena that is unexplained by conventional theory. The hot, gas orb in the heart of the Milky Way stretches 150 light-years perpendicular to the remainder of the galaxy. This physical arrangement consists of a shell-like structure similar to an atom, the planets, the Sun, the Solar System, and protostellar and protoplanetary nebulas, as well as other systems throughout the Universe.

Differential rotation displays a structure with the inner regions moving faster than the outer regions. As is often the case, gravity cannot account for this differential rotation, while such an arrangement is necessary for accelerating particles, plasma, and ionized gases toward and away from the core. A redshift field is also found in the plane of the Milky Way with what conventional theory would call, an "expansion," known as the Hubble constant, but ten times higher than the Universe as a whole. Either the constant is not very constant, or again, the redshifts observed are non-cosmological and the result of a helical field (*i.e., chirowaveguide*).

The distribution of gamma-ray bursts (GRBs) reveals a uniform distribution, but are slightly concentrated toward the galactic center and along the disk. A local anomaly (*anistropy*) occurs between the 30° and 45° latitudes at 0° longitude, which is noted by the gamma-ray burst locations.[325] This could be evidence that either GRBs originated in or near the galactic plane, and then spread throughout the galaxy, or they are the traces of a field system along which there is ejection. GRBs are accelerated along or are the result of a system of fields that displays core, polar, mid-latitude and ring (disk) features, typical of the Field-dynamical Model.

A local magnetic field component exists in the Milky Way roughly in the direction of the spiral arms and the North Polar Spur.[761] The general pattern of the galactic magnetic field(s) can be seen in the optical polarization of 7,000 stars, which disclose a structural component along the galactic equator.[208] A ring-like structure in the magnetic field exists.

Radio emissions emanate from the spiral-arm magnetic fields due to relativistic electrons. The origin of the electrons is unknown in terms of standard interpretation, but are routinely produced by thermonuclear fusion. At the extremities of the Milky Way are quasars and radio galaxies.[216] These facts support a model with characteristics of thermonuclear fusion and ejection.

A ring (*toroidal*) field runs away from us above the equator and toward us below the equator. A twisted semi-helical polar (*poloidal*) field is the only place where there is a bright polarized feature. This can be explained as a magnetized jet issuing from the galactic center, along with relativistic electrons, where the helical emission is most intense.[688]

The Milky Way's star clusters are concentrated along a flat plane, something like the planets around a star or a planetary ring. Globular clusters exist in a sphere mostly on the outer edge of the Milky Way, something like the moons around planets. X-ray sources, indicative of confinement or acceleration along field lines, radiate from the globular clusters and the central bulge.[407]

The galactic magnetic field appears to play a primary role in large-scale galaxy dynamics (*the flat rotation curves seen in H I in the outer regions of many galaxies*). The flattened rotation has led conventional theorists to imply that large amounts of hidden matter exist outside the optical disk. However, arguments against such an interpretation include a magnetically driven wind along the galactic plane (which is an aspect of the ring-like structure).[550] X-rays and gamma-rays, typical of acceleration, are most intense along the equatorial region.

Planetary nebulas (PN) are also concentrated along the galactic plane, even for the largest nebulas. The smaller nebulas are strongly concentrated in the direction of the galactic center. However, this is not the case for the larger nebulas. Longitude restrictions show a concentration for the smaller nebulas around 0° longitude, the GRB anomaly region. Larger nebulas display a distribution with no clustering, but are less numerous in the direction away from the center (anti-center), and around the 0° longitude.[605] This is also the case for GRBs and stars. Again, this is evidence of ejection within a field system.

The galactic distribution of PN forms a ring around the equator from about 0° to 10° latitudes. There is also a density increase around the 0° longitude, and a thinning around the 180° longitude.[491] The radial velocities of PN in the galactic bulge are good evidence for rotation of the bulge.[267] More will be discussed about protoplanetary nebulas in Chapter 10.

There is a brightness minimum in the radio emission near the North Pole or Polaris. One of the most impressive hydrogen (*H I*) shells persists near the equatorial north pole. It forms a huge arch that extends from 120° to 150° Galactic Longitude and about 25° to 45° in latitude. At 30° galactic latitudes there are lower temperatures around 240° Galactic Longitude, where the plane radio emission slopes down to a minimum. The feature suggests a site of ejection like that responsible for cratering on the planets. In fact, an astrophysicist says it looks "more like a flattened meteor crater instead of exhibiting a loop structure."[615]; p.122. Pulsars are observed in the arm or disk, and the 30° latitudes of the Milky Way.[436] As observed on the planets there are latitude, longitude and ring-like phenomena, and a structure that looks like a meteor crater that is likely to be a site of previous ejection (see discussions on cratering in planets and ejection phenomena in galaxies; Chapter 2, 3 and 6); supernovas also reveal this structure.

CHAPTER 9

Exploding Stars
(Supernovas and Their Occurrence)

Supernovas are stars that explode spewing out heavier elements that are important to galaxy evolution, and the existence of life. Theory and observations are in conflict, particularly with the fairly recent Supernova 1987A. An examination of the facts discloses that galactic magnetic fields in the spiral arms, and the molecular clouds they push are partly responsible for causing supernovas. The overall picture confirms the Field-dynamical Model of galaxies, as well as stars. The techniques used to find supernova remnants include visual observations, photography, and radio and X-ray astronomy.

9.1 Theory

Two types of "exploding stars" or supernovas are known. The Type I are two to three times brighter, on the average, than Type II. The Type I stay bright for a short time, and drop-off quickly in about a month or so. Type II supernovas' initial drop off is less rapid, but after about 100 days or so they start dropping off more rapidly. Type I fall less rapidly after the initial rapid drop. Most Type I supernovas behave much more like each other than the Type II supernovas who differ from one another to a much greater degree. The spectra of Type II supernovas show lines of hydrogen, while the Type I do not. These hydrogen lines are one of the most important ways of distinguishing the two types from each other.

Supernova locations in galaxies, and the types of galaxies in which each type is found, reveal other characteristics, both of supernovas and galaxies. Two types of star groups exist in galaxies, and are referred to as Population I and Population II stars. Population I stars are younger stars that include stars like our Sun, and stars which are currently being formed. All different star masses are included in this category, including very massive stars. They are generally found in the arms of spiral galaxies and in all parts of irregular galaxies. In contrast, Population II stars are older stars that may have formed shortly after the galaxy in which they are found was formed, and well before Population I stars. The central bulge and the halo in spiral galaxies are typically where Population II stars are located. And, they also include stars in the globular clusters. Type II supernovas seem to be closely associated with Population I stars, while Type I supernovas seem to be found in both population types.

Supernovas appear to leave behind gaseous nebulas and pulsars. A helical field (synchrotron) causes a rapid variation of accelerated electrons in a strong magnetic field. This time-varying nature has lead to the label "pulsar."

Present theory claims that this pulsing is the result of a rapidly rotating neutron star, which is the proposed massive black hole supposedly produced by a collapsed neutron star. Observations can be described by the "lighthouse" model where the neutron star rotates sending a beam of radiation that is swept past the observer as the star rotates, thereby creating a pulse. The pulse is said to be the motion of electrons and their positively charged counterparts, positrons, in an intense magnetic field along the concentrated fields at the magnetic poles. The pulsing is observed in the optical, radio and X-ray portions of the spectrum.

Over 400 pulsars are known, but only five pulsars (1.25%) have been found associated with supernova remnants. That is, the gaseous nebulas left behind in the explosion are absent from 395 pulsars (98.75%), and this is said to be due to the remnants becoming undetectable. Meanwhile, the pulses may instead originate from the interaction of fields with a time-varying mechanism, not a massive black hole or neutron star.

9.2 Observations

Supernova remnants are either "shells," that appear like rings of gas, or "plerions," which appear bright over their central regions. Supernova 1987A, a supernova that took place in 1987, has a bright ring or shell of gas surrounding its center, as photographed by the Hubble telescope.[245] Some supernovas, such as the Crab Nebula produced by the supernova that occurred in AD 1065, have "waves" and other brightenings in the central regions of the visible nebula. Others, such as the supernova remnant of the supernova that occurred in AD 1181, shows a small hot star in the center of the radio source 3C 58, but all attempts to find a pulsar have failed. This may be due to a pole that is not in our line of sight, and therefore, an estimated 90% of the pulsars are not seen, because the beam misses us. However, pulsars may form only under certain unknown conditions. This may be why, to date, astronomers have found 150 supernova remnants for which only five (3.33%) are associated with a pulsar.[86,299]

Field dynamics are quite evident in supernova phenomena. The overwhelming majority of spiral galaxies for which supernovas have been discovered are viewed almost pole on. Supernovas occur mostly in rich star clusters and are located at the bottom of vertical chimney-like, or steep-sided pit, structures that are embedded in the equatorial gas and dust layers. The high supernova discovery rate is due to our line of sight of the source, which is dynamically related to the galaxy. This kind of observation could be expected from the interaction of stellar and galactic fields in supernova occurrence. These galaxies are of the spiral types (*Sc and SBc*) whose structure allows for either the production or the viewing of supernovas, or both. The highest supernova rates occur in the most active star-forming, late-type galaxies with a pole-on viewing angle.[727,728]

About 50% of the approximately 150 supernova remnants (SNR) are detected as very structured sources (i.e., linearly polarized). The filled-center SNR or plerions are normally highly structured (polarized). Meanwhile, the shell type are weakly structured. The filled-center SNRs appear to be sources with uniform magnetic field structure. SNRs in this class display common properties, such as filled-center radio emission, flat radio spectra, and central energy sources. The shell-type display a stretching of the magnetic field lines, accompanied by expanding ejected matter. The young shell-type SNRs present a spherical (radial) magnetic field orientation, while the old SNRs have a tangential orientation. Common properties are found in linear (tangential) field orientations in old SNRs and radial fields in young SNRs, but the picture is confusing elsewhere. This inhomogenity may be due to the individual conditions at the time of the explosion, such as magnetic field interactions and/or the interstellar medium.[288]

Supernova shells appear to contain relativistic electrons (and/or positrons) at high overpressure, which indicates ordered magnetic fields and extremely efficient power sources. Some supernovas (*Type Ib*) are radio emitters within one week of the explosion, others within several months (*Type II L*), and others possibly not at all (*Type Ia, II P*). This makes it questionable that an efficient engine is causing the (*in-situ*) acceleration if only the star's magnetic field is considered.

In order to explain why young supernovas' shells tend to have radial magnetic fields and old supernovas' shells tangential (*parallel or circumferential*) fields it is necessary to look at their structure in more detail. By radial it is meant that the magnetic field diverges uniformly around the center in a shell. Tangential is on the edge of the SNR's outer spherical shell.

Few scientists have theorized magnetic fields, because such models are too difficult, even for today's supercomputers. Filaments which move relative to the weakly magnetized environments tend to become strongly magnetized, because they are good conductors. Hence, in the early stages the magnetic fields would tend to stretch radially as in young shells. After the filaments slow down in space, the interior will expand explosively overtaking the filamentary shell. This could cause a tangential magnetic field depending on whether the interior or exterior fields dominate.[459]

The interior field dominates if the SNR is no longer in contact with an interstellar spiral-arm magnetic field (ISMF) of the Galaxy's arms, thereby, producing the radial shell. When still in contact with the ISMF, it becomes tangential. This stellar-galactic magnetic field interaction explains many unresolved problems in supernova theory.

This SNR-ISMF linkage appears to be the case in the triple radio source in the center of SNR G179.0 + 2.7, within a double-sided radio galaxy. The peculiar properties include its radially projected magnetic field and the triple radio source. The triple radio source is steep from its center towards the outer component. The SNR and the galaxy are aligned in such a way that a physical relationship between both objects is responsible.[288]

9.3 Supernova 1987A

The recently observed supernova 1987A (SN 1987A) threw many contradictions in the face of present theory. A type II supernova, SN 1987A should have been a star that was a red supergiant before it exploded. Yet, observations indicate that the star was more like a blue supergiant. A star on the blue side of the supergiant sequence is in an earlier evolutionary state that theoretically should not undergo a supernova explosion. Only one other supernova is known to have had a blue supergiant progenitor. Furthermore, SN 1987A was not as bright as the typical Type II, and after the initial rise it should have leveled off and decreased, but instead it continued to increase for a couple of months. SN 1987A brings into question our present theories on stellar evolution, and the factors behind the occurrence of supernovas.[292,299]

SN 1987A was the first supernova for which we were able to detect neutrinos. However, a big surprise was that the neutrinos came not in one, but two bursts. SNRs are believed to give rise to neutron stars or black holes. However, the earlier neutrino detection requires that more than the rest mass energy—conversion of all mass to energy—of the neutron star would have to have been converted into neutrinos. Therefore, no neutron star could have been formed, because all the mass was lost to energy.

Because this neutrino flux was not in accord with accepted theory, it originally made astrophysicists doubt its origin was in the same location as the supernova (*i.e., the Large Magellanic Cloud*). The initial burst (detected at Mount Blanc), along with the later burst, produced far more neutrinos than could be expected from neutron star or black hole formation. Black hole formation could not have a burst lasting seconds. Confirming the data, concordant gravitational wave noise was observed in Italy that matched the time of the initial burst observed at Mount Blanc, Maryland.[649]

The evidence demonstrates that the supernova occurred in two stages, which is also strongly supported by luminosity and spectral evolution. The presence of a bright companion star, after the supernova explosion, is not explained by present models based on spherical symmetric collapse without distortion (*angular momentum*) and magnetic fields.[6] These observations suggest stellar models with stronger magnetic fields and/or rotation (as proposed in this book), bringing into question the validity of presently accepted stellar models.

The following statement exposes the necessary conclusion: "The Mt. Blanc burst would necessitate an initial collapse event that is quite different from standard models. Models with a large magnetic field and/or rotation have low temperatures, but it is hard to imagine an event which radiates a *minimum* of several neutron star rest masses in neutrinos, or has a very nonthermal distribution. However, the alternative of some new physics cannot be trivially excluded."[649]; pp.113 & 117

Interestingly, from the (*isotropic rate*) background and angular resolution the second event is given an angle of 40°, typical of the Field-dynamical Model. Likewise, the early ultraviolet burst from the supernova was of a brightness that can only be explained by extremely energetic processes.[649] Such extremely energetic processes, as well as other observations, can be accounted for by the interaction of the stellar magnetic field with a magnetic field of the galaxy's spiral arm.

Other observations confirm the unexpected character of supernova 1987A. One set of data (intrinsic polarization) implies a "lopsided" (asymmetric) shape of the outer edge or envelope. In contrast, theory predicts a spherical (symmetric) shape. The lopsided shape could be due to the galactic field distorting the envelope.

These data also present evidence for a strong rotation of the pre-supernova's surrounding layers (*stratification*) of the envelope. Furthermore, the asymmetries are different for the in and out lines. The prolonged brightness implies an additional energy source. A possible jump of position angle takes place in the long wavelength (*infrared*) range of the spectrum.[308] About 50 days after the supernova, a secondary hotspot was noted near the supernova. And X-rays and gamma rays were noted far too early for present models as if some "punch through" took place.[649,737,738] This punch through was caused by the spiral arm magnetic field puncturing the magnetic envelope of the star (see Plate 4.8).

Records of supernovas dating back as far as 1885 exhibit the fact that SNRs occur in steep-sided pits in the galactic plane of spiral galaxies. This "chimney" theory states that barrages of supernovas occur in certain regions of the galactic disk. There they affect the shape and rotational speed of the galaxy, while redistributing its chemical elements. In two classes of spiral galaxies (Sc and Sd) there is a sharp rise in supernova frequency in those that are seen almost face on (a tilt no greater than 25°).[728] This is an observation that could be expected if it were the spiral arm magnetic field that punches through into the magnetic envelope of stars, causing supernovas. Another indication of this interrelationship is observed in the end products of supernovas: pulsars.

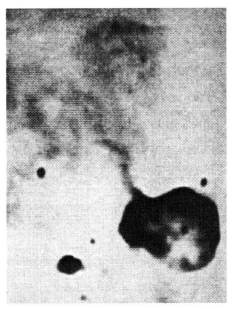

Plate 4.8. Supernovas and Supernova Remnants. This picture exhibits a vivid look at how the outer magnetic envelope of a star was punched through. [Reprinted by Permission from Nature; Roger, R.S., et al. (1985) Nature 316:44-46, Macmillan Magazines Ltd.]

9.4 <u>Pulsars</u>

The resulting pulsars also suggest this scenario. To date, astronomers have found 150 supernovas remnants for which only five are associated with a pulsar.[86,299] High supernova rates occur in the most active starforming galaxies that have a myriad of stars along the spiral arms. The dipole field of the star interacts at nearly right angles with the spiral-arm magnetic field. Occasionally, an interstellar or spiral-arm magnetic field punches through into the magnetic surface of these stars causing the supernova with its lopsidedness (asymmetries). This leads to pulsars only if the star's field-system is still in contact with the spiral-arm magnetic field.

Other recent supernovas show similar phenomena. Supernova 1984e, in the galaxy NGC 3169, was first reported in 29 March 1984, and reached a maximum in brightness around 1 April 1984. It displayed the largest hydrogen (*Lyman-alpha*) luminosity ever seen from a single star to the extent that it was comparable to the luminosity of a entire Seyfert or spiral galaxy (*narrow line region*). This hydrogen "spike" was much too narrow for what is expected from present models of supernova.[292] Again, this suggests an interstellar spiral-arm magnetic field (ISMF) that punches through into the star's magnetic field producing the narrow spike and intensified emission (i.e., gas, plasma and/or dust, possibly in double or more complex layers along the ISMF in conjunction with supernova emissions).

9.5 Supernova 1987A Pulsar

One of the biggest surprises was the pulsar that followed Supernova 1987A. The pulsar that emerged had a period of half a millisecond, making it the fastest known. In addition, an 8-hour variation suggests that the pulsar is a member of a binary system, a system with two or more stars orbiting each other.

The surprises were many. Pulsars were thought to give rise to a rotation of perhaps ten times a second, not the 5,000 times observed. Its rotational energy is about the same as its gravitational energy, which means it should have been physically flung apart. Dumping about half a solar mass of material onto a fledgling neutron star could spin it up to a millisecond period. However, the energy generated would be far above the (*Eddington*) limit where the radiation pressure would be enough to repel the in-falling material. This limit could be evaded if the in-falling gas were entering the equatorial region and the radiation were to emerge at the poles. However, this is not in accord with present models, but could be explained by the new model with its polar fields and ring phenomena (see section 10.2 on star formation). Notwithstanding, the ISMF-SNR linkage is all that is required to produce a pulsar.

Likewise, the high power of a typical pulsar points to its being some form of coherent radiation that flows along magnetic field lines at the poles. Yet, a self-consistent theory of the fields and currents for pulsars has not been found in the scenario of the proposed neutron star. An elementary assumption is that the pulsar's luminosity originates from electrons accelerated by the rotating dipole magnetic field of a massive object. However, the brightness of the pulsar suggests a different magnetic field than envisioned or an unknown process that interacts with the magnetic field.

A companion, if present at the time of the explosion, would have been sitting *inside* the outer layers of the giant progenitor star. Such a binary system is incredibly impossible, because if this were the case then it should have been destroyed long ago, or at least at the time of the supernova. All in all, these observations call for the wholesale revision of ideas, because the notion that pulsars are born slow may turn out to be an unfounded guess.[86,482]

X-ray sources also shed a different picture. Unpulsed sources of X-rays occur in the central bulge of the galaxy and clusters. Pulsed sources occur in the spiral arms where they still interact with the ISMF. Again, these details indicate the new model and the sudden development of the supernova as the result of interaction with an ISMF and a long-range force (e.g., plasma filaments along the spiral arms).

This can be seen in a number of observations. The polarization of SN 1987A implies an asymmetric shape to its envelope and gives evidence for a strong rotation of the pre-supernova star. These asymmetries are different on the in and out lines. The prolonged increase of brightness at all spectral ranges implies an additional energy source. One possibility is that these observations are due to the ejection or acceleration of relativistic particles.[308,622] Again, a gravitational perspective fails to explain the data, while a ISMF-SNR linkage can be expected to produce what has been observed.

9.6 Pulsar-Galaxy Linkages

Other observations are also inconsistent with present models. The magnetic field model of pulsars is admittedly "grossly incomplete." Intensely rotating magnetic fields generate strong electric fields that can rip material from the surface of a star. Yet, here is an intensely "rotating" pulsar and theory claims that material is collapsing onto it, forming a neutron star. As a result, theories about magnetic fields (and neutron stars) show "that our understanding of this important phenomenon is poor and incomplete."[86]; p.31.

The origin of pulsars in globular clusters is not well understood either. Theory indicates that pulsars should last about ten million years. However, the facts indicate some are about ten *billion* years old.

Meanwhile, there are two classes of galactic X-ray objects. These are pulsating sources near the galactic plane, and unpulsed sources in the central bulge of the galaxy, and in globular clusters. The 150 globular clusters in our galaxy are ancient self-gravitating clusters of 10,000 to 100,000 stars, which are disturbed in a sphere mostly outside the plane of the Milky Way. The total number of neutron stars in clusters (10^5) is too high to support conventional models of neutron star formation. These facts also indicate the importance of galactic magnetic fields, because the pulsed sources are in the region of the interstellar spiral-arm magnetic field (ISMF), and the unpulsed sources are outside of that field in the central bulge and halo.

The first millisecond-period pulsar (1937 + 21) has no companion, which is purported to give rise to the fast spin. The pulses could be due to a dense wind along the galactic magnetic field in conjunction with the stellar magnetic field. Or, the pulses may originate from the stellar magnetic field "bouncing off" (accelerate and decelerate along) the galactic magnetic field's contours. This is also suggested by radio pulsars.[86]

One of the youngest galactic pulsars (PSR 1800-21) seems to be in line with an extension of a supernova remnant (G8.7-0.1). A physical relationship between the two seems apparent. Yet, this creates a dynamical problem of having a simultaneously born pulsar travel to a position near the edge of a supernova remnant (SNR) within its visible lifetime. This would require a great velocity to reach its present position. However, the hydrogen (*H II*) regions show a finger emission extending towards the pulsar, as could be predicted if a magnetic field connected the two. Of the six young (<20,000 years) pulsars half have been associated with a SNR.[407] All of these details reveal an ISMF interacts with the stellar magnetic field that produces the supernova and its subsequent pulsar.

Conformation of this relationship is a pulsar (PSR 0833-45) and the proposed Vela supernova remnant. An absence of observed motion for a star (star M) is in sharp disagreement with the predictions. This brings into question whether the two objects are associated at all, as convention would dictate.[107]

Finally, the distribution of pulsars in the Milky Way confirm that galactic magnetic fields produce supernovas and pulsars. More than two dozen X-ray pulsars have been discovered. The Ginga satellite discovered four X-ray pulsars and three hard X-ray sources in the galactic plane at the galactic latitude of about 30º, as expected from the Field-dynamical Model. They are in the region of the (*5-Kpc*) spiral arm, indicating that the arm is a site of vigorous formation of X-ray pulsars. Six are along the 30º latitudes, while the others are in the galactic disk (ring).[436]

Again, evidence of the new model comes in the form of active regions at the 30º latitudes and along the equatorial region. There is an inflow of stars towards the central bulge and an outflow along the spiral arms, typical of the Field-dynamical Model. Hence, supernovas and pulsars occur in the core or bulge, at the 30° latitudes, and along a ring. These observations support both the stellar (as will be seen) and galactic models presented in this volume.

CHAPTER 10

Star and Planetary System Formation (*Protostellar and Protoplanetary Nebulas*)

The formation of stars and planetary systems confirm the universal nature of the Field-dynamical Model in producing the variety of celestial objects. Plasma clouds collide along the spiral arms of galaxies and generate mass. The mass then rotates, producing a dipole field due to the centrifugal and Coriolis forces. Winds flow in and out of the dipole field. The spinning also generates the mid-latitude fields as the result of the Coriolis force. This Field-dynamical Model, whose fields interact along the equator, then creates a ring around the star. The ring eventually forms the planetary system in much the same way as the galaxy forms stars. The locations of star and planetary systems also confirm that the Field-dynamical Model of galaxies is a reality.

10.1 Star Formation

Three main phases of stellar evolution are theorized. The pre-main sequence phase involves the gravitational contraction of the star's mass. The main sequence phase entails the thermonuclear fusion of hydrogen to helium in the central core. The post-main sequence phase consists of the burning of hydrogen, helium, carbon, and the heavier elements formed.

The process of star formation involves cloud-cloud collisions produced by the spiral arms of galaxies. At times the collisions are enhanced by shocks generated by supernova explosions. Both the collisions and shocks cause mass to aggregate, forming a whirling plasma cloud.

Angular momentum is a term used to measure the energy of a rotating object, such as the spinning nebula that eventually forms a star. Meanwhile, the observed angular momentum of protostars is too great for their contraction solely by gravitational forces. If the matter is closely coupled to the magnetic fields, however, the twisting of field lines can generate (*Alfven*) waves that would transfer angular momentum from inside the cloud to the external medium.

Even if the cloud density is initially uniform, a density gradient is soon formed producing shells of matter that are concentrated towards the center. Various stars have been observed to lose mass, but the physics and mechanism of the ejection process are not known. Such stellar oscillations can provide substantial information on the structure and evolution of stars. However, the physical mechanism producing the oscillations is not known in the theoretical framework of conventional theory.[122] The oscillations are the result of the time-varying aspects of the Fields of the Field-dynamical Model, and as with other celestial objects, there is ejection.

A detailed theory of convection in stellar interiors is lacking, while it is clear that other mechanisms must be at work.[122] In the protogalaxy, vorticity is generated along with shocks that generate a strong magnetic field. Strong magnetic fields in galactic halos, which would also lead to star formation, were inferred from recent measurements (*Faraday rotation toward radio-loud quasars*).[608,744]

This scenario led to the theory of "magnetic-breaking," which was suggested in order to solve the problem of angular momentum in star formation. However, the early Universe is theorized to be an unlikely source of strong magnetic fields. Instabilities occur in waves perpendicular to the rotational axis of the gaseous disks, and this could generate dipole magnetic fields in parallel waves. Magnetic field generation is the epoch of galaxy formation, as well as the onset of star formation.[608,744]

Angular momentum may be lost during the mixing (*convective*) phase as the result of the twisting force (*torque*) exerted by stellar winds in the presence of magnetic fields. The slow rotation of (*Ap*) stars is likely to occur in an early stage when there is an exchange of gas with the interstellar medium. The most effective braking mechanism involves magnetic stellar winds.[61,642]

With a large enough magnetic field the outwardly directed magnetic pressure force can dominate over the inwardly directed gravitational force. This is why cool shells have been detected (*by IRAS*) around (*A, F and G*) main-sequence stars and normal (*B*) stars.[61,642] According to relativity theories the sources of this non-gravitational force are particle mass, magnetic energy, and also the magnetic pressure itself (acceleration). The Field-dynamical Model develops and brings into existence the space-time curvature that produces the non-gravitational force (due to polarization).

Star forming activity and magnetic fields are intimately connected.[642] In fact, levels of molecular hydrogen gas in galaxies and magnetic field strengths are linked. Elongated dark, molecular clouds lie along the magnetic fields. This arrangement produces an additional magnetic pressure on the gas perpendicular to the magnetic fields. A star-forming wave results from cloud-cloud collisions and the generated vorticity, which brings about the field system.[270]

The angular momentum of a star in its diffuse protostellar state is several orders of magnitude too large to allow its collapse to the dimensions of a star. Rotational forces should halt protostellar gravitational collapse in the equatorial plane, making star formation impossible. Stars slow to a rotation with velocities that are one-third of the equatorial breakup velocity. This precision is evidence of a dynamic mechanism that has non-gravitational forces, and transcends the effects of angular momentum. A combination of fragmentation, magnetic fields, disk formation, gravity (i.e., accretion), and planetary system formation and evolution helps to solve the problem of angular momentum.[409]

Condensation of low mass stars is driven by self-gravity, but it is tightly controlled by an ordered large-scale magnetic field. The magnetic field exerts a twisting motion (*torque*) that reduces angular momentum.[248,486] However, most of one type of (*T Tauri*) star rotates slowly, while a number of higher mass pre-main sequence stars are rotating more rapidly. Therefore, like gravity, magnetic braking cannot be the whole story behind the birth of stars.[409,763]

Circumstellar disks play an important role in the process of star formation. These disks are especially necessary in order to redistribute angular momentum. Disks may produce significant luminosity, and have moderately sized masses. Since the mass fluctuates in the star's spiral disk this would force the star to move from the center of mass. Thereby, angular momentum is transferred to the star's orbit. This coupling may lead to the formation of giant planets and/or a binary companion within the disk.[4]

Observations of the formation of stars reveal that the Field-dynamical Model is at work. Energetic molecular outflows, stellar winds and jets are important to star formation.[286,462,701] Observations depart from the conventional framework, and support the Field-dynamical Model: "Neither gravity nor magnetic fields can confine these high velocities to the localized regions where they occur."[463]

The details disclose that there are molecular outflows as the star begins to form. Outflows result from a combination of magnetic fields and electrostatic fields (plasma in double or more complex layers) in the structure of the Field-dynamical Model. One of these outflows is like the outflows at the Polar Field on Earth, dubbed the polar wind, which, along with the inflowing solar wind, produce auroras. Likewise, both inflowing (*redshifted*) and outflowing (*blueshifted*) winds (*lobes of emission*) are involved in forming stars. Conventional theory is at a loss for an explanation: "However, the precise nature of the engine that drives the intense neutral wind remains uncertain. How can an object simultaneously undergo both inflow and outflow? Alternatively, how can a star form by losing mass? The answer to this question holds the key to understanding the basic physics of star formation."[463]

The collapse of molecular clouds into stars is ordered by a stabile magnetic structure that resists any random turbulence in the gas. Conclusions by the researchers tell the story: "Indeed, if we have learned anything, it is that star formation is much more mysterious process than anyone had expected even as late as 15 years ago. We do not understand how they are generated, when they are ignited, and how long they last. We now have reason to believe that, as we progress toward a greater understanding of star formation, we will also begin to unlock the secrets of the origin of planetary bodies."[463] A fitting remark this is, for both stars and planets originate in much the same way on different scales.

The sudden appearance of nebulous structures is not associated with the violent ejection of matter, but with an increase in the flux of ionizing radiation (photons) from the central hot core. A bipolar flow, along the two poles, and jet-like structures in the form of disks or streams in the equatorial plane reveal the new model.[731] Observations such as these are very much like phenomena seen on the planets and galaxies, including their rings and aurora, and disks and jets.

Many examples of bipolar outflow in protostars are known. In the Milky Way, bipolar outflow is observed from objects believed to be associated with star formation. An example is one star, SS 433, that displayed ejecting luminous material near the speed of light in a bipolar outflow.[66] Likewise, there is the bipolar outflow of the polar wind in the auroral zones on Earth, as well as jets and ejections from galaxies. Stars can be stretched to many times their size and bounce back in a few hours, indicating a highly stable structural confinement. They can also form in less than a century, such as the new star NWC 349, for which gravity could not possibly be solely responsible on such a short timescale.

Magnetic fields play a major role in cloud morphology. Certain observations (*range of forbidden line profiles*) are best accounted for by a latitude dependence of the wind's velocity and density structure. The wind leaves the stellar poles at a higher speed than at the equator, and a high degree of structure (*collimation*) is required to account for the observations.[701]

Generally speaking, outflows tend to be aligned perpendicular to the flattened filamentary-like structures present in the molecular clouds. A central issue for the understanding of the formation of bipolar outflow is how these flattened clouds and disks manage to form in the first place.[607] The Field-dynamical Model involves Field relationships that produce dipole and disk (or ring) phenomena usually perpendicular to each other.

Gravitational forces cannot explain the incredible range of the flattened or filamentary structures. Many filamentary objects and cores or "fragments" can be seen throughout the cloud. Strong magnetic fields form perpendicular to these large-scale structures, and cannot be static, or the filaments would align parallel to it (i.e., it must be time-varying). Plasma (*hydromagnetic*) processes are responsible for much of the structure involved.[607] A time-varying acceleration along the equatorial region of the protostar is unmistakable.

A bipolar nebula S106 has a line of sight magnetic field that is the strongest (*137 +/- 16 uG*) ever detected in the interstellar medium. Its origin and the nature of its driving force are uncertain when using conventional, gravitational interpretations.[410] Magnetic fields of this intensity require a structured field system and plasma layers.

Star formation does not have to occur as the result of gravitational forces involving an accretion disk. "However, a possible lack of axisymmetry due to, e.g., density waves, time variations, and associated dynamic changes, and possibly many other effects, which we have not even been thought of, may determine the structure of the disk and leave their signature behind. The question still remains how the disk can respond so quickly."[533]

Accretion, the gravitational collapse of mass aggregates (e.g., the molecular clouds of the disks), is inadequate when explaining the observations. As diagramed in Volume One, and in this volume discussed under planetary rings, the disk is the result of the Fields descending limbs coming into conjunction along the equator. The magnetic field at the equator appears to be closed for this reason (i.e., it turns inward towards the star).[284]

Additionally, energetic winds are only found in young stellar objects surrounded by massive disks. "The formation of disks of solar system mass and dimension appears to be common, if not inevitable consequence of star formation processes."[701] There are very narrow jets of more tenuous gas than the disk of gas and dust around the equator of young stars. And, this can be explained as the result of large-scale magnetic fields.[749]

The final stages of star formation are accompanied by violent instabilities (i.e., time-varying acceleration processes). Usually, the outer material has too much angular momentum to fall directly onto the star and so a disk forms (due to the Field-dynamical Model's curvature of space-time). Magnetic stresses allow material to spiral into the central star. Much of the gravitational energy and angular momentum released by the inward-spiraling material is carried away by an outflow of matter, and by (*hydromagnetic*) waves along the polar regions at right angles to the disk. A jet often occurs along with this outflow.[759]

Dynamical phenomena produce time-varying light, velocity, color and spectral variations typical of the Field-dynamical Model. These data provide important constraints on the nature of the system involved. Rotation in some stars enhances and alters the character of stellar winds. The mechanism for mass loss is uncertain in orthodox perspectives, and involves different speeds in the mass motions.[763]

These observations can be understood as the time-varying acceleration of different populations of particles and/or plasma by the Fields.

Because this model is not known, an issue in star formation is the on-off coupling between the gas and a magnetic field pervading the gas. The coupling had to be on in the early stages to reduce the angular momentum of the spinning gaseous nebula. Meanwhile, it had to go off in later stages to prevent stellar magnetic fields from reaching unrealistic intensities. Furthermore, organic dust, which is often superconducting and aligns in magnetic fields, could have aided in this coupling.[359]

Radio emission arises from continuously injected particles in the magnetosphere along closed field lines in the outer magnetic fields. X-ray emission is the result of accelerated stellar winds. These phenomena are not observed in some stars, which could mean that some stars are different than others, the phenomena switches on and off (time-varying), or that we see them from different viewing angles (i.e., whether or not they are discrete sources).[61]

The behavior of "discrete absorption components" is different for each star. Yet, for a given star the behavior is rather similar over a timescale of years. There are large variations in the absorption parts of unsaturated lines (Si IV) and at the steep edge of saturated lines (C IV and N V). These two types of absorption variations are correlated, indicating a unified mechanism(s). Timescale variability is hours to days without substantial differences for a given star. The implications of the observations are clear: "This implies a rather constant controlling mechanism responsible for the rapid structural changes in the winds of early-type stars. The nature of this mechanism, however, remains unknown."[405] That is, until now.

10.2 Planetary System Formation

What follows star formation is the formation of protoplanetary (PPN) and planetary nebulas (PN). The evidence confirms that the Field-dynamical Model is operating. Highly complex and highly symmetric shapes of PN are due to interacting winds under the control of the model. There is a central cavity with a bright rim, an inner halo, and an outer halo something like the layered interior of the planets, and the clusters and stars of galaxies. The PN are round, elliptical and butterfly (i.e., toroidal) shaped, as could be expected of the Field-dynamical Model.[90]

Many stars possess disks of hot gas and dust that can evolve into planets. Infrared studies of 83 solar-type stars in Taurus-Auriga indicate that half of the stars younger than three million years old emit larger than expected amounts of infrared radiation. Infrared observations of 33 stars surrounded by disks suggest three of them have holes or gaps near their center, again reminiscent of planetary rings. Young star clusters, such as the Pleiades, emit double the expected amount of infrared radiation.[158,699] This indicates the presence of grainy, glowing disks that extend close to the stellar surfaces, like planetary rings and galactic disks.

Multiple-shell phenomenon are prevalent in PN. Most outer shells are filled with material. "Nevertheless, it remains to be explained how the shells manage to fight diffusion and maintain a steep density gradient at the interface."[175] The break between shells could be formed by an abrupt change in the mass loss rate of a super wind and/or ionization products from confinement (see discussion in section 10.3).

Young PN have radio continuum, a very small but dense shell, and the presence of a neutral envelope, with the largest envelopes in the youngest PN. Young PN often have a stellar appearance and the nebulas are ionized, preventing their dispersion. Radio observations of young PN display rapid changes in the radio characteristics of the nebulas, indicating a time-varying acceleration mechanism.[293]

The geometry of shell structure (bubble-nebula interface) indicates that mass ejection prior to the hot wind-phase is the most dense in the equatorial plane. Some astrophysicists infer that a portion of the heating arises from a source other than stellar ultraviolet radiation. Evidently, this heating is due to confinement and acceleration by the Field-dynamical Model. A fast wind that blows into denser material that was ejected previously along an equatorial disk can explain the spherical geometries. As the "hot bubble" grows it moves fastest toward the low density poles.[90] These observations can be explained by a Field-dynamical Model with field lines along the equator accelerating particles into a disk, as well as acceleration towards the poles in something like a self-induced aurora (this also occurs with the Sun; see Chapter 13).

The mapping of some reflection nebulas reveals parallel bands (*of polarization vectors*) across pre-main sequence stars. These bands can be explained by dust disks whose grains are being aligned by a donut-shaped (*toroidal*) magnetic field as noted in the Field-dynamical Model (i.e., it is similar to magnetospheres). The parallel bands are essentially perpendicular to the direction of the major axis (poles) of the proto-planetary nebulas (PPN). The peculiar shapes of PPN and PN are a direct result of an internal donut-shaped (*toroidal*) magnetic field.[574]

Winds add very little mass to nebulas' shells, but could affect internal motions (*kinematics*). The shapes of PN are only partially explained by an initial pressure gradient and radiation pressure acting on dust. Winds (*from progenitors*) produce extended halos that are found in about two-thirds of all PN.[326] These observations demonstrate that PN are controlled by much more than random processes and gravity.

Time-varying turbulence could lead to dust sedimentation within the disk and subsequent planet formation. Short timescale phenomena, such as electrostatic discharges, appear necessary to produce local rapid heating events (*i.e., to produce the observed chondrules in meteorites and other chemical properties*). Turbulent gas motions produce collisions, which can lead to greater mass, and this appears to be an essential pro-tostellar and protoplanetary process.[733]

Planetary nebulas display short-term heating events, such as lightning-like discharges, or rapid magnetic reconnection events (i.e., electrostatic time-varying characteristics). A great deal of evidence indicates that not only is there rapid heating, but also rapid cooling of meteoritic material. Dust models do not explain rapid heating and cooling, because the actual process takes place much too fast (by a factor of about 1,000).[533] Bipolar planetaries exhibit very strong emission lines of low ionization particles (*O I, N I, N II and S II*) due to photoionization by the central star and shock heating.[317] The electrostatic time-varying effects of the Field-dynamical Model appear to be necessary in order to explain these observations.

Molecular hydrogen ions (H_2^+) are observed in PN. Investigations of winds that occur in the center disclosed that planetaries appear like stars and are more dynamic than assumed: "The results of this survey are that planetaries have flux distributions like what you would expect from hot central stars (i.e., fluxes increasing steadily toward shorter wavelengths), but a couple of planetaries had odd flux distributions. Clearly, there is some source of absorption at the shorter wavelengths."[346] One object (NGC 6210) shows that wavelengths drop off dramatically (*shorter than about 1500 angstrom units*). Molecular hydrogen ions should not be in a steady-state nebulas, but could be in an evolving nebula under the control of the Field-dynamical Model.

In addition to gravity there must have been field-dynamical— "hydrodynamic and electromagnetic"—effects in the formation of the Solar System. "Either the orderly arrangement of the planets is a coincidence or some as yet unknown physical mechanism has operated to organize the Solar System in this way."[214]; p.130. The following section will address this topic.

10.3 Critical Velocity and Magnetic Materials

The new model can be seen in observations of our own solar system where there is a local origin for cosmic rays. These local cosmic rays are noted during solar minimums, a time when it is least expected. Most activity in the outer heliosphere beyond the planets (*about 25 to 30 AU*) ceases in the ecliptic plane and the 30° ecliptic latitudes. The planets are like a ring system, including Pluto's orbit, which is inclined 16° to the ecliptic plane. Rings are sometimes wider on their outer edges where the Fields deflect from each other, and back towards the planet. Typical of the Field-dynamical Model there are "ring" (ecliptic plane) and 30° latitude phenomena in the Solar System.[355]

Comets' composition of submicron crystals show little indication of a long sojourn in interstellar space.[184] Therefore, comets might just be considered the outer ring or halo who, like planetary rings or galactic halos, are younger than what present theory would suggest. Comets are more in a dynamic relationship with the Solar System as a whole in maintaining mass relationships (see Chapter 4 on comets).

When a falling non-ionized gas has obtained the same energy that is necessary for ionization, it will become ionized, and hence, stopped by electromagnetic forces. The different ionization potentials cause various compositions to fall into certain bands from which planets form. This is known as critical velocity, which produces band structure, and it is an example of action at a distance by the Fields.

The shells around forming stars and PN are the result of critical velocity. The band structure of the Solar System explains why there are certain distances to the planets around the Sun, and moons around the planets. Their different compositions can be explained by the different ionization species of each band. That is, the orderly arrangement of the planets (*gravitational potential distribution of secondary bodies*) around the center of mass is the result of band structure, due to critical velocity.[11,66,73]

Another formative process is the Cosmogonic Shadow Effect. It is manifested as a two-thirds relationship between the orbital radius of rings and gaps in the Saturn ring system. An analogous effect is noted in the Solar System's asteroid belt.[11,66,73]

Band structure is the result of the interaction between a thin magnetized plasma in the region around the Sun (*circumsolar cavity*), the magnetic field of the central body or bodies, and the neutral gas and dust falling toward them. The partially ionized protostellar source cloud gives rise to the Sun by gravitational settling, and the field-system's acceleration and confinement processes. Neutral gas continues to move in toward the proto-Sun from the remaining protostellar cloud, which is partially ionized and magnetized. The ionized component of the cloud is suspended by the magnetic field, while the neutral component is free to settle gravitationally toward the proto-Sun. As the neutral gas is accelerated in free fall, it eventually reaches a critical velocity. Ionization takes place at this critical velocity, which is different for different elements. The ratios of ionization potential to atomic mass would divide the elements into discrete bands, separated by gaps (*this is not photoionization*).[11,66,73] In the formation of planetary systems we do see gaps form in the nebulas before any planets form, and this process can explain that observation.

The distribution of the elements in the Solar System has been a problem for purely gravitational theories. However, metal is easily ionized and would therefore concentrate near the Sun. Hence, the terrestrial planets with metals formed closer to the Sun. High ionization potentials would be generated in hydrogen, helium and the noble gases further out. As a result, the gaseous planets formed further from the Sun.[162]

This band structure, arising from critical velocity, can explain the formation of planets, moons (satellites) and planetary rings at certain distances from the central body. The moons or satellites, and rings of the planets would have developed when the planets had established their magnetic fields.[73] Rings are very likely to be the beginning stages of the formation of moons.

For example, the Earth's moon displays a good deal of evidence for an origin by fission. That is, they separated from a common mass. This is because they are not isolated entities, but are an interconnected dynamic system.[108] Analogous to ejections from galaxies, moons are ejected at an early, primitive plasma-phase and develop in relation to the main body's magnetic field. Likewise, the other planets would have developed a mass and substantial magnetic field before their moons formed.

The typical gravitational—accretion—theories of mass distributions for the planets poses a problem when considering the outer planets. Accretion is the gravitational clustering of mass into planets, which is often referred to as an era of violent bombardment that is purported to have caused most of the cratering on planets. Two scientists discuss the unsolved problem: "In the case of the outermost planets with their satellite systems, the much longer timescale required, comparable to the age of the Solar System, poses a problem in all accretion theories which depend solely on classical collision dynamics for accretional evolution. It is likely that the solution to this general problem in early solar system formation theories lies in the accelerating effects of accretion in dusty plasma which follows from gravitoelectrodynamic considerations."[73]; p.363

Critical velocity phenomena appears to operate in space today. However, there are some anomalies, such as the Martian satellites appear to violate critical-velocity band structure systematics, and mass discrepancies appear to arise in some of Saturn's rings.[73] Meanwhile, the Martian satellites may be captured objects, such as asteroids, and the Saturn rings may be still developing and could eventually become satellites.

In the origin of the Solar System, plasma processes were decisive in its early formation and evolution. The importance of plasma is believed to have decreased when planets were formed and celestial mechanics then dominated. At present, plasma (*magnetohydrodynamic*) effects are considered unimportant. Yet, a great deal of evidence demonstrates a solar-planet-lunar linkage due to plasma dynamics and the electrostatic forces they create. In Tome Five, the full scope of the solar-terrestrial-lunar linkage of FEM will be illustrated. Furthermore, critical velocity and band structure are important for understanding the Solar System, though this is not widely known.[11] The Fields were formed through plasma processes, dominate planet formation, and even celestial mechanics to a certain extent (especially the non-gravitational forces).

Plasma processes can be seen in the Solar System today where the Interplanetary Magnetic Field (IMF) is rather complex with an inner boundary in the vicinity of the Earth, and an additional structure extending from the distance of most asteroids (*about 3 AU*). At that point the fast solar wind interacts strongly with the pre-existing shock in the range from the asteroids to about Jupiter (*about 2.5 to 5.0 AU*). This structural transition may be the reason why asteroids formed at this distance instead of a planet forming. That is, the IMF is involved in generating the dipole magnetic fields of the planets, and at this distance there are two independent components to the IMF. The inner heliosphere is severely disturbed by several interacting shock fronts containing peculiar magnetic field configurations, while the structure beyond Saturn (*about 10 AU*) remains relatively undisturbed. It has been observed that the energizing of ions is a pervasive phenomena in space-time, even low energies (*about 30 keV*) are noted over an entire solar cycle to at least the distance of Neptune (*about 30 AU*).[441-444]

"Cosmic rays" at the highest energies are arriving from the interstellar medium, as well. However, they are moving into the heliosphere mostly in the ecliptic plane where plasma and the IMF interact most strongly.[441-444] Plasma processes and the IMF are accelerating cosmic rays toward and away from the Sun, and these observations display the long-range force of the fields involved. In addition, a similarity can be noted here between the ecliptic plane with the planets, and the spiral arms of galaxies where particles are accelerated in and out of a ring-like structure with secondary bodies (stars in the case of galaxies).

Strong gravity is analogous to the strong nuclear force. Gravity is not only attractive, but possesses a torsion analogous to the spin of nuclear particles. Matter is continually created near existing matter and is spinning outward due to the creation of angular momentum and plasma processes. Furthermore, this rotational mass produces a magnetic field that has been observed in most celestial objects.[7,8] All of these phenomena result from the space-time curvature of the Fields (including vacuum polarization).

The usual models of dynamo processes of celestial bodies encounters difficulties and alternatives must be considered. The protoplanetary nebula may have contained active plasma regions in which most of the energy exchange took place. This would have converted gravitational and kinetic energy into electromagnetic energy so that they were and are now self-supporting.[11,391] Furthermore, the magnetic dipole moment and angular momentum are proportionally related and this implies that there is an unknown electric field. Electric fields have a very long-range force and are an aspect of the Field-dynamical Model.

All materials in Nature are magnetic, and many features are the result of the magnetic properties of atomic and/or molecular bodies and dust (particularly if it is organic, which it appears to be). Materials are attracted or repelled by magnetic fields, but in most cases the forces are extremely small. Another force exerted on minerals is electrostatic, particularly if the force changes with time (time-varying; i.e., *Faraday's Law applies*). Electrostatic, magnetic and electric forces can orient plasma and the subsequent formation of minerals (*diamagnetic, paramagnetic, antiferromagnetic, ferromagnetic and ferrimagnetic constraints*). The magnetic properties of minerals and ionized mass allow for the composition and structural development of planets and their satellites. Electrostatic forces are known to overcome gravitational forces and create unique structural arrangements of matter.

Through the evolution of plasma processes to atomic species and finally minerals, Field dynamics guide the formation of the wide variety of celestial objects. The magnetic properties of minerals and ionized matter under the influence of the Fields can explain a number of physical features of the planets, including their main dipole magnetic fields. The materials in terrestrial-planet interiors are of the class (*ferromagnetic*) that acquire a large magnetization in relatively small magnetic fields. The molecular field in such a material always forms parallel to the magnetization. If the magnetization direction rotates under the action of an applied field—produced by a rotating mass of a protostar or PN, and the IMF—the direction of the molecular field rotates with it. An applied field is necessary for maintaining the strength of the magnetic field (*due to ferromagnetic saturation*), dynamo action only reinforces it.

The initial applied field was of stellar origin and similar to the Sun's Interplanetary Magnetic Field (IMF). This field "churned" the protoplanetary nebulas perpendicular to the IMF, and through rotating the nebula, produced the main dipole fields (through the Coriolis and centrifugal forces, and dynamo action in the plasma). Herein is the basis for the formation of dipole magnetic fields. For the gaseous planets, hydrogen under high pressure and low temperatures produces a metallic form of hydrogen.[601] The resulting field system is a product of the condensing planetary nebula in conjunction with the IMF, and thereby, solar linkages would be, and are apparent, in various observations.

As a result, gravitational effects are not the only influence, and electrostatic time-varying effects also play a role in Solar System and planetary dynamics. Gravitational forces are indistinguishable from the mechanical forces as a result (i.e., the Einstein Equivalence Principle applies). Furthermore, gravitational mass is identical with inertial mass and mass is equivalent to energy. The spiraling planetary nebula produced the main (dipole) field, and on further condensation the Coriolis Force produced the other 30° to 40° latitude fields. The forces were present during the formation of the planets and guided the alignment of minerals. It is a case of the electromagnetic and weak forces—the electroweak force—controlling gravitational forces, leading to the Field-dynamical Model, which brings about further structural dynamics.

A modification of the distance dependence of gravity is inadequate, and there must be a long-range force that couples to angular momentum originating with the planetary nebula. An external-field-effect (EFE) consists of a microdynamic system (plasma) placed in an external field of dominant acceleration (the Field-dynamical Model) that is affected above the tidal or gravitational effects. The EFE leads to the structural arrangements of the planets (or any other celestial object). Furthermore, the EFE can explain the inconsistencies in conventional gravitational theory.[526]

In further support of this scenario, there is the fact that angular momentum versus mass relationship for the full range of astronomical objects—from asteroids to superclusters—shows a remarkably tight clustering around a specific power law (*of index 2*). This relationship shows that there is a saturation of the electromagnetic force that is unaccounted for theoretically. Most classes of systems are grouped more tightly around the mean relation than the stability argument would require, indicating the necessity for a dynamic component(s).[715]

Furthermore, a new constant that equates angular momentum and mass must exist on theoretical grounds. There is a symmetrical property (self-similarity) which most astronomical systems share. Most systems have density and rotation laws that are scale-free (i.e., not restricted by size, composition and space-time).[144,752-754] This is strong support for the existence of the Field-dynamical Model and Unified Field producing the various objects as the result of curving space-time (see the sections "The Cellular Cosmos" and "Churning Up Creation" in the **Conclusions** for further discussion and references).

10.4 Distribution

Star formation occurs under much the same conditions, only the galactic magnetic field, particularly spiral arm fields, produces the initial "churning" that yields the main magnetic field of the protostar. Consider, as an example, how radio galaxies have a strong tendency for the radio jets to be aligned with the observed optical emission. This alignment indicates that massive star formation is triggered by the jet.

A central bar is a basic feature of disk-shaped clouds as the star's magnetic field develops. The rising blobs experience the Coriolis Force due to the vorticity that is generated, and this causes the star to spin. Gas motions such as these are known to produce magnetic fields.[608] As the star spins its convection brings the Coriolis Force into existence at the 30° to 40° latitudes forming the Fields whose interaction causes the stellar disk to form.

This interaction of galactic magnetic fields with star formation is why PN are concentrated along the galactic plane. Furthermore, this relationship is noted even for the largest nebulas. The smaller nebulas are strongly concentrated in the direction of the galactic center. Such a distribution is not the case for the larger nebulas who grow larger as the result of plasma clouds being accelerated away from the galactic center. Further out, plasma clouds experience more cloud-cloud collisions, which generates larger nebulas.

Additional support for this scenario is the longitude and latitude constraints of the nebulas. A concentration for the smaller nebulas is noted around 0° longitude. This is not the case for the larger nebulas whose distribution shows no distinct clustering. However, fewer of the larger nebulas are found near the center and around the 0° longitude.[605] This scenario suggests ejection has sent the nebulas away from the core along the 0° longitude, eventually forming larger nebulas away form the site of the ejection.

Another indication of this ejection is the galactic distribution of PN. The distribution forms a ring around the equator from about 0° to 10° latitudes. There is also a density increase around the 0° longitude and a thinning around the 180° longitude, showing a mutual relationship with the core (AGN).[491]

Furthermore, radial velocities of PN in the galactic bulge are good evidence for rotation of the bulge. The rotation is an acceleration process that ejects plasma, which leads to plasma cloud collisions. These collisions generate protostars that produce planetary nebulas.[267]

Studies of stellar rotational velocity in clusters have shown that there is a distribution of angular momentum among young stars. Angular momentum is larger for the more massive stars. Conventional theories are at a loss to explain this relationship, while it is important for understanding star formation. Slow rotation in lower-mass stars is typical, but in some clusters there is an excess of rapid rotation on the lower main sequence. For the younger clusters, such as the Pleiades, the low-mass stars also display strong stellar halos (*coronas*). This scenario suggests that enhanced X-ray emission (greater acceleration and confinement) is connected with the higher rotation rate of stars.[667] Again, the evidence suggests acceleration processes along large-scale fields in clusters.

Moreover, in some clusters (like the Pleiades and NGC 752) the evidence reveals that there are several epochs of star formation. Blue Stragglers, stars that lie above the main-sequence turnoff points of the massive stars, also present evidence of epochs of star formation. Time-varying acceleration and ejection are evidently responsible for these epochs.[667]

Young clusters (like NGC 6530 and NGC 2264) are still imbedded within their parent clouds. Many of the stars in these young clusters still have dust shells and are contracting toward hydrogen-core burning. Flare stars and T Tauri stars, the more active and "dynamo" stars, are associated with the younger clusters.[667]

About ten clusters show diffuse X-ray emission centered on their cores. Like galaxies, the concentration of stars greatly increases towards the center of the cluster. Each core contains tens of thousands to millions of stars, and a few appear to be flattened (ring-like). Globular clusters move in giant, highly eccentric, elliptical orbits around the galactic center, forming something of a halo. Open clusters, which are younger star systems, form circular orbits and tend to reside in the spiral arms. Considerable evidence exists for several epochs of star formation in clusters; again, time-varying acceleration and ejection are involved.[305,667] Clusters, like galaxies, display all of the constraints expected of the Field-dynamical Model.

Star formation is observed in giant molecular clouds in the spiral arms of galaxies.[732] Galaxies' boundaries are associated with self-regulated star formation due to dynamic structural arrangements.[573] The Field-dynamical Model is evident in that there is a law governing star formation in galactic disks; a ring-like phenomenon.[413]

Moreover, fewer cloud-cloud collisions occur in the outer Galaxy than in the inner Galaxy, because the spiral arms are less tightly wound in the outer region. For the same reason, inner star formation is more frequent and massive than outer star formation. However, this does not occur in proportion to the density of molecular clouds, and therefore, cannot be explained gravitationally. It is, however, explained by the spiral arms being instrumental in star formation, because most of the material is confined to the narrow arms.[511] All of these observations support the star formation scenario discussed above, as will Tome Five's discussion of the nearest star, our Sun (Chapter 13).

CHAPTER 11

A New Physics Contrasts Conventional Thought
(New Paradigms and Physics are Needed)

Overall the paradoxes between what is observed and present theoretical perspectives leaves us with a necessity to form new paradigms and require new physics. There are serious flaws in a number of theories in terms of observational support. A fresh perspective for deciphering the facts is essential if we are to be truly scientific about our views.

11.1 Theoretical Problems

AGN could be identical except that they are viewed from different angles. Each is time-varying and have the same physical properties, but differ in some way (central mass flow rate, or stronger acceleration and/or ejection). It is quite possible that they are evolutionary stages of the same physical objects and/or each has its own time-varying phenomena (i.e., each may appear exactly the same if observed long enough). The basic problem is that "what we do see is likely to be misleading," because of our relative viewing angle, preconceived ideas, and the limits of our instrumentation.[623]

Blazars, and Seyfert I galaxies could be the final products of far-infrared evolution. The nucleus may use or blow away the dust covering the AGN, exposing the nucleus, thereby causing luminosity shifts from the infrared to the ultraviolet and optical.[379,683] The dust appears to be organic or dry bacteria, which would align in the AGN's magnetic field, and enhance conductivity and electric field generation (due to super-conductivity).

Magnetic field strengths are smaller in the AGN of spiral galaxies than in elliptical galaxies. Compact radio sources are located in quasars. Blazars have radio spectra cut-offs, while Seyfert and Markarian galaxies have steep spectra.[74] These differences could arise from the evolution of quasars and Blazars into spirals, which then evolve into ellipticals.

One general mechanism has been shown to operate: the Field-dynamical Model. The observations made of all objects support this model in a number of ways, some more than others. In contrast, a growing number of objects provide evidence that super massive black holes (or collapsed neutron stars) and dark matter cannot be responsible.

Another observational stumbling block is that direct measurements of electric currents are very difficult or impossible, and therefore, would not be directly observed. Often the ionization of cosmic plasma is produced by a hydromagnetic conversion of kinetic or gravitational energy into ionizing electric currents. When an electric current is sent through a plasma it is changed (becomes noisy), which is not considered in classical theory. Yet, electric currents are essential for understanding double layers, transfer of energy from one region to another, the cellular structure of space and current shifts (*i.e., current sheet discontinuities*). An understanding of cosmic hydromagnetic phenomena is impossible without describing them explicitly by electric currents.[11] Or, in other words, we must look for electric currents indirectly and then explain the dynamics, or we are missing a very important part of the story.

This limited observational perspective also applies to cosmic magnetic fields. Celestial bodies' magnetization must be due to electric currents flowing through their interiors. Plasma theories indicate a need for the Field-dynamical Model, which involves primordial fields that curve space-time. A new model is needed purely on theoretical grounds alone, as the most notable and knowledgeable plasma physicist states: "Even the usual models of dynamo processes in the interiors of celestial bodies seem to encounter difficulties so that alternatives to those should also be considered."[11]; p.87.

When an electric current is passed through a "quiet" (*quiescent*) plasma a number of complicated phenomena are produced that present theory has not explained, and appear to require new physics. Sometimes the plasma becomes more noisy than expected, while the energy distribution produces an extremely large excess of high-energy particles (*strongly non-Maxwellian*). The electron temperature is orders of magnitude larger than the ion temperature, which is larger than the neutral gas temperature. Meanwhile, classical theory assumes all are the same temperature.

Instabilities can occur, such as a large current density causing the plasma to contract into filaments. If the plasma is a gas mixture, then components often separate, and when electron drift velocity exceeds the thermal velocity, double layers may be produced that could explode, producing violent current surges of plasma acceleration. The importance of boundary conditions (i.e., plasma layers flowing along Field contours) is often neglected, which can give and has given completely erroneous results when interpreting the observations.

In many theories it is thought that the behavior of the plasma depends on local conditions (density, temperature, magnetic field, etc.), which can also be misleading. For example, in a current-carrying (*non-curlfree*) plasma the local conditions are only part of the effect. The outer circuit also plays a part to the extent that the violence of the plasma explosion is determined largely by the circuit.

Field contours determine the circuit producing electric fields. Electric fields have long-range force, or to put that another way, one event leads to another, which leads to another and so on. Our only knowledge of plasmas are from the laboratory, magnetosphere and heliosphere. It may never be possible to obtain an equally reliable knowledge of plasma in extremely distant regions of space and time.[11] If we are to be truly scientific, we must admit that we are left with critical gaps in our observational perspectives and present theoretical framework, and new interpretations are not only called for, but inevitable.

A few of the generally accepted astrophysical concepts can no longer be valid, because more than 99.99% of the Universe consists of plasma. The plasma-planetesimal transition was a series of small-scale transitions of individual cloudlets in a rapid and time-varying process, like auroras in a long sequence through time. Motions were largely regulated by the magnetic field produced by dipole spin, and Coriolis and centrifugal forces in the plasma. The evolution of an interstellar cloud to the present planets and satellites was governed by mechanical and electromagnetic effects, not the often mentioned accretion of gravitationally bound "clusters" of matter.[12]

Until the middle of this century magnetic fields were only known to exist within the Solar System, but since have been evoked for almost every phenomena. Magnetic fields would be expected of the Field-dynamical Model. Magnetic fields are certain in gas cloud support, star formation, stellar activity, interstellar gas dynamics, cosmic ray dynamics, and jet formation. They are highly probable in supernova expansion, interstellar dust dynamics, spiral structure, density waves, galactic winds, and nuclear activity. Magnetic fields are also possible in stellar evolution and supernovas, and galaxy interaction, rotation, evolution and formation.[99,285,632]

Here the gaps in theory are quite evident. "The origin of magnetic fields, however, is still dubious."[99; p.3.] Magnetic fields can also solve the so-called missing mass problem (i.e., the Dark Matter Hypothesis is not needed).[550] Furthermore, the existence of magnetic fields means large-scale electric fields also exist, which have long-range force.

Even black holes, regardless of whether super massive or not, work better with a magnetized medium. The electromagnetic fields far from the black hole are analogous to the fields created by a rotating magnetic star. In order for black holes to remain charged, extract energy and maintain accretion disks, a magnetized medium is superior.[76] Again the gaps are noted: "the role of magnetic viscosity in accretion disks appears primordial, although yet unclear."[76]; p.201.

The obstacle involves a theoretical mind-set.

> "In order to keep their models simple theorists usually tend to neglect the influence of magnetic fields on the dynamics of astrophysical systems, and only after observations revealed the importance of magnetic fields were they willing to take them into consideration. The physics of the outer layers of our Sun is a first and typical example, and the dynamics of the inter-stellar matter has turned out to be another clear case. But, as usual, answers to certain questions raise new ones. The origin of the seed fields, for example, is still subject to considerable controversies."[350]; p.264-265

The reason for the considerable controversies is that it may indicate a purposeful Universe, and therefore, some scientists are immediately opposed to the idea of primordial fields.[350,351]

The timescale for diffusion of plasma across field lines is not a problem if a hierarchy of vortices exists. The ordering of small-scale fields can be brought about by small-scale vorticity in relation to large-scale vorticity.[146] The Field-dynamical Model is a system of vortex fields that accelerate layers of particles in vortex motions, which lead to other Field-dynamical Models on smaller scales and so on; a hierarchy of vortices.

Quasars are the result of an interaction effect and not due to expansion. The levels of hierarchy indicate a higher strength of redshift within systems than between them.[388] The higher redshift quasars are more luminous than low redshift ones. Moreover, the predicted angular-size/redshift relationship disagrees with the radio observations of distant sources. At high redshift there is a dominance of flat spectrum radio sources. Again, the evidence shows that the redshifts are not due to expansion, but involve a system of fields with a hierarchy of vortices.[740] The different redshifts result from the handedness, known as chirality, of the vortices.

Another bit of observational support for this scenario is that helium abundance is in uniformly high levels in galaxies. Helium is so uniform that it is difficult to suppose it came from ordinary stars. The conclusion reached from this scenario is that the Universe had a singular origin or it is "oscillatory" (i.e., time-varying).[361]

The gamma-ray bursters and the X-ray background radiation can only be explained by ad hoc assumptions in the conventional framework. More than enough energy is available to produce the microwave background. So, the essential problem is to find a mechanism which can accomplish this conversion and give a rather high degree of uniformity or "isotropy." It could be due to some shell or cocoon-like structure, as with supernovas and shells around galaxies.[11] The Field-dynamical Model is an arrangement of interconnected fields that form somewhat spherical "shells."

> "An important and puzzling problem in astrophysics is how the energetic, sometimes extremely energetic radiation is produced and how matter is accelerated to velocities sometimes approaching the velocity of light. Half a century ago it was discovered that the Earth receives an (almost) isotropic radiation consisting mainly of protons and other nuclei, with very large energies. This radiation is referred to as cosmic radiation, nowadays often 'galactic' cosmic radiation, although it has not been proved that the radiation we receive on Earth is necessarily generated in the galaxy."[11]; p.104

The Fields, through hydrogen fusion and particle acceleration, produce mainly protons and other nuclei with very high energies, and therefore, the source involves various astrophysical objects, and this is why it is isotropic, or in other words, uniform.

X-ray and gamma-ray astronomy has revealed that a large number of astrophysical objects (possibly all) emit high energy light-particles (photons). Some radiation is time constant or shows slow time variations. Others show rapid variations in fractions of a second. Time variations of magnetic fields pump the energy of a particle up and down. However, if particles are scattered by magnetic field irregularities (time-varying properties), it results in a net energy gain. When particles pass a double layer with the necessary voltage, they also gain energy. The maximum energy a particle can gain is limited only by the magnetic field(s).[11] Structural dynamics beyond conventional theory are required.

11.2 <u>Big Bang?</u>

The explosive vigor of the Big Bang had to be extreme enough to prevent gravity from overwhelming the dispersing material. If it had not, then the expansion would have been reversed engulfing the entire cosmos in an implosion, producing something like a huge black hole. It would have to have been fined tuned so that the material did not disperse too much or collapse on itself. Furthermore, the Universe is extraordinarily uniform on the large scale, which continues to be preserved through time. All excited quantum systems, such as the false vacuum comprising much of space, are unstable and will tend to decay. When that happens then the repulsive force disappears. This would put a stop to the "inflation" or "expansion" bringing the system under the control of gravity.

According to gravitational theories a small degree of clumping would had to have been necessary in the early Universe or there would be no galaxies and galactic clusters. Astronomers hoped that gravity could explain the aggregation since the Big Bang. A cloud of gas will tend to contract under its own gravity and then fragment into smaller clouds, which would fragment into even smaller clouds. The original Big Bang could have been uniform, and purely by chance accumulations became over dense here and there, and under dense elsewhere.

The deficiency of these assumptions is immediately apparent in that the growth of galaxies by this mechanism would have taken longer than the age of the Universe. There had to be something that brought about structure at the onset, and it had to be of just the right magnitude. With the Big Bang scenario, we are at a loss in explaining the large-scale structure of the Universe, why the galaxies are the size that they are, or why the clusters contain the numbers of galaxies that they do.[213]

New maps on refined measurements still show no obvious signs of any fluctuations in background radiation, which is believed to be the remnant of the Big Bang.[686] Discordant redshifts of many galaxies in the Local Supercluster indicate that there is no (*Hubble*) expansion in some objects. Meanwhile, other objects show an "expansion" ten times greater than the Universe as a whole. Therefore, the evidence that has been used to claim expansion is contrary to the Big Bang scenario, which depicts all expansion at the same rate. The large fluxes of helical (synchrotron) radiation maintained in the very small volumes also contrasts with the predicted observations of the Big Bang. Quasars may well be local objects that have been ejected along a system of fields from the nucleus of our galaxy or the Local Supercluster.[150] This seems especially evident when considering that a "young" quasar was recently discovered, when it was thought that quasars are some of the oldest celestial objects.

The center or core of a rich cluster called Abel 2029, extends six million light-years in diameter, which is more than sixty times the width of the Milky Way. It emits more than a quarter of all the light produced by the entire cluster. The central galaxy is remarkable for its huge size and its uniformity of structure, so much so that it could be labeled the core.

The most distant "normal" galaxy G 0102-190 resembles the Milky Way. The galaxy is surrounded by an extensive gaseous halo that is not usually found around nearby "normal" galaxies. Its discovery provides valuable clues about the formation and evolution of galaxies in the early universe. The great distance and order conflict with Big Bang predictions and massive black hole scenarios.[720] Conventional theories are at a loss when attempting to explain clusters, which, like galaxies, stars and planets, show the structure and time-varying aspects of the Field-dynamical Model.

The microwave radiation was immediately assumed to be relict radiation by those who believe in the Big Bang. It also requires that the Universe must be a closed system, which is an ad hoc assumption. This raises a number of questions that have been relatively ignored, such as indicated in this comment: "Suppose there was no initial dense state? Suppose that matter and radiation were never strongly coupled together? Suppose the laws of physics have evolved, as has everything else? The current climate of opinion requires that these questions *not* be asked."[151]

The large-scale structure of the Universe discloses an extreme lumpiness as far as luminous objects are concerned. This is contrasted with the extreme smoothness of the microwave background. Attempts to explain this by hidden (*baryonic or non-baryonic*) matter have reached a point of crisis. If fluctuations in the microwave background are not found it could mean that matter and radiation were never coupled. The microwave background has no imprints to mark the occurrence of events of a non-thermodynamic character, such as the condensation of galaxies, clusters and the large-scale structure of the Universe. It may be that evolution has occurred, while no conventional theory predicts evolution (which is like life).

The properties of all discrete objects now observed can be explained by having been generated by other sources. An obvious weakness in such a universe is that there is no reason that we should be moving at such a slow speed with respect to the microwave background.[72,150-152] The chaotic inflationary scenario is also not realistic and is flawed in many essential points.[422] The major problem can be stated thus: "The root of the matter, it seems to us, is one of time ordering. In the Big Bang model, the microwave background came first and the galaxies second, whereas the observations suggest (almost to the point of compelling) the opposite."[72]

The microwave background departs from the black-body form, which indicates that either extra components are present or the radiation has a quite different origin. In both instances the answer must be that the radiation or total flux arises in specific, discrete sources. The sources are the nuclei of galaxies, which radiate largely at the infrared and microwave frequencies.[148] Other sources are the planets and stars, which contribute much less than that of the galaxies. This could be predicted of the Field-dynamical Model of these sources, as well as a large-scale system of fields.

The discovery of a highest redshift quasar ($z = 4.7$) would make its formation back near the age of the earliest formation of galaxies in the Big Bang scenario. Yet, it shows no spectral differences from the lower redshift quasars that are supposed to be much nearer and younger. Moreover, quasars are said to be the oldest and farthest objects, but the discovery in 1998 of a young quasar, about a million years old, contradicts both the Big Bang scenario, and conventional theories on quasars. This is direct evidence that the presumed evolution of the Universe after the Big Bang did not take place.[72]

Another embarrassment for the Big Bang is the presence of compact objects, sometimes stellar in appearance, displaying gaseous emission lines. Generally not associated with other galaxies, they are usually referred to as "isolated extragalactic H II regions." The embarrassments are that if their redshifts are taken at face value they are completely isolated, and they seem to be very young objects.

The Universe is not supposed to be forming young, isolated galaxies. A question emerges about where the material came from, because the remnants of the Big Bang are supposed to have been consolidated into galaxies long ago. However, when these regions are plotted they fall into the connecting lines between low-redshift quasars and high-redshift companion galaxies.[66] Likewise, recent infrared surveys reveal considerable numbers of young galaxies, because they appear to contain mostly atomic and molecular gas, and high-mass young stars.[72] All of these observations contradict the Big Bang scenario, but support the idea that an intergalactic system of fields exists in a still growing and evolving universe.

11.3 Large-scale Structure of the Universe

A new map of the distribution of galaxies brings into question the presently accepted theories on how galaxies form, as well as the Big Bang. A markedly uneven arrangement calls into question the validity of the missing mass or dark-matter model of galaxies, as well. Huge galactic clusters separated by vast intervening voids do not fit the standard cosmologies. "There is more structure on large scales than is predicted by the standard cold-dark-matter theory."[638]

Surveys of the Universe's galaxies shows "bubbles," "filaments," "sheets," and "Great Walls" separated by gigantic voids. In one survey, galaxies were noted to be distributed on the surfaces of shells tightly packed together, next to each other. Several large regions void of galaxies are surrounded by galaxies in elongated, "sharp" boundaries. They appear something like a slice through the bubbles in a kitchen sink with the galaxies on the edges of the bubbles, but with a more orderly arrangement. Some voids are nearly circular. The observed and expected redshifts are different with two marked peaks. These observations are not explainable by gravitational models nor the Big Bang theory.[218,651,652]

A dipole structure (*anistropy*) within the microwave background also exists.[320] Along with this dipole structure there is also a quadropole structure. The quadropole structure is centered along the equator and displays the sign shape—more or less like an S on its side—noted in galactic warps.[217,226,290] This arrangement appears to indicate that those fields which produce the quadropole structure are along the 30° latitude of the Universe (see Figure 4.6).

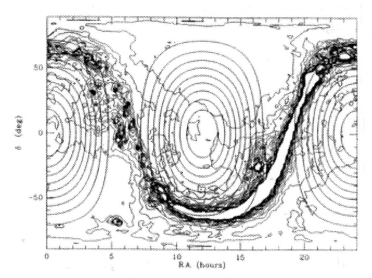

Figure 4.6. Structure in the Universe. The sine curve, typical of the Field-dynamical Model, is also displayed in the quadropole structure of the large-scale structure of the Universe. [from reference 217].

Scientists insist that the large-scale structure requires "the existence of primordial magnetic fields."[567] Over a distance of seven billion light-years galaxies are regularly spaced in clumps about 420 million light-years apart. This particular line of sight pierced a sequence of thirteen evenly spaced walls of galaxies. All galaxies are arranged in a large-scale, regular pattern that is beyond the random chance gravitational accumulation of matter and the explosive effects of the Big Bang.[460]

The Universe is also known as the Metagalaxy, the observable part of the Universe. The concept allows for the possibility of other universes, or an infinite universe with other characteristics that we may never observe. It includes an expansion mechanism relying on the presence of a magnetic field, not on a Big Bang.

The metagalactic plasma is penetrated by a highly stable magnetic field. As gravity contracts the initial metagalactic cloud's radius, it strengthens the magnetic field. Hence, part of the gravitational potential energy released during the contraction is transferred into magnetic energy. If the metagalactic magnetic field is sufficiently strong a good part of the potential energy will be transferred into magnetic energy. Furthermore, with a large enough magnetic field the outwardly directed magnetic pressure force may dominate over the inwardly directed gravitational force. According to Einstein's theories the sources of this force are particle mass, magnetic energy, and also the magnetic pressure itself. The result will be an expansion of the metagalaxy without a Big Bang.[470]

The expansion or growth of the metagalaxy can also be caused by the pressure of very energetic electrons and positrons formed during the annihilation phase or by thermonuclear fusion. A multi-stage, multi-step acceleration, such as the time-varying acceleration of the Fields, would produce much higher velocities. Observations of galaxies and quasars make it necessary to envision a number of separate expansion mechanisms or time-varying acceleration, and a time-varying gravitational field. If the highest observed redshifts are representative for the metagalaxy then a homogeneous or Big Bang scenario encounters serious difficulties. This is further evidence of an non-Big-Bang inhomogeneous metagalaxy.[470]

Galaxies seem to be receding from one another, with the more distant galaxies receding more rapidly. At the same time an astonishingly uniform cosmic microwave background endures. This uniformity is difficult to reconcile with the clumping of matter into galaxies, as well as the even larger features extending across vast regions of the Universe, such as "walls" and "bubbles." Large groups of galaxies seem to be streaming in particular directions. These observations indicate that there must be a structure-forming mechanism that creates features as large as five percent or more of the size of the Universe.[66,72,298,320,483]

The Big Bang allows too little time for the force of gravity by itself to gather matter into the patterns of galaxies and vast areas of "nothingness" observed. With the Big Bang model and the physical constants, there is no way that a very homogeneous Big Bang could produce the structure of the Universe in its lifetime. A mechanism must produce fluctuations in the geometry of space-time, or in the density of primordial matter early in the history of the Universe. The character of the mechanism would also have to be small enough not to disturb the microwave background or produce it in a later stage.[66,72,298,320,483]

A structural arrangement is obvious:"Also very impressive, however, is the fact that we see many examples of chains of galaxies throughout extragalactic space. In many cases these chains are aligned with radio ejections. There seem to be abundant precedents for the configurations to actually represent some sort of line or filament of objects."[66]; p.132

Because chaining is a phenomena of galaxies it may mean that it is a short-lived phenomena, and therefore, the timescale presently used is wrong.[71] Likewise, the outer planets and satellites have methane. However, hydrogen and helium ions, abundant within the ecliptic plane, erode or polymerize methane. Time, much less than the proposed age of the planets, should have caused methane's absence, if no mechanism exists to replenish it.[466] Likewise, the rings of planets should have disappeared long ago. The Universe may have formed much more recently than we have imagined, and has and will continue to form in epochs.

When quasars are projected on a two-dimensional sphere of the Universe definite features become observable, as well. The equatorial region is void of quasars, and the 30° to 40° latitudes show a greater concentration.[581] Typical of the Field-dynamical Model, a ring effect and time-varying ejection at the 30° to 40° latitudes are noted. This is in addition to the dipole and quadropole structure observed in the microwave background.[217] The Universe itself displays all the attributes of the Field-dynamical Model.

Acceleration processes are also evident. For example, a large-scale streaming motion, at speeds of about 640 kilometers (400 miles) per second, exists for our galaxy. No theory has predicted this motion, and it is unexplained by gravitational forces. Other galaxies are also moving. Typical of conventional thought, this motion has been labeled the "Great Attractor," even though gravity cannot possibly be involved.[232] Like everything in the Universe, the Universe itself rotates, which is expected of the Field-dynamical Model curving space-time, but not of strong field gravity.

Unconventional wisdom about magnetic fields is that there were weak "seed fields" present in the early Universe. These seed fields were amplified by shocks, and arose in the current systems present in plasma flowing along the curvature of space-time. Meanwhile, magnetic fields and electromagnetic phenomena are not included in conventional theories about the early Universe.

One unsolved problem involves the strengths of magnetic fields in extended radio sources outside galaxies. These fields are as strong as the fields in the galactic disks. However, the densities of the gas in these two different regions are vastly different. Primordial fields with a mechanism directing mass flow must be responsible (warping time-space; i.e., the Field-dynamical Model).[151]

The mind-set, molded by conventional, is what deters the truth of the matter. "In the 1980s we all talk as though the only detectable objects in the Universe will be galaxies and QSOs. Are observers thereby foreclosing the possibility of finding something else? The answer is probably 'Yes.'"[151]; [p.227]. Many observations cast doubt on "established" theories, while some new ideas are discouraged nearly every step of the way.[66,151]

As stated, the angular momentum versus mass relationship for the full range of astronomical objects (from asteroids to superclusters) shows a remarkably tight clustering around a specific value (*a power law of index 2*). This demonstrates that a saturation of the electromagnetic force is involved. Something still remains to be explained about the predominance of the electromagnetic force.[715] In contrast, conventional perspectives have not allowed electromagnetism to be studied fully: "All four forces in physics—gravitation, strong nuclear, weak nuclear and electromagnetism—have been used to explain the Universe, electromagnetic effects have been less explored than the others."[327]

Magnetic monopoles would complete the symmetry between electric and magnetic phenomena, satisfying a demand of Grand Unification Theories (GUTs). Magnetic monopoles would interact in a way that is like an electric point charge. Two like monopoles would repel each other, while two unlike monopoles would attract each other. The electric current produced by the motion of the electric point charges generates a magnetic field. A magnetic current formed by moving monopoles would induce an electric field. Plasma flow in conjunction with the electric field would induce a stronger magnetic field. A monopole model accounts for the intensity of the particle (*positron-electron pair*) annihilation line, and the high energy photons in the galactic center (*monopole catalytic reactions in the center*).[741]

Cosmic vortex lines are a consequence of symmetry breaking in many grand unified gauge theories.[320,567] All that is required is a field system that polarizes the vacuum producing (*virtual*) particles, which are then accelerated in right- and left-handed vortices (helixes). Anomalous gravity measurements are now known to arise from polarization of the vacuum for this reason.

Furthermore, a new constant that equates angular momentum and mass must exist on theoretical grounds. A symmetrical property most astronomical systems share is called self-similarity. In addition, most systems have density and rotation laws that are scale-free. Planets, asteroids, the Solar System, double stars, spiral galaxies and superclusters (probably even the large-scale structure of the Universe) share this property.[144,752-754]

Rotational mass produces a magnetic field which can be observed in objects ranging at least from the Moon to pulsars.[7,8] The magnetic dipole moment and angular momentum are proportionally related, and this implies an unknown electric field. Electric fields have long-range force. Again, the facts conform with the Field-dynamical Model.

CHAPTER 12

A Fundamentally Biological Universe
(A Teleological Cosmos)

The Universe is fundamentally biological. Even the Urey-Miller experiment that simulated the theorized early pre-life conditions on Earth, and produced amino acids, suggests this. The ammonia used was obtained by a process involving hydrogen of bio-origin, and the methane was also biological in origin. Non-biological catalysts would be poisoned almost instantaneously by sulfur gases under pre-life conditions. What this means is that most of the material in interstellar grains must be organic or life itself would have been impossible.

12.1 Widespread Organic Constituents

The spectrum for all grains along the line of sight from the galactic center to the Earth is very much like that of dry bacteria. Either the grains are bacteria or are organic grains in proportions like bacteria (amino acids, nucleic acids, lipids and polysaccharides).[359] Therefore, both theoretically and observationally, organic constituents fit the observations.

Organic materials or bacteria would easily align in magnetic fields, and could produce superconducting surfaces that would generate filaments. Organic materials or bacteria could more easily produce the variety of objects in the Universe than inorganic or non-biological materials. As with so much of its constituents, the Universe itself is fundamentally biological.[348,359,362,563,760] In fact, so much is this the case that life constitutes a physical law; it had to arise, it was an inevitable result of the laws of physics as they exist. The difficulty with accepting such a perspective in scientific circles is described:

> "If this were an entirely scientific matter, there is little doubt from the evidence that the case for a fundamentally biological universe would be regarded as substantially proven. The reason why the scientific community passionately resists this conclusion is that biological systems are teleological, which is to say purposeful. And if we admit the Universe to be inhabited by a vast number of purposeful components then the thought cannot be far away that perhaps the Universe itself might be purposive, a conclusion that not only stands astronomy immediately on its head, by making a large fraction of our efforts futile, but which offends the tenets on which modern science presents itself to the public. It is here where the biggest issue ultimately for science may lie."[359]; p.6

Bacterial membranes act as one-way systems with respect to electric charges and chemical flow. Some bacteria contain an ordered (ferromagnetic) domain that enables orientation in a external magnetic field (see Chapters 5 and 15 of Volume One). Wavelength dependence of polarization produced by alignment with the magnetic field agree "excellently" with the data curve of iron-bacteria.[359] This also evokes the idea of primordial magnetic fields, which in turn suggests the idea that the Universe itself is teleological.

12.2 Thermodynamics

Like life, some evidence of activity which does not conform well with thermodynamic law, as it is usually envisioned, can be found in the

Universe itself. Self-gravitating systems become hotter and are progressively self-organizing. In fact, the structural tendency of self-gravitating systems, including the Universe itself, is a fundamental principle. Similar to life these observations run counter to the usual interpretation of the Second Law of Thermodynamics, which indicates that systems should become less ordered and lose energy as time passes.[214]

If the Universe collapses after having reached a maximum radius then hydrogen would be regenerated. Thereby, the Second Law of Thermodynamics is contradicted again, because by definition hydrogen should only decay. This is among the other difficulties of the standard cosmologies.[581]

Black holes were once though to overcome the Second Law of Thermodynamics because of their ability to "swallow" entropy.[214] Then a physicist theoretically described that a black hole possesses entropy, and it is measured in terms of the size of the boundary surrounding the hole. Known as the "event horizon," this boundary can be passed on the way into the hole, but nothing can ever cross it if it is trying to move outwards. Yet, as discussed, ejection phenomenon streams from the region, even the seeds of new galaxies.

Applying this theoretical perspective to the Universe, which should have its own event horizon, shows the Second Law fails at least in some situations. A physicist comments: "In a perfectly normal region of space-time the generalized Second Law of Thermodynamics fails."[215]; p.1349. Like life, the Universe seems to be exceptionally good at overcoming its own entropy.

Because nothing that happens within a black hole could ever influence anything outside, the presence of a singularity there does not matter. A singularity is a point in space where physical qualities, such as density and gravitational force, become infinite. The Cosmic Censorship Hypothesis states that singularities will always be clothed by event horizons, and can never be visible from the outside. However, a recent model using a supercomputer uncovered evidence that the gravitational collapse of certain distributions of matter leads to the formation of a "naked" singularity. As a result, one cannot say anything precise about the future evolution of any region of space containing the singularity, because new information could emerge from it in a completely arbitrary way.[661] This brings into question our theories on black holes, the Big Bang, and even the gravitational aspects of general relativity itself—all of which are gravitationally focused theories.

12.3 <u>Semantics</u> <u>of</u> <u>the</u> <u>Mind-Set</u>

When it comes to the order we observe it should make us think more philosophically about why. Einstein's equations can just as easily describe emptiness, as well as galaxies and quasars. Mass and energy are evenly distributed, as occurs with the microwave radiation in the Universe, which varies no more than a few parts in 50,000. Why is there anything in the Universe at all? Why should widely separated regions of the Universe be so similar when there are so many ways to be different?

In extragalactic astronomy, part of the problem concerns the fact that it is an observationally dominated science. Therefore, it tends to be descriptive rather than quantitative. This "orderliness" of the Universe has been repeatedly commented upon, and the following quotes will serve only as examples which reflect the current mind-set's contrast.

"Contemporary science when interpreting the existing reality does not like to ask the question 'why,' and it has completely dropped the question 'what for' from its vocabulary. The Anthropic Principle in cosmology has again focused our attention on the question 'what for.' I think we should not be ashamed of the return to that old question, which seemed so old fashioned for a long time. What was pure philosophy is now science."[624] "It is clear that I espouse here that Baconian principle of induction of general laws from a body of facts. It would seem obviously that if a scientist only reasons deductively from known laws then he or she can never do more than recover those laws, and will never discover anything fundamentally new."[66; p.178.] "Indeed, if the history of science is any guide, theory and observations have never exactly been in phase."[546]; p.244. "Cosmology has always been—and will by definition always remain—a borderland between science and philosophy—some would say a religion."[11]; p.123

After more than three centuries of Newtonian gravitational theory, built into even our observational perspectives and relativity theory, our observations will yield only gravitational descriptions of the observed phenomena. The heart of the dilemma of today's theoretical panorama emerges from such a mind-set. The problem is that we see what we have been taught to see, but is it *really* reality?

The expansion of the Universe's radius occurs at a constant rate, known as Hubble's Constant. Evidence also shows that the Earth has expanded at the same constant.[427,729] When "expand" or "expansion" are used it is merely a philosophical perspective, because grow or growth could explain the observations just as well. In fact, one can find these words are synonyms in any thesaurus. Other terms used are "inflation," and an "accelerating" or "runaway" Universe.

The formation of very bright (*ultraluminous infrared*) galaxies involves the interaction or merger—a symbiosis or consuming—of two molecular gas-rich spiral galaxies. The funneling of molecular clouds toward a merger nucleus accounts for nuclear starburst and provides fuel—food—in the form of gas and stellar remnants.[120,683] A possible evolutionary connection between these galaxies and quasars is that consuming the dust causes it to evolve into a quasar.[120] Conversely, the dust could be ejected by the quasar as part of the evolution towards forming a galaxy.[66] Furthermore, the dust could very well be organic grains or even bacteria.[359] Again, all of the terms commonly used could be replaced with words that give the Universe a more life-like character.

Observations of quasars and these galaxies suggest that the majority of quasars are formed through galaxies consuming each other, or as usually described, by collisions or coalescence.[735] Again, the use of certain words is a matter of semantics, emerging from a philosophical perspective. Consuming or eating (even cannibalism) are biological terms, and therefore, their use is discouraged. The other possibility is that they are ejected or given birth, if you will, as discussed. Large galaxies are believed to consume smaller galaxies and then grow in size. Likewise, twin or double stars can lead to the larger stars consuming smaller stars. Stellar evolution is the predominant mass loss mechanism for the smaller nuclei, while collisions dominate the larger nuclei.[538] One could describe this as the smaller stars or nuclei growing, while the larger stars or nuclei are reproducing or consuming other stars or galaxies. One could say that the Universe is a continually fruitful, huge buffet.

All quasars begin in the ultraluminous infrared galaxy phase. Blazars and Seyfert I's could also arise as the final product of this evolution.[683] Ontogeny, the development of an individual organism, could also be just as descriptive. Moreover, in the evolution of the biological world the development of an individual organism is said to show its evolutionary stages (ontogeny recapitulates phylogeny).

Double nuclei and multinuclei galaxies confirm that the Universe goes from a dense condition to a rarefied one, and is again like life. Some galaxies (i.e., Markarians, etc.) have similar brightness, and about the same strong emission line nuclei or knots in a common envelope. Approximately one hundred multinuclei galaxies have been noted that are not the result of the interaction or merger of two independent galaxies. Close-by galaxies with a single nucleus may have originated from multinuclei galaxies. One astrophysicist notes their life-like character, because they "are formed from a single maternal body, as a result of its divisions, in analogy to what takes place in the biological cell."[421]; p.118. It is rare to find a scientist who is willing to use biological terms when describing anything outside of the Earth's biosphere.

The galaxy Arp 220 shines brighter than 100 Milky Ways in the infrared. Two closely spaced sources of intense radiation buried deep within the galaxy's core are present. Astronomers have found that four out of ten known ultraluminous galaxies have what appear to be close pairs of nuclei. Evidence suggests that this arrangement occurs in all ten.[321] Such an observation has suggested to some astronomers that it is in the late stages of a galaxy merger, or in other words, the consuming of one galaxy by another. Other interpretations are that they are ejecting matter or splitting into two galaxies, analogous to a biological cell, giving birth to new galaxies.

How special is the Universe?

> "An examination of the delicacy of advanced life forms and the requirement of highly special conditions for their existence would seem to support the long-standing tradition that the Universe has been created in a singularly convenient form for the presence of intelligent life. For example, it is sometimes pointed out how various naturally occurring numbers must be restricted in their values in order to be consistent with the existence of living matter. [One scientist] has remarked that an increase in the coupling constant of the strong interaction by only a few percent would have resulted in a catastrophic synthesis of most of the hydrogen in the Universe during the early stages, thereby depriving stable stars such as the Sun of their energy source, making life seemingly impossible. Another important naturally occurring number is the ratio between the electrical and gravitational forces between an electron and a proton (a staggering 10^{39}). Strangely, this number coincides with the age of the Universe measured in atomic units of time. Thus the approximate equality of the two numbers is not a coincidence, but a direct result of our own existence."[212]

Some of us would not even marvel at the idea that these naturally occurring constants exist, and that without them the Universe would not allow for life's existence. Some are even trying to explain them away as coincidences. This mind-set shows how much of a religion science can be for some people.

Furthermore, more objects are alive in the Universe than what we have already accepted as living. The planets, Solar System, galaxies and the Universe also display characteristics that can be considered living, as well as harbor the life forms we accept as living. And then, here we are literally the self-consciousness of the Universe, and able to marvel at our own existence. We have not, however, completely accepted the purposefulness or teleological nature of the Universe. **Life itself is the highest physical *law*!**

In Tome Five we will look at the unity of the Sun, Moon and Earth, along with the life on it. Their intimate connections will very likely surprise you. They are, in fact, a single unit.

TOME FIVE

The Unity of the Sun, Earth and Moon
The Solar-Lunar-Terrestrial Linkage
In the Historical Process

"Science is the organized attempt of mankind to discover how things work as causal systems." Conrad Hal Waddington (*The Scientific Attitude*. London, Penguin, 1941)

"In our universe, matter is arranged in a hierarchy of structures by successive integrations." Francois Jacob (*The Possible and the Actual*. NY, Pantheon, 1982, p.30)

"Our business is with the causes of sensible effects." Sir Isaac Newton (in *Physics Bulletin*, Sept 1977)

CHAPTER 13

Celestial Orb of Light, From Where Comes Thy Might? *(Solar Structure, Solar Activity, and the History of Solar Activity)*

Being the dominant body of our Solar System, the Sun exerts a tremendous influence on the Earth, and the other planets. Within the Sun, thermonuclear fusion brings helium to a temperature of about 15,000,000° Kelvin (nearly the same in Celsius, or 27,000,000° F). Deep within the interior of this fiery globe the pressure reaches many thousand million times that of the Earth's atmosphere at sea level.

Above the visible surface of the Sun (photosphere) is the inner atmosphere (chromosphere), a layer 5,000 kilometers (3,100 miles) deep. Extending above this layer lies an exceedingly thin and obscure gaseous envelope, known as the corona, that stretches far beyond the Earth. Normally, the Earth intercepts only one part in 2,200,000,000 of the enormous output of the Sun, yet such a consistent flow has been sufficient to perpetuate life on Earth since its inception.

13.1 <u>Solar</u> <u>Structure</u>

In spite of a great deal of recent research, and satellite observatories launched to study the Sun, the actual way in which the Sun functions has remained a mystery.[332,469] Our purely physical perspectives on the origin of the Sun's energy have proven to be inadequate (especially with regard to the neutrino flux; as will be discussed). A number of the observations have made it evident that the Sun has more structural dynamics that what was expected.

Skylab missions in 1973 and 1974, and early 1970s observations made by the Naval Research Laboratory, revealed intriguing facts about the Sun's structure and operation. A film was taken of the Sun in June 1973, revealing "a transient blob" the size of the Sun itself moving outward through the corona at speeds of about 400 kilometers (250 miles) per second. A few months later, a study of the outer gaseous corona (*coronal spectrography*) showed an eruption of helium rising about 800,000 kilometers (500,000 miles) above the Sun's surface. This blob then stopped as if it was blocked by an invisible wall, which was described as a "total mystery." The Solar Maximum Mission satellite also took photos of a huge bubble-like structure that extends from the solar limb to several solar radii. These observations revealed that the Sun has an outer magnetic sphere that confines the solar surface. More recently, these observations have been attributed to a magnetic sphere that confines the Sun's corona, in particular.[36,428] Yet, originally no scientist envisioned magnetic fields of this nature for the Sun.

An unsolved riddle is that nobody has been able to explain why the regions along the Sun's equator rotate faster than any other region. A point near the equator takes about 25 days to make a round trip, while the polar regions take 33 days to complete a rotation. These observations have been called the most "surprising and puzzling solar features."[469]

This differential rotation is the result of the Field-dynamical Model of the Sun. The entire Sun is more visibly involved in the acceleration process than are most of the planets. However, Jupiter, Saturn and Neptune all show a similar differential rotation on their surfaces. This differential rotation is the surface manifestations of the gradient essential for the time-varying, linear acceleration of hydrogen plasma towards the core.

Fairly recent observations of the Sun have disclosed this model, and phenomena at the poles present some of the evidence. For one, there are spiral patterns at the poles that have been referred to as polar "topknots." In fact, sunspot numbers during solar maximums are correlated with the strength of the Polar Field during the preceding minimum.[644] A group of scientists comment on these polar topknots:

> "The maintenance of such a concentrated polar field requires the presence of a poleward bulk flow. Around the time of sunspot minimum, most of the interplanetary magnetic flux observed at Earth originates from the Sun's polar regions. A strong polar field is required to account for the observed strength of the Interplanetary Magnetic Field around sunspot minimum."[644]; p.714

The poleward bulk flow is something like a self-induced aurora (see Figure 5.1).

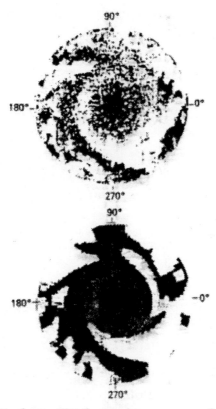

Figure 5.1. Polar Topknot. Similar to the patterns of aurora, the stacked polar plates of Mars, the pole of comet Halley and spiral galaxies, the Sun's Northern Hemisphere polar region displays the spiral pattern typical of the Field-dynamical Model. [from reference 644; Figure 2]

During solar maximums the polar streams (*faculae*) disappear, and energy is being ejected at mid- to low-latitudes in sunspots, solar flares, coronal mass ejections, and solar prominences.[330] The greatest impact on the Earth occurs with coronal mass ejections. Like ejections from active galactic nuclei, and ejections from the planets that caused some cratering, the Sun also has ejections. In order to make the following comments more uniform, and to adhere to some researchers original comments that it was solar flares, not necessarily coronal mass ejections, that caused great influences on the Earth, they will be referred to as solar flares.

The polar regions are typically shifted upward in brightness, which, unlike other features, is not dependent on the solar cycle. During solar maximum, however, the polar streams (*faculae*) disappear, but not the polar brightness.[330] The polar streams and brightness are something like a self-induced aurora, and are like the planets with their polar-wind induced aurora, while like the Sun, there is ejection along mid-latitudes.

This Field-dynamical Solar Model is also revealed by other observations along the mid-latitudes and the equator. During solar minimum, and at about the 35° latitude, new sunspot activity begins, and advances towards the equator, as the cycle moves towards maximum.[516] Sunspots decrease in latitude as the cycle progresses, which indicates acceleration along latitude bands on the solar surface.

Sector regions in the magnetic field extend up to the 35° latitude, and also spreads about 90° in longitude. These regions are very stable, and are associated with the interior, not the solar surface. Regions of opposite magnetic field polarity are separated in interplanetary space by a thin boundary layer that is inclined approximately 15° to the Sun's equator. Also large-scale electric fields coexist with the magnetic fields, whose magnetic lines form a spiral.[603] Moreover, surface temperatures at equatorial and mid-latitudes are displayed as more active (see Plate 5.1).[330] These observations are some of the evidence that reveals mid-latitude and ring-like structures on the Sun, but there is more.

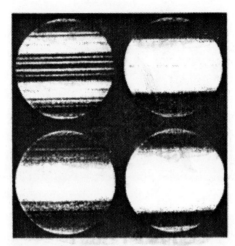

Plate 5.1. Solar Latitude Temperature Bands. Solar temperature distributions show distinct latitude bands similar to the bands on Saturn and Jupiter. From the upper left and moving down, then to the right, the images are of the summer Sun in 1983, 1984, 1985 and 1987. [from reference 330]

The Sun's (heliospheric) magnetic field has an underlying spiral or helical pattern that closely resembles a stellerator used in laboratories as a plasma device designed to cancel out drifts.[199] This field is known as the Interplanetary Magnetic Field, abbreviated IMF, which reaches beyond the planets. It is a field that issues off the poles, but ends up in the ecliptic plane, something like a huge planetary ring, but composed of solar plasma and a large-scale magnetic field.[199,516]

The Sun also has a ring of globular matter around the equatorial region. A latitude dependence displays speeds approximately 10% lower, and particle density 20% higher, along the equatorial solar-ecliptic plane. High-speed solar winds are associated with coronal holes near the equator, and produce the solar wind in the Earth's vicinity. The IMF forms a tight spiral near the Sun's equator, with the most active regions typically between 10° and 30° latitudes.[603]

Interconnections between the solar wind and the IMF can be seen in observations of solar particle events. Acceleration processes in the solar corona, and in interplanetary space as the result of the IMF, are what drives these particle events.[311,635] Ions and electrons are accelerated along the plane of the IMF (partially due to shockwaves), as has been observed from inside Mercury's orbit out to Neptune's orbit. Field-aligned particle flow, especially electrons, takes place throughout this distance, with additional acceleration as the wind reaches each of the planets.[395] Meanwhile, IMF loops stretch from the Sun's surface to beyond the planets. The IMF's strength increases with solar activity, and its polarity reverses every 11.1 years, along with the Polar Fields, during solar maximums.[516,603]

Observations of the formation of Sun-like stars confirms that the Field-dynamical Model is at work. The details disclose that there are molecular outflows as a star forms. These outflows are like the outflows in the Polar Field on Earth, dubbed the polar wind. Likewise, there are both inflowing and outflowing winds at the poles of the Sun and other stars, as with the solar and polar winds in the Earth's auroral zones, as well as the other planets. Ring-like disks form along the star's equatorial region, as do planetary rings, galactic disks, and so forth. The Field-dynamical Model can explain the collapse of molecular clouds into stars, because the process requires a stable, structured magnetic field that resists any random turbulence in the gas (see Chapter 10 for further discussions on star formation).

13.2 <u>Flare</u> Processes <u>and</u> <u>Formation</u>

Active solar flare regions are known to occur along certain longitudes, as well as latitudes. Longitudes of flares were noted for 216° to 246° in the Northern Hemisphere, and 60° to 150°, and 260° to 320° in the Southern Hemisphere. Typical of magnetic field interactions, the angles in relation to the core are 30°, 60°, and 90°.[45]

Others see the longitudes of flares at 0° and 180° as significant centers of activity. However, for approximately three years these areas did not show much activity. This conclusion was based on 32 flares and subflares for 1972.[360] These regions show a slow drift against solar rotation, which demonstrates the stability of Field position in spite of other physical processes.[684] The longitudes are similar to the jets and anitjets, and the distributions of planetary and protoplanetary nebulas, in galaxies (see sections 5.1, 5.7, 8.1 and 10.4).

Flares in both hemispheres show a 152-day to 158-day period, with an average of 155 days, independently (both within and outside active zones). Other periods are one-third and one-half this cycle, and are 51 days and 77 days.[45,598] Flares also rotate with a 26.75-day cycle. These observations are surprising in the conventional framework, because they suggest that there is a regular time-varying component that is dynamically coupled to internal processes.

It is clear that there is magnetic stability in flares, which is not what was expected from a turbulent and abrupt event on the Sun. Likewise, flare build-up displays magnetic complexity. Sunspots of opposite polarities are embedded in a common structure (*penumbra*). Flow, otherwise known as flux emergence, and motions at the Sun's surface display magnetic shear with complex configurations that are essential to flare formation. A fast realignment of magnetic structures takes place, even on a fine scale.[587,588] For example, on 22-27 May 1980 flare conditions were preserved for many hours, even days, despite the eruption of flares, and the eroding effects of sunspot groups. Flares also erupted about 2.5 days after the first sign that a small amount of new flux was emerging.[205] This shows that there is something that stabilizes the flare regions that is unaccounted for in present theoretical perspectives.

Flare loops are formed through a magnetic reconnection in a "local" outwardly active field, and a short-lived transient that is enhanced by an *external* magnetic field.[152,480] New structures (*filaments*) form again after the flares take place due to a heating mechanism that was active before flare maximum. Upward motions occurred without disturbing the initial down flows.[152,370] A process is at work that is not the result of an interruption of a circuit on the solar surface, and the particle heating occurs in discrete bursts (i.e., time-varying acceleration).[152]

These observations show how stable the flaring region is itself, because great bursts of solar activity do not disturb the formation of more flares in the same location. In fact, a good place to look for more flare activity is in the same location in which it occurred previously. Experts find it puzzling when describing flare events, and magnetic field structure.[9,674] The vortex structure requires a differential rotation of the plasma that is accelerated along the same line, which, according to scientists, is very much like the magnetic reconnection required for quasars and jets in galaxies.[82] A scientist discusses the stability of flare formation:

> "The chief lesson to be learned from this survey is that the essence of the flare process, whatever it may be, is surprisingly durable. At one extreme we have situations where magnetic patterns evolve slowly; minor subflares recur with similar shapes at the same, but somewhat modified, locations for many successive days. At another extreme, in rapidly evolving magnetic situations, we see increasingly powerful flares repeat the same patterns at precisely the same locations in just a few hours. As in the case of emerging flux, photospheric motions seem to be an essential but still obscure ingredient in triggering flares."[205]; pp.26 & 29

The photosphere is the Sun's surface, and the obscure ingredient in triggering flares is the effects of remote field(s) in conjunction with the IMF, as will be discussed.

13.3 Remote Field Effects

This remote field effect can be seen in observations of magnetic forces. Surface features, known as coronal transients, start upward before a flare's impulsive phase. This implies that there was a magnetic field instability in the flaring region before particle acceleration, and plasma heating took place.[205,526] Large motions of sunspots and magnetic fields indicate that the magnetic field was stressed prior to flare buildup.[249] Indicative of a remote field "pulling" at the solar surface, magnetic arcades or arcs form above the Sun's surface in the pre-flare structure.[271]

Studying the process from the development of sunspots to flares reveals what could be expected of the effects of a remote field. There are high-velocity motions among sunspots of opposite polarity within the group (*i.e.*, *magnetic shearing develops*), and the leader sunspot of the group moves rapidly in the direction of solar rotation.[163] From growing flux to the formation of flares, an intimate interaction occurs with adjacent old flux-flare sites. The center of activity is the new/old flux boundary, where the new flux erupts when the surface magnetic field becomes to weak to prevent its eruption.[382]

As could be predicted from a remote influence, the shear of the magnetic field in a solar flare is partly due to sunspot motions, but "other mechanisms" are required. Likewise, the observations show that the magnetic field is parallel, and the potential field is perpendicular, to the solar surface.[235] These manifestations indicate that a remote field with a helical contour is interacting with the solar surface, producing flares.

For the same reason, the centers of sunspots (*umbrae*) are the centers of the magnetic fields. In large flares there are fast motions, magnetic shear, large gradients, and strong curvature (*of the zero line*).[296] Those theories that attempt to describe flares as an internal process are known to be flawed, while flare formation is best described by an "exploding" current.[12,13,152] A scientist's comment reveals what could be expected from a remote influence: "The likelihood of remote magnetic interconnections was certainly high."[205] These and other facts indicate that a remote magnetic influence from a remote field(s) is responsible for flares.

For the first time, solar astronomers have observed bursts of coherent radio emission coming from a large sunspot. The bursts emanated from a maser, which is the radio equivalent of a laser, in an intense magnetic field. The radio waves are generated by high-speed electrons in a region with converging magnetic fields, and produced short pulses (i.e., time-varying acceleration). These discharges are billions of times brighter than other radiation of the same wavelength produced on the Sun by any other process.[206] Furthermore, electron beams are noted in solar flares.[326] These observations suggest that more than just solar processes are at work, because the region would not otherwise be so magnetically stable to produce a maser and electron beams.

One of the first mechanisms proposed for the acceleration of flare particles is a time-varying electric field. Electric fields have long-range force, and can produce both the perpendicular and parallel features on the solar surface. This is what could also be expected of a remote field.

A number of observations indicate that a field(s) outside of the Sun effects the formation of flares. Solar plasma emits x-rays, and additional "run-away" electrons, that are produced by electric field acceleration.[326] Magnetic fields are about 75%, and plasma streams about 25%, of solar-wind outflow.

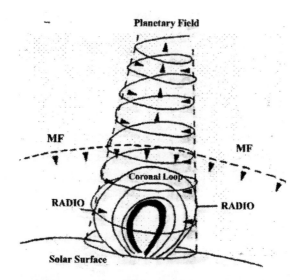

Figure 5.2. Mechanisms in Solar Activity. This figure displays the mechanisms that cause solar flares, coronal mass ejections, sunspots, and prominences. At the bottom is the solar surface. A Field from a planet is accelerated towards the Sun's surface after passing the outer magnetic field (marked MF). The planet's Field then uncouples the magnetic confinement at the surface and accelerates solar plasma, forming solar flares, sunspots, and other forms of solar activity. Radio signals are observed on the outer edge of the prominence (referred to as the coronal loop) and are due to the Field's acceleration processes. The arrows represent the direction of the magnetic lines of force. What is referred to here as the coronal loop is a site of coronal mass ejections, solar flares, and sunspots.

Likewise, when compared to quiet times, solar flares' magnetic fields are enhanced during the formation of flares. During solar maximum, shocks have the strongest magnetic fields. Pre-flare mass motions include "surging arches" of an *overlying* magnetic field structure. Hard x-ray brightening indicates that an acceleration mechanism is at work (see Figure 5.2).

Furthermore, mass ejections and x-rays are the primary particle acceleration event of the flare. Present theories do not offer the answer, as can be seen in this statement: "In each case, some 'speed-up' mechanism is needed to provide the impulsive timescale and particle acceleration."[623]; p.102. Moreover, magnetic fields increase the energy output of flares.[561]

Initial energy release occurs in a compact magnetic loop that later develops into a larger structure.[245] Radio emissions and hard x-rays are directly related to the coronal loop, while electron beams hit the "footpoints" of the loops. The footpoints are those portions of the loops that make contact with the solar surface. The production of energetic electrons plays a major role during the impulsive phase. Also indicative of a remote field are white-light flares, which are produced from energy *above* the surface (*photosphere*).[258,468] Bright features (*i.e., plages, granulation boundaries, etc.*) correspond to strong vertical magnetic fields. Dark features (*i.e., fibrals and filaments*) are associated with strong horizontal magnetic fields. The heating shows that there are considerable differences between the horizontal and vertical magnetic fields.[563] As could be expected from an external, time-varying field, coronal mass ejections show "multiple spikes," as does the magnetically disconnected outer corona.[282,402]

Further support for this understanding is in the twists, vortices, spirals, and other features observed in flares. An upward vortex motion is often related to flare formation. Typical of a time-varying particle accelerator, the energy stored in the emerging flux region, close to a sunspot, is periodically released as elementary flares.[366,400] Flare structures twist and spiral, and the answer to their formation does not lie with present mind-sets: "Particle acceleration remains a big problem, the whole process can be much more complicated and dynamic then we thought before!"[488]; p.43

Observations of large flares disclose a high degree of magnetic complexity, shear (*near the polarity inversion as indicated by a plage filament*), and rapid motions that are either vertical (emerging flux) or horizontal (satellite sunspots). Some flares with no emerging flux show strong twisting.[488] Minute to minute changes are not mass motions of ions, but a time-varying magnetic field accelerating high-energy electrons in waves. Coronal mass ejections form in loops, or most often, bubbles caused by an overlying (i.e., remote) field. The electrons are accompanied by radio-generating shocks typical of time-varying acceleration.[641] There is also a bright ring whose southeast corner ascends, and northwest corner descends, as could be expected from a remote, helically contoured field.

Likewise, observations of filaments disclose that the ejection of matter has a "rolling" or "twist" in the active region (prominence).[290] Eruptions, called prominences, are generated by a huge magnetic field system. Often magnetic filaments creating the eruptive prominences are twisted and intertwined.[515] Again, this indicates a remote, helically contoured field. The unresolved question is where does the remote field originate?

13.4 Earth Triggers on Solar Activity

When considering all of the observations, they call for a spiraling electromagnetic helix typical of the Fields on the planets, like the Polar Fields' and their helical auroras. These fields are accelerated towards the Sun's surface by a coupling with the Interplanetary Magnetic Field; abbreviated IMF. A solar physicist's comments strongly indicate that the Field-dynamical Model of the planets are at work: "The similarity of auroral particle acceleration to the solar flare case has suggested to several authors that the same acceleration mechanisms may explain both phenomena. It seems likely that more than one acceleration mechanism may be necessary to explain all observations."[109] A planetary correlation with sunspots has been claimed by many scientists throughout the years, but has been denied by others, because no mechanism was known, and gravitational effects are far too weak.

Notwithstanding, a number of observations indicate that the Earth and other planets trigger solar activity. Flares do not form at the brightest x-ray point in sunspots, and other types of solar activity, which indicates that the trigger is far from the site. Also the magnetic fields are continuous, but would not be if they were the result of internal processes. Internal motions would disrupt an internal field. In fact, some flares affect different depths of the solar surface (i.e., corona, chromosphere or photosphere), which is another indication of a remote field.[537]

The Earth's magnetic field undergoes changes of intensity that reflect the magnitude of changes in solar activity *before* they take place on the Sun. This study took into account eight solar cycles totaling 89 years (1884 to 1972). The observations disclosed that magnetic data for the Earth at sunspot minimum indicates the "depth" of the following maximum.[81]

Furthermore, the majority of solar maximums take place in a period around the vernal equinox.[399] The scientist that conducted the study states: "Of the 21 maximums occurring during the period 1750-1970 thirteen take place during the 4-month period February-May, while only four occur during each of the periods June-September and October-January."[399]; p.205. Flare formation around the time of the March or vernal equinox unveils the effects of the Earth. After all, months are relative to the Earth's motion around the Sun, not the Sun itself.

Consider an example. Just following the equinox (April 1980) the similarity of flare structure (*homology*) was caused by the overall consistency of the magnetic field.[367] On 12 April 1980, hours before the flare, there were slow changes in coronal structure, accompanied by radio emissions. This suggests that there were large-scale magnetic field changes, and the release of energetic electrons, which are typical effects of a time-varying accelerator.[367,534]

The long-lived x-ray loops require sustained energy for, at least, an hour prior to the flare.[367,534] At this time there was an interplanetary disturbance near the Earth, which was "probably related to this limb flare, although the (unexpected) absence of a shock makes identification uncertain."[534] Afterwards there was a coronal mass ejection overlying the decaying outer magnetic field. What is not even being considered is the possibility that the disturbance near the Earth could be the result of the Earth triggering the flare, due to a coupling of the Earth's magnetic field with that of the Interplanetary Magnetic Field (IMF), and this is why there was no shock.

The radio burst at the low corona, the formation of a sunspot of opposite polarity four to five days prior to the flare, and the x-ray loops during the main and post-flare phases, indicate the effects of a remote particle accelerator. Most of the activity was to the south of the leader sunspot in the region of the decaying magnetic field, not the emerging magnetic flux to the north.[534] This is what the shifting of the Earth's axis after the vernal equinox would cause as the Southern Hemisphere moved away from the Sun in fall (decaying magnetic field in the south), and the Northern Hemisphere moved toward the Sun in spring (emerging magnetic flux to the north).

Evolving magnetic fields of flares on 16 to 23 June 1980 showed that four of the five major flares occurred within one to three days of the two largest flux regions. Around the Earth's June solstice there were 17 emerging flux regions in seven days.[381] These are only examples of what is fairly typical of solar activity.

The Earth's influence is immediately understood in the fact that months are relative to the Earth's rotation around the Sun. These seasonal variations are also noted in naked-eye Oriental sunspot sightings, which show a peak in March, and a secondary peak in April. A third peak is noted around the December solstice. This effect is not random, and it has been suggested that late winter-early spring dust storms obscured sightings in other seasons, bringing about this effect.[110,665]

However, this does not explain the December peak, a time when there were winter thunderstorms, especially during solar maximums.[147,642] Furthermore, dust storms have more often been an aid, rather than a hindrance, to naked eye viewing of the Sun.

More recent solar events also show these peaks, as well. Major solar activity was noted for June 1988, March 1989, March 1990, March and June 1991, and so on. Some of the strongest solar proton events also occur around these times. The strongest proton events for certain years are 24 September 1978; 13 October 1981; 13 July 1982; 26 April 1984; 13 and 18 March 1989; 13 August 1989; 30 September 1989; 20 October 1989; 1 December 1989; 24 March 1991; 11 June 1991; 8 July 1991; 9 May 1992; 31 October 1991; 21 February 1994; and 21 April 1998 (*all are 1,700 pfu @ > 10 MeV or greater*). All of these events are within five weeks, some are within a few weeks or days, of the equinoxes or solstices. In fact, the strongest was 24 March 1991, just a few days following the vernal equinox (*43,000 pfu @ > 10 MeV*).*** Moreover, in 1991 there were very strong solar activity, solar flares, and geomagnetic storms that revealed the FEM-solar linkage.[71] Other facts confirm the FEM-solar linkage (as will be shown throughout this tome).

One such fact is the 1913 and 1969 jerks of the geomagnetic field, which are correlated with the 11-year solar cycle.[680] Shifts in the geomagnetic field peak *first*, then solar activity.[18] Again, this phenomenon indicates that the Earth is responsible for the solar transformations; it is a matter of cause and effect. Further support for this linkage is evident in theories that attribute the polar jerks to both internal and solar variations. Impulses were first attributed to internal processes, but were later correlated to solar activity.[18,19,120,127] The reason they are attributed to both is that FEM and the Sun are a single unit. Polar jerks, reversals and wander will be discussed in Chapter 19. The entire solar-FEM

*** Data from the NOAA/NASA Space Environment Services Center (http://umbra.gsfc.nasa.gov/SEP/seps.html); accessed August 1998

linkage can be noted in the fact that solar activity, geomagnetic activity, changes in the length of day (i.e., the Earth's rotation), and various geophysical phenomena, including climate, earthquakes, and volcanic eruptions are correlated.

The vernal equinox is a time when the Earth has both poles oriented perpendicular to the ecliptic plane, and is moving toward the Sun, making it an ideal time for interaction with the solar wind and IMF. Geomagnetic activity reaches a maximum around the time of the equinoxes, with the largest variations in spring and fall, as well as daily and annual variations, and this effect is controlled by the Earth's interaction with the IMF.[434,525]

Due to the interactions between the IMF and geomagnetic field (GMF), there is a twelve-month wave in geomagnetic activity with a maximum at the March (vernal) and September (autumnal) equinoxes. The largest sunspot average effect on the Sun's Northern Hemisphere switched to the largest effect on the Southern Hemisphere in 1913, the year of a polar jerk. This occurred along with the deepest solar activity minimum since 1811. Also, it was a year when solar motion around the center of mass (*barycenter*) attained a minimum. The conclusion is inescapable: "It seems, however, that these phenomena are the result of a common cause."[608]

One of the most established connections between the Earth and Sun is the close correlations between changes in the Earth's magnetic field, and fluctuations in the coming and going of sunspots.[434] So interconnected are these fields that the polarity of the IMF can be inferred by observations of the Earth's magnetic field.[584,585] An Earthly influence on solar activity has been commented upon by one scientist who proposes that the GMF could be a strange attractor due to some type of oscillations, and thereby, creates turbulent flow.[523] Others claim that different mechanisms are responsible for an Earthly influence on solar activity (see Plate 5.2).

Plate 5.2. Earth's Sunward Field. This photo, taken by the IMAGE spacecraft, shows a sunward feature in the Earth's magnetosphere at the upper left. Theoretically, all features of the magnetosphere should flow away from the Sun due to solar wind pressure. Yet, here we see a feature that is directed toward the Sun, which is the Field of FEM responsible for triggering solar activity. [NASA/JPL].

Observations near the Earth also confirm a solar-FEM linkage in solar activity. The most energetic solar flare phenomena are associated with what are called Ground Level Events (GLE). All GLE between 1942 and 1978 displayed no delay in onset times; they occurred on the Sun and Earth simultaneously.[468] In order to explain this, the IMF must intersect "open field lines connected to Earth."[114] Such events are only partly accelerated on site, but also in the heliosphere, which extends beyond the planets.

The number of protons and high-energy electrons were accelerated during the impulse phase of the 7 and 21 June 1980 flares. This impulse produced low-intensity proton events near the Earth, while some flares with no high-energy (gamma-ray) acceleration still created important events near Earth. Some events were not related to any identifiable flare or persistent, gradual electron acceleration.[262] This suggests that the acceleration involves the Earth's Field(s).

This can be seen in other ways. An AU is an astronomical unit, which represents the distance between the Earth and the Sun, and particle acceleration demonstrates the Earth's influence. Lunar-based samples indicate that the greatest acceleration of the solar wind occurs between the Earth and Sun (up to 1 AU). This was considered very unlikely in light of present theory, as the greatest acceleration should occur near Jupiter, just past 5 AU, due to the much larger magnetosphere. Low-energy protons were discovered between 0.7 and 1.0 AU in 1966 and 1978, with approximately a 200% increase per AU. Between 1973 and 1976 an increase of approximately 350% per AU was noted between 0.3 and 1.0 AU, or up to the Earth, but only around 100% per AU between 1.0 and 4.0 AU, or after the Earth's orbit. In the period of 1978 to 1980, near solar maximum, the intensity of anomalous helium and oxygen was a factor of ten lower near Earth's orbit.[603] The only way in which these observations can be explained is that something at the distance of the Earth is controlling the events, and that something is the IMF-FEM linkage.

Moreover, a negative gradient in the solar wind of -40% to -100% per AU occurs between 4 and 9 AU. An astrophysicist notes that, "these data may be explained by some particle acceleration which acts mainly in the region between 1 and 5 AU; from this region particles diffuse into the outer and inner heliosphere."[603]; p.196. This region is between the Earth (1 AU) and Jupiter (5 AU).

It was noted that an active boundary of the IMF (*current sheet*), and plasma layers in the geomagnetic field (GMF) were surrounding a strong rotation of the GMF that contained compressed and heated solar wind plasma. This observation reveals one of the IMF-GMF interactions that take place. An enhanced field strength and considerable wave activity direction is noted, but not in the direction predicted by the magnetic stresses across the boundary. The source of the energy and momentum were unknown.[543] However, the unknown is the Polar Field of FEM interacting with the IMF, and accelerating the solar wind earthward. This is why solar wind streams display a pronounced local depletion of ion concentrations near the Earth.[597] In contrast, during quiet periods the Earth is one of the main sources of protons.[603]

The Moon, which influences Earthly phenomena (as will be discussed in this tome), is also observed in solar activity. A lunar effect is indicated in proton counts (1952-1963), and neutron counts (1958-1963) near the Earth (*i.e., the lunar mean synodic rotation of 27.3 days, and the synodic month of 29.5 days, which is especially beyond random distribution*).[170] There is also a lunar influence on the occurrence of aurora.[259] The Moon influences the dynamics of FEM by triggering particle flow, which in turn may influence the Sun due to the solar-FEM linkage.

Other indications of an Earthly influence on solar activity have been noted. Only 25% of all solar flares cause magnetic disturbances at the Earth (geomagnetic storms). However, as stated, the Earth's magnetic field always reveals solar activity prior to that activity.[399,407]

Even flare formation and aurora formation are very similar. Hydrogen (-alpha) flares develop in about 20 to 30 minutes, and auroral substorms in about 30 minutes. In x-ray and microwaves the flare occurs in less than a minute, and auroral intensity increases two to three orders of magnitude in about a minute. Both have x-ray emissions and electron spectra that are similar. An astrophysicist makes the inevitable statement: "It is fascinating to infer that an auroral 'curtain' and a flare 'ribbon' are produced by similar processes."[9]; p.309

Large-scale fluctuations of the solar wind occur in the Earth's vicinity (1 AU), and display a period of four days. The intermediate-frequency fluctuations occur in periods from approximately nine hours to four days, and low-frequency from about four days to the solar rotational period.[86] In addition, there are periods of three to four days in Interplanetary Magnetic Field phenomena.[677] All these phenomena have a common period of four days.

A number of phenomena on Earth have a period of four days, and this is associated with the time-varying aspects of the Fields. There is a 4-day period in mid-latitude electron zone flow induced by lightning. Thunderstorms increase about four days after cosmic-ray events (e.g., flares). The electrical potential change in the lower troposphere and radioactive elements (*radionuclides; e.g.,* Be^7) increase about four days after solar eruptions. Likewise, when the IMF sector structure passes the Earth there is an increase in the size of cyclones (VAI; Vorticity Area Index), which peaks in about four days. Weather events, earthquakes, and other geophysical phenomena also display a 4-day period (this will be discussed). Putting it all into perspective, the life connection of FEM can be seen in the fact that all life on Earth has a "preperception," four days *prior* to magnetic changes on Earth that are later caused by solar events.[172] To better state this, life on Earth contributes to the functioning of FEM, which in turn influences the Sun, and this then affects FEM.

Further support for this understanding is afforded by observations of the interrelationships between the geomagnetic field (GMF) and the Interplanetary Magnetic Field (IMF). The aurora known as the Theta Arc configuration occurs along a certain structure of the IMF, which is the north-directed IMF (*i.e., the y-component*). Dynamics of the Earth's polar region, where auroras take place (*i.e., polar cusp*), are related to IMF solar wind particles when they are injected into the region where the northward IMF and GMF lines merge. The IMF carries shock waves that cause geomagnetic storms, and corresponds with other polar phenomena (*such as, the electrostatic potential*).[294,304] The active regions of the IMF, at 0° to 40° latitudes, are ideal for interaction with the Earth's orbital plane, which is inclined 7° to the solar equator.[559,603] The IMF's annual change form toward or away form the Sun occurs when the heliographic latitude of the Earth is 0°, and this IMF shift depends on which solar pole is tipped toward the Earth. So interconnected are the GMF and IMF that the polarity of the IMF can be deduced from observations of the GMF.[658]

As discussed, the IMF and GMF interact more during the equinoxes, and one pole during the solstice. As a result, there is a maximum in geomagnetic activity during the equinoxes due to the Earth's inclination on it axis relative to the IMF. The strongest activity occurs when open field lines merge, particularly when the IMF is southward with maximums just following the equinoxes on about 5 April and 6 October.[525,608]

This linkage can be noted in that the majority of solar maximums take place in those months around the vernal equinox.[399] For example, a period from 10 April to 1 July 1981 was a time when the Sun was observed for flare formation (*i.e., STIP Interval XIII*). The first part of this period, during April and early May, there were several x-ray flares, particle events and shocks.[213] In fact, there were new small sunspots, and more complex magnetic field structure before the flares. In February 1981, a part of the solar surface was empty, but in March an active region developed until flares occurred.[296] A flare on 9 April 1980 produced a (*Type-II*) radio signal moving towards the Sun, opposing velocities (*dichotomy*) in two regions of the Sun's surface (*chromosphere and lower corona*), and a flare-associated transient before the flare started. All of this indicates a source acting on the solar surface that is not internally driven, and occurs more often around the vernal equinox.[641]

Other flares displayed the influence of the solstice, as sunspots increased and the flare region became more complex on 21 June. Less than a day later another substantial flare occurred at the boundary of the old-new flux region. The gradient of the magnetic field increased until midday on 22 June, the day of the solstice, and then the active area disappeared.[382]

Processes other than the Sun and the IMF are required to accelerate the solar wind in interplanetary space.[311,603] Not realizing that the Earth's and other planets' Fields influence solar activity and the solar wind, scientists make comments like these: "The nature of the sources of accelerated particles is still unclear."[603]; p.194. "The nature of the Sun's magnetic field and the processes that cause sunspot-belt flux to erupt as observed remain enigmatic."[644]; p.717. In contrast, another scientist comments on the Earth's influence: "The phenomena observed indicate that the Earth also apparently repels sunspots to the farther side of the Sun."[531] Other studies also indicate the Earth has an influence on solar activity.[389]

Numerous flare observations confirm that a mechanism(s) outside the Sun triggers flares and sunspots. Studies of flares indicate that there are mechanisms that are *overlying* the Sun's surface.[235,247,262,465] For example, the charge interchange with atomic and ionized hydrogen at particle sources appears to require two different acceleration mechanisms.[465] The preexisting field on the Sun must be moved "aside" for the emerging flux to develop. Flare loops form through a magnetic reconnection in a local, outwardly extended and external magnetic field.[480] Magnetic disturbances at the Earth and corotating, high-speed streams are so co-fluctuating that the magnetic disturbances have been used to determine solar activity, including seasonal peaks. A team of scientists explain this phenomenon: "That such similar structure is observed in the same phase of different cycles over such a long time (up to tens of years) suggests a quite stable source structure for the geomagnetic disturbances generated by the corotating structure in the solar wind."[684]; p.1242. The Sun and Earth are a single unit when it comes to solar activity.

Aside from the solar maximums occurring more around the vernal equinox, and the Earth's magnetic field indicating the depth of the following maximum, there are other connections with FEM. Solar cycles fluctuations in phenomena on Earth and the life upon it are well known. For example, a double solar cycle of 22 years is apparent in the heavy hydrogen (*deuterium*) to hydrogen (*D/H*) ratios in fossilized tree wood.[164] This shows that there are cyclic variations in air or water temperature that are in accord with solar cycles.[180] The thickness of annual sediment layers also display a 22-year cycle.[192] There are many similar cyclic correlations that will be discussed in Chapter 15, and cannot be explained solely by the relatively minor fluctuations in solar output.

Life and the electrical potential of the Earth vary with solar activity. There is a minimum (*1.0 GV; gigavolt*) at sunspot minimum, and a maximum (*2.7 GV*) at solar maximum, which is also reflected in magnetic disturbances on Earth. The air and Earth potentials fluctuate with life. For example, tree potentials (*L-fields*) shift along with the Earth's magnetic and electric activity, the light-dark cycles, and lunar cycles.[87,88]

Chemical reactions (*Piccardi Tests*) on Earth peak *first*, then solar activity. These chemical reactions are also correlated with magnetic storms, sudden cosmic-ray bombardments, solar flares, and have an annual cycle with a peak, like solar activity, around the March (vernal) equinox.[94,473,474] This chemical-reaction effect is also apparent in living things.[474] All life on Earth, even the simplest creatures, such as bacteria, display behavioral and physiological changes four days *prior* to magnetic disturbances on Earth that are later caused by solar activity.[172] In turn, alternating electromagnetic fields, like those that accompany geomagnetic storms, are biologically active, changing animals (*systolic*) rhythms, bioelectricity, and blood dynamics.[634]

In accord with life's stabilizing control, the growing season is longest approximately one year after solar maximum, and if it is an unusually high maximum, a correspondingly longer growing-season peak occurs. For example, the growing season was 200 days long up to the mid-1930s, then three solar maximums produced 220 days, 225 days and 235 days, as each maximum increased in intensity.[225] Likewise, life's response can be seen in the fact that very early and extended growing seasons were noted in 1938, 1948, 1957 and 1959, all years of high solar activity.[306,307] The fluctuations of solar output that accompany the solar cycle are not sufficient to account for these facts, and it is life, coupled with FEM, that affects solar activity.

A highly significant statistical test indicates that the economy (GNP and Price Index), which is actually a measure of the life destroyed, has an influence on solar activity, and *not* the other way around.[545] Finally, it is a well-known fact that when cycles of the same length here on Earth coincide with solar cycles, it is the Earth's that peaks *first*.[159,172,473,474,545] Chapter 14 will illustrate this further.

Other observations bring conformation of this relationship. Polar faculae of the Sun have been observed more since 1951, during minimums. Faculae are produced by solar plasma that is accelerated into the poles of the Sun. The 1950s were a time when the world economy became more industrialized (post-World War II) and solar activity began a new maximum. Since that time these faculae first appeared and continue to reappear.[521] The Sun is maintaining its fusion reactions in response to more losses in the solar wind than previously.

The Northern Hemisphere on Earth is more inhabited by humans. As a result, the North Polar Field is more disturbed due to the destruction of natural life systems by humans. As a result of the IMF-FEM interaction, this should cause the Sun's Northern Hemisphere to be more active, and it is more active.[285]

The northern IMF entering the summer polar cap on Earth cases a reverse GMF disturbance, which is most frequent in summer and rare in winter. Its occurrence is dependent upon solar activity almost (linearly) proportional to the sunspot number. The interaction occurs along with a two-vortex structure in the electric field in the ionosphere near the Earth's magnetic pole.[285] Again, this reflects the intimate connection between the IMF and GMF.

An Earthly influence on sunspots was discussed as far back as 1907, and has continually reappeared in scientific literature since then. The 1907 article was called "An Apparent Influence of the Earth on the Number and Areas of Sunspots in the Cycle 1889-1901." A recent review of these data leads a present-day scientist to conclude that the Earth triggers sunspots.[531]

13.5 Planetary Influences on Solar Activity

The Earth is not the only planet to display an influence on the formation of sunspots, though it is the most influential. Electron acceleration has been observed near Jupiter, not only along magnetic field lines, but across the regular magnetic field and toward the Sun. In fact, some of these electrons have been noted in the Earth's orbit. An astrophysicist's comment reveals what could be expected of a time-varying accelerator, "the source acts for some days and is then switched off."[603]; p.198

This is also true of the other planets, and this is one of the reasons why we observe planetary periods in solar activity. It has been known for a long time that planetary positions that create angles of 0°, 90°, 180°, 270°, or two planets at 180° with the third at 90° affect short-wave radio reception, and radio signals due to increased solar activity. These angles are typical of the dynamics of electric and magnetic field interactions. From 1952 onwards, forecasting based on this understanding alone has been 80% effective.[422]

The Earth in an angular relationship with any of the other planets, such as Venus-Earth configurations, influences the formation of sunspots. Configurations involving Mercury, Earth and Venus show some of the greatest effects.[531] A 110-day cycle of angular acceleration between Venus, Earth and Jupiter is correlated to energetic x-ray bursts.[340] The sources can be noted in the observation that the solar wind is associated with the local depletion of ion concentrations in the Venus and Earth ionospheres.[597]

The four outer, largest planets (Jupiter, Saturn, Neptune and Uranus) are the most important for determining the position of the Sun, and the center of mass in the Solar System. The three closest (Mercury, Venus and Earth) are the most important for causing the jerk or change of acceleration of the Sun.[673] These influences are the greatest gravitational effects, which are still far too weak to cause the observed shifts in solar activity.

The terms used to describe the angles between the planets are conjunction (0º), square (90º), and opposition (180º), all of which are noted in effects on solar activity. When Venus and Earth are in opposition there are 60% more sunspots than during conjunction.[531] When the Earth, Venus and Jupiter are in conjunction, there are even more sunspots.[446,671] A study covering a 300-year period disclosed that sunspots increase when Jupiter and Saturn are in conjunction, square and opposition.[77] Uranus and Neptune are in square during maximums, and in conjunction or opposition during minimums. The positions of Mercury, Venus, Earth and Jupiter are correlated with solar proton events.[85,291,460] Mercury's revolution around the Sun is also a solar cycle (87.976 days). When Venus, Earth and Jupiter are on the same side of the Sun with Mercury at closest approach (*perihelion*) the effect doubles.[62] That is, Mercury's orbital period in sunspot data also depends on the phases of Venus, Earth and Jupiter.[671] The conclusions from the data are clear: "There is a close link between various planetary alignments and the dates of sunspot maxima and minima."[224]

The influence is a combination of electromagnetic fields in dynamic interaction that overcome the gravitational (tidal) forces. This is why studies that claim gravitational forces are responsible have been put into the skeptics corner, so to speak, and have somewhat discredited this whole area of study. Gravitational forces are far too weak to be responsible for the effects.

This electromagnetic long-range force is also why there are planetary periods in sunspots. What particularly illustrates this is that Pluto, the farthest (most of the time) and smallest of the planets, has a period that shows up in sunspot data. The influence cannot possibly be gravitational, as a scientist exclaims: "These planetary influences cannot be gravitational, but must be magnetic or electrical in character."[291]; p.56. Planetary positions have been used to predict solar flares, which is not explainable by gravitational effects. A scientist studying planetary positions and solar activity makes a comment that reflects the limitations of present perspectives: "There was as yet no understanding of why this should be."[460]; p.35

The five outer planets' orbital (synodic) periods display close associations with sunspot periods, with the exception of Neptune. A scientist expresses concern over this enigma: "Note that, in spite of its size and great distance from the Sun, Pluto is included as one of the planets involved in a synodic period associated with a sunspot period. This is most surprising. Pluto is the most eccentric of the planets; Neptune, next to Venus, is the least eccentric. In fact, Neptune's orbit is almost exactly circular."[160]

The reason Neptune's synodic period does not show up in sunspot periods is due to the way in which Neptune's magnetic field is offset in relation to the ecliptic plane and the IMF. It is not anywhere near the angular interactions that the other planets have with the IMF. This is also evident in lower radio emissions and an offset auroral zone that rotates with Neptune away from the Sun and the IMF.

Not only do planetary synodic periods correlate with solar activity, but also variations from orbital eccentricity are found in sunspot periods.[672] Conjunctions of Jupiter and the center of mass of the Solar System (barycenter) have been used to predict energetic x-ray flares.[193,340] Planetary effects on solar activity have been shown in numerous other studies.[20,22,40,62,435,446,556,557,659] Many scientists acknowledge that the effect is not gravitational (i.e., tidal).[40,291,435,460]

A local origin for cosmic rays, especially during solar minimums, reveals these linkages, as well. There are two types of shocks. One is a somewhat (quasi-) parallel shock where the angle between the shock (*normal*), and the up-steam magnetic field is approximately equal to or less than 45⁰. The other is a somewhat (quasi-) perpendicular shock that is approximately equal to or more than 45⁰. The parallel type display more magnetic and electric field fluctuations, and occur mostly in the inner heliosphere, generally up to distances just beyond the Earth (less than 1.5 AU). The perpendicular type occur most often at distances greater than this (2.0 AU) where the angle between the IMF and the shock (*normal*) is closer to 90⁰.

Most "cosmic-ray" electrons are not cosmic, but originate from the planets and are accelerated downstream. The Sun is also a source. Cosmic rays accelerated into the heliosphere occur mostly in the ecliptic plane (the Solar System's "ring plane").[329] Meanwhile, most activity in the outer heliosphere beyond the planets (about 25 to 30 AU) ceases in the ecliptic plane and the 30⁰ ecliptic latitudes. The planets are along the ecliptic plane, as well as the IMF, and their interaction affects solar activity, and cosmic electron flux.

The Interplanetary Magnetic Field (IMF) is rather complex with an inner boundary that extends up to the vicinity of the Earth, and an additional structure extending to past most asteroids (about 3 AU). At that point the fast solar wind strongly interacts with the pre-existing shock in the range from the asteroids to about Jupiter (about 2.5 to 5.0 AU). The inner heliosphere is severely disturbed with several interacting shock fronts containing "peculiar" magnetic field configurations, while the structure beyond Saturn (about 10 AU) remains relatively undisturbed. It has been observed that the energizing of ions is a pervasive phenomena in space-time, and low energies (*about 30 keV*) are noted over an entire solar cycle to, at least, the distance of Neptune (about 30 AU).[309]

Presently, there is the general agreement that there are two components of cosmic rays, the galactic and the solar.[329] However, cosmic rays at the highest energies are arriving from the interstellar medium, but are moving into the heliosphere in the ecliptic plane. In contrast, present models predict a density gradient of cosmic rays with cosmic-ray intensity increasing at higher heliographic latitudes.[329] This restriction to the lower ecliptic plane indicates that the planets and IMF in combination are accelerating cosmic rays along the ecliptic plane.

Ground Level Events (GLE) are cosmic-ray events that reach ground level, while showing no delay in onset time. A GLE was observed during the Voyager missions, and the protons were noted to be of a different type than the expected flare-associated enhancement. The low-energy ion and proton channels of Voyagers 1 and 2 displayed that there were clear fluctuations at the times of the equinoxes and solstices, suggesting that they originate from Earthly influences. The enhancements dropped abruptly following the encounter with Jupiter in July 1979.[329] It is the planets from Jupiter towards the Sun that affect solar activity the most.

The most remarkable aspect of the entire set of observations, according to conventional (*modulation*) theory, is the absence of low-energy protons in the outer heliosphere as Voyager 2 was between Uranus and Neptune (25 AU). However, with the Field-dynamical Model, protons would be accelerated into the planets' cores and would have undergone hydrogen fusion. Hence, an absence of protons would be expected (except for those produced by neutron decay).

13.6 Solar Neutrinos

Observations of solar neutrinos confirm both the structure of the Sun, and the influence of planetary Fields on solar activity. Solar neutrinos are subatomic particles produced by the nuclear reactions that take place within the Sun. A direct test of how the Sun produces its luminosity is to observe the quantity of neutrinos that are emitted.

Recent observations show that the Sun is discharging only one-third of the expected number of neutrinos. This greater solar efficiency is in disagreement with the supposedly well-established theory of stellar evolution. Therefore, the Sun's mechanisms are not understood or the classical theory of neutrinos is wrong. Both possibilities are unattractive to conventional theorists. Admittedly, a scientist relates: "Something else in the Sun has to be efficiently transferring energy out of the center."[600] This greater efficiency is also suggested by the long-term (*secular*) variability of solar activity.[179,555]

One possibility is that the Sun has a stronger magnetic field in its interior than is theorized. If the magnetic field decreases towards the surface, then this would increase the pressure gradient. A primordial magnetic field could survive (*in the Cowling mode*), and several scientists have considered a strong magnetic field confined to the core.[51,61,106,405] An expert on magnetic fields in the cosmos indicates that there would have to have been primordial fields that were later reinforced.[455] A specialist in neutrino and solar physics comments on this possibility: "The decay lifetime for such fields would be expected to be much less than the age of the Sun, and therefore, some unspecified mechanism must be involved that continuously regenerates the field."[43]; p.120. This statement is descriptive of the Field-dynamical Model, which involves the "unspecified mechanism."

Furthermore, relativistic effects in the Solar System disagree slightly with the predictions of general relativity. Unaccounted for are such observations as the redshifts of radio sources during solar eclipses, the limb-shift near the solar limb, and the angular displacement of objects situated behind the Sun. These observations call for new models of the Sun.

A central black hole is another suggestion, but according to conventional physics a black hole is composed of very dense matter and strongly gravitational. Within this framework it is questionable that such a massive object exists due to the observed mass-luminosity relationship. The Sun would not have the observed luminosity if this greater mass were present. However, if the black hole were a magnetic confinement chamber without the hypothesized mass (as discussed for active galactic nuclei in Tome Four), it could fit the observations.[42-44,111,112]

Other suggested non-standard models of the Sun also implicate the Field-dynamical Model. One is that the Sun has a small burnt-out core in which hydrogen has been exhausted. A rapid rotation in the solar core would reduce the thermal pressure required to support the star against gravity, and therefore, reduce the emitted neutrinos. However, the rotation must be 1,000 times faster in the core than on the surface. Hydrodynamic phenomena, causing internal gravity waves, is another unorthodox proposal. Other theories of non-standard solar models have emerged, but with little observational support. Less neutrinos indicate that models of the Sun, and the Earth, and/or the "laws" of physics need to be revised.[42-44,111,112]

The biggest surprise was that neutrinos fluctuate in relation to the solar cycle. Less neutrinos are emitted during solar maximum, but they should not vary, except on a timescale of billions of years. A neutrino with a large magnetic-moment can interact with strong magnetic fields, causing the spin of the neutrino to flip from left-handed to a right-handed helicity, and thereby, go unobserved in the experiments. That is, the weak interaction coupling (constant) occurs mostly with left-handed spins, and the experiments were designed with this perspective in mind. The spin conversion would have to occur at or above the solar surface, or it would not be correlated to the solar cycle. Therefore, a strong magnetic interaction that originates from *outside* the Sun is required, because sunspots are bubbles of less dense material coming off the solar surface (*e.g., magnetic flux tubes*). This sunspot-cycle effect by itself could not bring about the required conditions with only the Sun, and therefore, remote fields are required.[42-44] The facts at this level also encourage the idea that the remote fields of the planets are involved in solar activity, and this correlates with fluctuations in neutrinos.

Another possibility is that there is an extremely large magnetic field deep in the interior of the Sun, surrounding the core. This could provide a restoring force for oscillations that change neutrino flux. However, this does not resolve whether the variation is due to deep-seated neutrino production, or to neutrino transmission at the surface or in space.[328] Regardless, the idea of an extremely large magnetic field in the core is not presently an aspect of solar models, and indicates that the Field-dynamical Solar Model and/or planetary influences are at work.

A serious shortfall of solar neutrinos was revealed in more recent data from the Soviet-American Gallium Experiment or SAGE. The gallium interacts with low-energy neutrinos created in the prime energy producing reactions of the Sun, but their detected levels were far too low. This statement hints at the Earth's influence: "Such a low rate provides strong evidence that electron-neutrinos somehow disappear between their creation and reaching the Earth."[34] Notwithstanding, it has also been suggested that the neutrino shortfall could be the result of an energy source in the Earth's core, as described for FEM.[243]

In fact, the present theory of flare formation relies on magnetic flux tubes beneath the surface, in the photosphere, while no photospheric process can produce the energy required. Before a flare, dramatic, relative motions of sunspots indicate force-free fields that produce a twisting or shear of the magnetic field lines. Force-free and field-aligned currents are essential to flare occurrence. Furthermore, vortex flows are essential for generating field-aligned currents, while electrons carrying the upward field-aligned currents must be streaming down into a deeper layer, the chromosphere.[8,9] One group of scientists summarizes:

> "Considering all the evidence, it seems that a full solution of the solar neutrino puzzle requires something in addition to the hypothesis of neutrino magnetic moments. The primary concern is the possibility that ordinary cosmic rays, by some unknown process, contaminate the $_{37}Ar$ data and lead to a correlation with solar activity through the modulation of galactic cosmic rays. Nonetheless, perhaps some other vector could mediate between the cosmic rays and the detector."[61]

Statements like this suggest planetary fields are interacting with the solar surface, causing the spin flip of the neutrino, as well as the other observations. This is especially true of FEM, because the detector is on Earth, and therefore, the spin flip must take place before reaching the Earth's surface. Then again, as indicated by scientists, this could be due to the fact that the Earth has an energy source deep within its interior, as noted with FEM.[243]

13.7 <u>Solar</u> <u>Activity</u> <u>and</u> <u>History</u>

When considering life's contribution to the stability of FEM's dynamics, and the efficient functioning of the Fields, there should be a correlation between life crises on Earth and solar activity. When life systems are not complex and stable, the Fields are not efficient accelerators, and their diameters enlarge (due to less electromagnetic contributions to the system). Through the solar-FEM linkage, solar activity increases as, for example, when humans created life crises as civilizations arose, devastating natural systems on a global basis. Confirmation that the Earth influences solar activity is evident in how solar activity fluctuates along with history.

The Maunder Minimum is a good example of times of minimums in history. It included 70 years (1645-1715) of sunspot and aurora minimums. For 60 years not a single sunspot was seen on the Northern Hemisphere of the Sun, and not more than one sunspot group was seen on the Sun at any one time. During the entire period the total number of sunspots was less than what we see in a single year, and were mostly single spots at low solar latitudes that lasted one revolution or less. For 37 years not a single aurora was reported anywhere, contrasting with today for which aurora are observed every day. The relative absence of aurora indicates that there were far less sunspots, flares and other solar eruptions.[175,176,178] It was only before and after the Maunder Minimum that the 11-year solar cycle is evident in the data.[38,540]

An anomalous solar rotation was noted for the early 17th Century, as well. For the period from 1625 to 1626 the rotation was like that of today. However, by 1642 to 1644 the equatorial velocity of the Sun had become 3% to 5% faster, and the Sun's differential rotation was enhanced by a factor of three.[177] The Sun had become more efficient, and this was due to the undisturbed surface with the relative absence of sunspots and other forms of solar activity.

A sharp angular shift in the geomagnetic field took place in 1375, and again in 1575.[595,596] The geomagnetic field intensity was lowest by about 1400 (0.89), increased somewhat by about 1500 (1.07), and declined by about 1650 (0.94; figures are relative to present-day mean strength).[452] Both the angular shift and the 1400 low intensity of the GMF occur around the time of the Late Medieval Maximum.

The Late Medieval Maximum took place as medieval civilizations were widespread. Following this, the Earth had become renewed with wilderness, as disease, particularly the bubonic plague or Black Death, other diseases, conquests, war and natural disasters had removed life-destroying settlements. Thereby, the Sun had not been disturbed by the dysfunctioning of FEM, as life was renewed, leading to the Maunder Minimum.

Furthermore, in regard to fast-spreading or new types of diseases, such as the bubonic plague, there is a relationship with increased solar activity. Alternating electromagnetic fields, like those accompanying geomagnetic storms, induced by increased solar activity, cause microorganisms to multiply, and undergo changes in morphological and cultural properties.[634] As the humans of that time again began their destruction of wilderness the Sun again became active, and it and FEM subdued civilization. However, from 1300 through to 1700 there was a decrease in solar activity with a succession of three minimums. Activity was well below that of the present except for the late 14th Century when the bubonic plague struck, and the early 17th Century when the toll of disease, war, natural disasters, and the conquests of the Americas had revitalized the wilderness. As the 18th Century began civilization again started to flourish, and this is noted in that there was a lasting ten-fold jump in the number of aurora, beginning in 1716. For the minimum there were only 77 aurora, but more than 6,000 marked the activity in the 18th and 19th Centuries.[175,176,178,358]

Other facts confirm this understanding. There was a long delay in the discovery of the chromosphere, and there were no ancient or historical descriptions of coronal streamers during eclipses. The chromosphere was discovered in 1706, and the first description of coronal streamers was made in 1715. A long-lasting ten-fold increase in the frequency of recorded aurora began in 1716. All of these discoveries came after the Maunder Minimum, and the return of civilization with its life destructiveness.[175,176,178,358]

Today these phenomena are as common as is the human destruction of wildlife on Earth. Likewise, polar faculae are more intense when sunspots are less, and there were none during the three solar maximums up to 1964. Before 1951 they were rare, sporadic, and bore no known relationship with the sunspot cycle.[546] It is no coincidence that these phenomena occur after the end of World War II, when industrialization began a sweeping impact on the biosphere.

The same understanding emerges when the Modern Maximum in solar activity is examined. The decay of the Earth's magnetic field was noted from the time of the Modern Maximum in the early 19th Century.[101,394] A scientist comments on a short-lived minimum just prior to the Modern Maximum: "The change in sunspot periodicity occurring about 1800 might be taken as a hint that the Sun was on the verge of lapsing into a passive state."[508] The Industrial Revolution began in the late 18th Century in Britain after a period of extreme cold and other factors depressed the economy, and the destruction of wilderness. With the return of the destruction of life, as the Industrial Revolution spread, there came the Modern Maximum.

This same pattern can be seen in the trend toward ever increasing solar activity during the second half of the 20th Century, which is a marked feature of coexistent indexes of the geomagnetic field.[179,197] It has led to the understanding that this "period was one of distinct, secular change in the character of the solar wind."[179]; p.126. Likewise, the Sun itself changed in response, because the "spectacular structured coronas seen at modern eclipses might be considered a modern phenomenon of the Sun."[179]; p.130. Life's response can be seen in the fact that very early and extended growing seasons were noted in 1938, 1948, 1957 and 1959, all years of high solar activity.[306,308] Again, a correlation exists between the destruction of life, the weakening of the Earth's magnetic field, and increases in solar activity. This fact is all the more evident in the present with the most active period in solar activity known to science, matched by the greatest destruction of life in history, and the ever decreasing intensity of the geomagnetic field (GMF).

Making it even more undeniable, a test of causation demonstrates that the Earth's biosphere, transformed by the economy, has an impact on solar activity. The Gross National Product or GNP is a measure of the life destroyed by economic activity. This is because components of the GNP include capital investments, such as expenses for new factories, machinery, and housing; consumption of goods and services; government expenditures for goods and services; and net exports (exports less imports). These activities destroy wilderness for mining, natural resource exploitation, the clearing of land to build, and so forth. Likewise, the Wholesale Price Index is also a measure of the demand for goods and services. Both the GNP and Wholesale Price Index peak before solar activity, demonstrating that it is not solar activity that influences the economy, as was thought. Those who verified this causal relationship state their surprise:

> "We are forced to conclude that the U.S. economy has a significant impact on the Sun, but that sunspots have no influence

on the economy. We were, of course, more than a little surprised to the find that the U.S. economy has any influence on the Sun. However, as [others have] said so well, one 'would have to be rather more pigheaded in order not to have the evidence change his views.' We agree that identifying the linkages constitutes an important topic of future research."[545]

The entire story can also be seen in a comparison of solar maximums and minimums, and the history of civilization. At the heights of civilizations there are maximums, and following their downfalls there are minimums. For example, there is the Sumerian Maximum of 2700 BC to 2550 BC, and the Pyramid Maximum of 2350 BC to 2000 BC, following a flourishing of the Mesopotamian, Indus, Egyptian, and Olmec civilizations. Likewise, following a height in the building of megalithic structures in Europe and elsewhere, there came the Stonehenge Maximum. After the Egyptians and other civilizations underwent a series of events marked by disorder and the destruction of many structures and cities, there was the Egyptian Minimum from 1400 BC to 1200 BC. After Troy fell and numerous other civilizations collapsed throughout Eurasia there was the Homeric Minimum. Following the fall of the Greek States and other civilizations, the Grecian Minimum commenced. As the Roman Empire reached its peak there was the Roman Maximum from AD 1 to AD 140, and just prior to the collapse there was the Late Roman Maximum in the 4th Century AD. Just before we see the destruction of many Byzantine cities, the Islamic or Arab conquests, the Mayan collapse, and more, there was the Byzantine Maximum of the 6th Century. As the Crusades were undertaken, the Toltecs collapsed, and numerous other events, the Medieval Maximum took place from 1140 to 1240, followed by the Wolf Minimum of 1280 to 1350. Then a short-lived, but strong, maximum accompanied the bubonic plague, the Prince of Destruction, and more, after which there was the Sporer Minimum of 1420 to 1570. The conquest of the

Americas and additional events marked a number of maximums that were followed by the Maunder Minimum. Then came the full-scale destruction of the living world with the Industrial Revolution and the Modern Maximum beginning in 1800, the greatest maximum in the known history of solar activity.[176, 178,179,540,579-581]

There are many other sources that reveal the same correlations. The historical records of the Chinese also display the same periods of maximums and minimums.[540,643,661,681] The aurora records of China and Europe from AD 500 to AD 1549 indicate peaks that conform to four of these solar maximums.[357,358] Another study of aurora for the period of AD 1001 to the 1900s also reinforces this understanding.[334,358,578-581] This relationship of the waxing and waning of civilization with solar activity will be throughly exemplified in *In Defense of Nature—The History Nobody Told You About*, which covers history and prehistory.

CHAPTER 14

Things Go Round and Round
(Cycles of the Solar-Lunar-FEM Linkage in the Biological World and Historical Process)

Certain phenomena recur at regular intervals, often referred to as cycles. There are solar and lunar cycles in weather, earthquakes, and volcanic eruptions. Other phenomena also show cycles that match lunar and solar cycles in biological, physical and socioeconomic phenomena. These studies reveal that there is a synchronicity shared by all of these phenomena so much that the solar-lunar linkage of FEM, including life, can be considered a single unit.

14.1 Biological Cycles

The lunar cycle and lunar phases are observable behaviorally and physiologically in numerous life forms, such as sea creatures. Shellfish (*crustaceans; e.g., Anchistiodes, etc.*) renew their shells, undergo regeneration and sexual activity in accord with the lunar tidal cycle. The Fiddler Crab (*Uca pugnax*), as well as other crabs (*Sesarma, etc.*) display lunar rhythms in behavior (*locomotor activity*).[614] Guppy-fish (*Lebistes reticulatus*) have a color sensitivity on their back that is most responsive during Full Moon, and least responsive at New Moon.[345,346] Eels react to lunar induced microseismic activity, or the lunar triggering forces responsible for the microseismic activity.[614]

Animals and insects reflect similar behavior. The Golden Hamster displays lunar rhythms in activity, and urinary volume and acidity. On the same lunar day of the month urinary volume reaches a maximum, while activity is at a minimum. As the peak in activity is reached, with a maximum during the Full Moon, minimums in acidity occur on approximately the same day.[314,315] Spontaneous activity of the Mongolian Gerbil displays a minimum around the Last Quarter, and several days before the Full Moon in summer, fall and spring. Meanwhile, during winter a mirror image of this effect takes place.[582] This effect is also found in the hamster[314,315], mealworm[91], planarian[80], and other animals. Even the flight of the Honey Bee displays a lunar period.[432]

Plants also respond to lunar phases and cycles. For example, the South African Iris (*Morea iridoides*) flowers twice per month with the first flowering on the first day of the First Quarter. The flowering ends the day before the Full Moon. No flowering occurs on the New and Full Moons.[614] Water uptake by bean seeds is regulated by extremely weak electromagnetic fields, such as those that fluctuate with lunar phases and cycles.[78,79] Lunar phases and plant growth are correlated for plants in the wilderness, as well as in the laboratory.[3,4] In fact, two different lunar cycles affect *all* plant growth.[60] Various life forms exhibit lunar periods in behavior and/or physiology.[161,251.477,614] These lunar effects on life are components that sustain and partially control FEM's dynamics.

Solar cycles of various lengths are also evident in many life forms and their activities.[218,219,629,633] Consider some examples. Alternative cycles of sunspots reverse with a period of four years. Also having a cycle of four years are plankton yields in Lake Michigan, and the abundance of Arctic Fox, Canadian Fox, salmon, Snowy Owls, field mice, and Ruffed Grouse, as well as various socioeconomic activities. A 5.9-year solar cycle is also manifested in Grouse abundance, and various socioeconomic activities. The 9.11-year solar cycle is found in the abundance of Lynx and the Woodcock, agricultural production, and various industrial activities. Grasshopper and partridge abundance, and tree-ring growth display a cycle similar to the 9.3-year sunspot cycle. At least twenty industrial activities also have a cycle of 9.3 years. The 9.93-year sunspot cycle is apparent in Lynx, Ruffed Grouse and Waxwing abundance. Cycles of diverse lengths are also evident in human activities and physiology, other life forms, and physical processes on the Earth.[156-159,161,633] In addition, the correlation between solar cycles and cycles in the biosphere makes it clear that there is an underlying common cause, and it bears the evidence of a unified whole involving life integrated with FEM, and its solar and lunar linkages.[156-157,161,449,614]

14.2 War and Peace

Human aggression, conspicuous in national and international wars, occurs in cycles, as well. A study of civil war from 600 BC to AD 1941, and international battles from 600 BC to AD 1943 revealed cycles of 6, 9.66, 11.25, 12.33, 17.75, 22, 53.5, 143 and 164 years.[158,548] Cycles of 5.98, 9.73, 11.26, 13.2, 17.2, 21.9, 29, 52.9, 59.8, 101 and 143.7 years were noted in a more recent study. For example, the 29-year cycle peaked in 1913, which was the beginning of World War I, again in 1942, during World War II's most violent battles, and again during the escalation of the Vietnam War in 1971. The next peak began in the year 2000. Civil war also displays cycles, which are 130, 91, 62.9, 27.4, 18, 10 and 8.9 years.[408]

The global political system shifts in cycles of 8.6, 17.4 and 34 years. Cycles in global economic transitions occur in cycles of 5.6, 9.4, 20.5 and 54 years, which often result from or contribute to human conflict.[11,121] One of these studies took in the political and economic histories of 58 nations.[11]

Instability in social change often occurs at the time the Sun's center, the center of mass and Jupiter are in alignment. Such are the times when solar, earthquake, volcanic and climatic transitions peak, as well.[341,558] When considering all of the cycles mentioned in this chapter, they are at or close to the lengths of solar and geomagnetic cycles.[456-458,604]

Solar activity, geomagnetic activity, and weather are correlated to changes in blood pressure, blood composition, and the physical and chemical state of humans.[215,219,552,611-613,633] It is very possible that these factors play a significant role in triggering political, international and civil conflict (i.e., wars, riots, etc.). Changes in human behavior are know to take place during periods of high or low solar activity, as well as during geomagnetic storms (also see Volume One, Chapters 14 and 15).[203,298,458] Solar and lunar effects are known to influence civil unrest, which can be triggered by air ionization due to particle flow along the Field lines of FEM.[458] The cycles of national and international conflict noted above are at or close to the lengths of solar and lunar cycles.

14.3 Economic Cycles

Cycles in the economy are probably the best known and researched due to attempts at predicting market trends (i.e., to make money). More than two decades ago the following industrial and economic cycles were recognized: seven 4-year, eight 5.9-year, thirty-five 6-year, twenty-three 8-year, twenty 9-year, five 9.4-year, three 9.9-year, eight 12.6-year, two 16.67-year, four 17.3-year, two 17.7-year, twelve 18.3-year, four 22-year, and thirty 54-year cycles.[157] All of these cycles, with the exception of the 54-year cycle, were recently reaffirmed a quarter century later (i.e., 54 years had not elapsed since the first study).[244]

In addition, there are cycles of economic lows and peace of 41, 49 and 53 years, while economic highs and war occur in a 60-year cycle.[244] Major economic and social instabilities are noted for the times when the Sun's center, the center of mass, and Jupiter are aligned.[341,343] Analyzing annual data from the 18th Century to the present on wholesale commodity prices, cotton prices, and the stock market revealed a number of cycles. Five long-term (*secular*) sunspot periods appear in all three measures of the economy.[202,508,545] As with other facts, this again reveals an Earthly influence on solar activity. Planetary configurations, which influence solar activity, and in turn geophysical phenomena, also have effects on heredity, the birth cycle, the birth of successful professionals, and personality.[207-210] These factors evidently have an effect on the economy.[244] Many of these cycles appear in data on solar activity, as well as in biological and geophysical phenomena.

14.4 Cycles of Disease and Pestilence

Disease, epidemics, pandemics, and pestilence are also cyclic. It was early in the 20th Century that a relationship between solar activity, and the onset of epidemics and widespread diseases that widely effect many animals (*epizootics*) was discovered. Also correlated were increases in nervous and mental diseases, and other biological and social phenomena.[107,108]

More recently, studies show that epidemics of plague-carrying rodents follow a cycle of 10 to 11 years. These epidemics take place at the transition from minimum to maximum of the solar cycle. Secondary and intermediate peaks also occur at times after solar activity fluctuates. These effects are only partially due to shifts in climate.[351] Epidemic disease fluctuates with solar activity regardless of weather conditions or the population level of disease-carrying hosts.

Correlations between disease and solar activity are particularly clear in data for the 19th and 20th Centuries, as the Modern Maximum in solar activity became effective, and more complete records were kept. Such correlations involve recurrent typhoid in European Russia (1883-1917), cholera in Russia and all of the former Soviet Union (1823-1923), cholera in Pakistan, and diphtheria in 50 regions including Asia, Europe and the United States. Also correlated are scarlet fever in Russia, malaria in the former Soviet Union, plague in Java, smallpox in India, polio worldwide, infectious hepatitis in the United States, and numerous others. The conclusion of the researchers was that a solar-terrestrial linkage was involved.[678]

Another measure of solar activity involves changes in the geomagnetic field (GMF), which are also apparent in disease data. Dysentery and polio the world over, tetanus in Australia, and scarlet fever in Leningrad (St. Petersburg) display a clear dependence on the state of the GMF.[678] The incidence of nervous system and mental disorders, and the intensity of glaucoma are correlated with the strength of the GMF.[448,686] Even road accidents have been known to increase on those days with disturbed geomagnetic activity.[383] Geomagnetic storms and other GMF fluctuations are most intense during the equinoxes, and at extremes in the lunar nodal cycle and lunar phase, as well as times of increased solar activity.

On days when there are geomagnetic storms there is an increase in the number of heart attacks, and in the number of deaths due to heart failure and hypertension.[371,568] Furthermore, increases in the intensity of GMF disturbances increase the frequency of various cardiovascular disorders, and the number of related deaths.[427] Similar observations have been made for the cerebrovascular diseases.[10] In fact, early warning systems for sudden death from cardiovascular diseases have been proposed on this basis.[217]

Animal populations and their migrations also fluctuate with solar activity. The onset of seasonal migrations for birds, and the population size and migration of insects, fish, fur animals and rodents, fluctuate with the rhythms of solar cycles, and the Earth's magnetic and electrical environment.[107,219] One of the earliest studies that linked cycles in animal populations with solar activity compared the material purchases of Canadian Hare skins over 100 years with sunspot numbers (*i.e., Wolf numbers*). Peaks were noted during solar minimums.[182] Water voles have also shown a similar cycle of 10 to 11 years, and unlike Canadian Hares, their habitats (i.e., swamps and lakes) were the least affected by human economic activity at the time. Other animals were noted to have had population shifts as their habitat changed with solar activity.[369] Numerous insects have been observed to fly in greater numbers on those days with geomagnetic storms, and the effect increases along with storm intensity.[104]

The epidemic process involves the simultaneous action of three factors: the source of the infection (the "germ"), the means of transmission, and the condition of the populations exposed, all of which are influenced by solar activity and the solar linkage.[678] This is why influenza occurrence displays seasonal and solar cycle relationships.[233,272-274] In general, cancer deaths in Turkmen and the former Soviet Union are correlated with solar activity.[214] Solar activity, as measured by cosmic radiation, is also correlated with tuberculosis[438-440,442-445], and cancer.[441,478] Likewise, polio has been correlated with sunspots, aurora and geomagnetic activity.[279,678]

Solar activity is not the actual cause, but instead it is the Earthly phenomena associated with an activated FEM. As discussed (see Volume One, Chapters 14 and 15), these effects are due to a combination of air ionization producing an excess of positive ions, which lowers immunity, offsets other biometeorological factors, and geomagnetic activity offsetting the immune system and biochemistry. These influences also increase the populations of disease-carrying organisms, and alters their genetic strains and expression.

This is why physical changes in organisms, including humans, have been shown to fluctuate with solar activity. Systems of the blood (*coagulation and anticoagulation, especially the fibrolytic system*) under go changes that can lead to both thromboses and hemorrhages at times of solar activity.[479,520] As solar activity declines certain components of the blood increase (*erythrocyte and leukocyte indexes*). There is also a correlation between the formation of blood and blood cells (*hematopoiesis*), and solar activity, in such a way that blood formation is inhibited when solar activity increases.[214] Human blood was observed to undergo noticeable fluctuations in strict correlation with solar activity indexes (*especially, Wolf numbers*), and particularly, large flares.[633] Furthermore, the solar-induced geomagnetic field background coincides with the threshold sensitivity of humans.[419] An increase in solar activity is matched by a higher rate of reproduction, toxigenicity and virulence, among other biophysical factors. A weakening of the functional vitality of humans is noted, increasing the likelihood of disease.[568]

Cosmic rays have been correlated with fluctuations in a number of indexes of the vital activity of organisms. However, for long intervals of time this correlation was replaced by an anti-correlation, which clearly illustrates that there is some other factor other than cosmic rays in the solar wind.[633] That factor is FEM, which embodies geoelectric, geomagnetic, bioelectric, and air ionization (biometeorological) effects on physiology and central nervous system function.

Other related phenomena display this influence on human functioning. Geomagnetic influences on immunological and enzymatic reactions are known.[519] Weather and climate alter blood pressure and composition, and the physical and chemical state of humans.[610] Changes in capillary patterns take place during planetary conjunctions, while under constant environmentally regulated temperature and humidity.[324] Disease organism and human immune physiology are both affected.

Human unrest and excitability are also influenced by these same factors. The number of epidemics of psychopaths, mass hysteria, mass hallucinations, and mental diseases fluctuate with solar activity. As soon as solar activity reaches a maximum, the number of important historical events, taken as a whole, increases.[108,156] An extraordinary chart spanning 2,400 years displaying all major wars, civil unrest, mass migrations, epidemics and pandemics, recorded in the histories of all people, as well as other research, uncovered regular cycles that are synchronized with those of the Sun.[11,108,186,298,341,408,456-458]

14.5 The Cause of Cycles

Looking more deeply into cycles it becomes even more evident that what we do here on Earth has effects that go beyond the Earth. Many major physical forces on Earth are interrelated with those of the Sun.[184,338-344,456-458,551] Other cycles that coincide include variations in the GNP, stock prices, interest rates, periods of general instability, economic turning points, animal populations, climate, geomagnetic storms, earthquakes, volcanic activity, changes in the length of day, and others. Cycles would not normally cluster unless they were related by cause and effect, or by a common cause.[156,614] The cause is the human destruction of life, which in turn weakens the Fields, resulting in terrestrial, then solar, and finally, solar-induced terrestrial effects.

This interrelationship has been observed in a number of ways. One of the earliest cycle researchers comments on this: "When cycles of seemingly the same length are found on both the Sun and the Earth, the solar waves crest *after* the Earthly ones."[159]; p.69. Putting that into a perspective of cause and effect, the conclusion would be that Earthly phenomena is the cause of the solar cycles.

This relationship is also apparent in the Gross National Product (GNP) and the Wholesale Price Index, both of which are a measure of the life destroyed in the manufacture of goods and the waste generated. A study compared these indexes over a time period from 1889 to 1978 with sunspot cycles, and disclosed the causal factor. In this article, "Sunspots and Cycles: A Test of Causation," in which the authors state: "In an attempt to identify the limitation of exogeneity testing, we gathered data on a 'classic' question in economics. Is there a causal relationship between sunspots and the business cycle? On the basis of the evidence, we conclude that economic activity has an important influence on the Sun."[545]

The stock market similarly reflects the degree to which life is destroyed, and also displays a cycle. Sunspots have a cycle of 9.15 years, which is "exactly the length of the dominate cycle in the stock market."[185] Furthermore, when other cycles around this same length are examined the majority of biological, industrial and economic cycles peak first, then solar, and finally, biological, social-unrest and geophysical cycles affected by the solar cycles.[185,614]

Included in the last phase of these interrelationships is the cycle apparent in the International War Battles Index, and the 11-year solar cycle.[158,548] That is, we humans begin to destroy the very industries and urban centers in war that were destroying life to begin with. Not only does war follow, but geophysical events, such as earthquakes, floods, droughts, and so on increase, which also eliminate life-destroying urban-industrial centers.[456-458] These relationships are particularly true for the dominant 11-year solar cycle.

Other observations confirm these interrelationships. Just as there are differences in the activity of plants and animals at night, as compared to day, there are differences in chemical reactions (*Piccardi Tests*) during the night and day.[552] A worldwide force affects both organic and inorganic compounds (*i.e., colloids*).[4,611-614] Definite correlations exist between chemical reaction rates (*precipitation curves*), and the curve of the daily relative sunspot number.[55] None of the meteorological factors can be said to be responsible.[610-614] Furthermore, there are differences in these reactions according to latitude that are in accord with the latitude phenomena of FEM.[94] These chemical reactions reflect solar activity in precipitation rates with sunspots peaking *after* the reaction times. They are also correlated with geomagnetic storms, sudden cosmic-ray bombardments, solar flares, and have an annual peak, like solar activity, at the March (vernal) equinox.[473] Moreover, these reactions are evident in living things, which are a major component of the functioning of FEM (particularly as a bioelectrostatic generator).[4,474]

The electrical potential of the Earth varies with solar activity showing a minimum value (*1.0 Gv*) at sunspot minimum, and a maximum (*2.7 Gv*) at solar maximums. The electrical potential of trees (*L-fields*) fluctuate with the Earth's magnetism, and light and dark cycles, lunar cycles and solar activity. Meanwhile, air and Earth potentials fluctuate exactly in phase with tree and plant potentials, though occasionally it may be that only a few trees fluctuate, or the air and Earth are 180° out of phase. Atmospheric electricity shows a double periodicity annually, with maximums at the time of the equinoxes and minima around the solstices.[87,88]

Alternating electromagnetic fields, like those of geomagnetic storms, are biologically active. Animals undergo changes in (*systolic*) rhythms, bioelectricity (i.e., heart, cortex, and other biophysical potentials), and blood physiochemistry. Microorganisms display non-hereditary changes in morphology, and the properties of their cultures.[633,634] Most Earthly cycles are apparent about 40° from the equator (i.e., the mid-latitude Fields).[159] *All life* on Earth displays the reactions at least four days *prior* to any solar influence that is exerted on the Earth.[172] The conclusion is inescapable: *What we do here on Earth controls the Sun!*

CHAPTER 15

The Sun and Moon in Whether Weather Changes
(The Solar-Lunar-FEM Linkage in Climate and Weather)

While the actual mechanisms have remained obscure, the topic of solar influences on weather has attained an unprecedented scientific respectability. One understanding which is shared by all scientists is that the solar influence could not be direct.[151,261] As occurs with most other geophysical phenomena, it is the varying levels of solar plasma and particle flow along Field lines that produces the effects.[457]

Because this is unknown, many unanswered questions remain. One is that the superrotation of the upper atmosphere has always been theoretically unexplained.[277] As is typical of the deficiencies involving present-day models, a scientist makes a comment that is still timely: "No physical mechanism for Sun-weather effects is generally accepted at the present time."[261] Yet, FEM is a model of the Earth that would be expected to create what has been observed.

The unity of the Earth, Sun and Moon is also clear in weather phenomena. Solar triggers and solar cycles have been uncovered in individual events, as well as long-term cycles. Again, the conclusion is that the Earth, Sun and Moon are an integrated single unit.

15.1 Solar Triggers on Weather

Solar cycle, sunspot and solar flare influences on weather phenomena have been known for a long time. However, such relationships have been denied any real attention by most meteorologists. The reason for this lack of attention is that no satisfactory explanation could be derived from classical physics, particularly with regard to gravity (lunar tidal forces) and mass (fluctuations of solar plasma).

Meanwhile, studies disclosing a solar influence on weather are quite extensive. A list of just a few is quite lengthy.[48,83,84,96,155,165,211,212,375, 406,436,437,457,509,539,554,567,653,655,656,662,676] Even sunspot structure and climate are correlated.[275] New studies on solar activity and weather can be found in the literature nearly every day.

One study disclosed that even the 27.5-day solar rotation was present in weather.[309] This study was criticized on the grounds that it resulted from a problem with filtering the data. However, the real difficulty was remarked upon by the critic, and reflects the actual dilemma, since there is "no physical mechanism to explain a 27.5-day solar rotation in weather."[663] Such a comment reflects the deficiencies of present theories, and calls for a new model of the Earth.

Aside from the 11-year and 22-year cycles, there is also the 45-year Double-Hale solar magnetic cycle reflected in storminess and high tide. The all-planet synod, when all the planets are on one side of the Sun, which can influence solar activity, creates a 178- to 179-year solar cycle that is reflected in Greenland ice cores, and Hudson Bay sediment anomalies.[190-192]

Solar cycles are correlated with sea level, atmospheric pressure, and surface air temperatures in summer, and especially over the oceans in winter.[68,301,619] Ozone varies with the long-term solar cycle, as does upper atmospheric airborne particles (*stratospheric aerosols*), and shifts in climate.[47,317,496] The extent of Newfoundland's ice cover for the period between 1860 and 1988 has been correlated with solar activity.[263] An influx of air from the uppermost (stratospheric) layer of the atmosphere into the layer below (troposphere) has been observed three to four days after solar flares.[501-503,42,628] What is particularly surprising to scientists is that observations indicate a source of ionization from *below*; a completely unexpected phenomenon.

Single energetic flares and periods of enhanced solar eruptions (e.g., coronal mass ejections, etc.) are related to weather, including the quality of forecasts, atmospheric circulation, rainfall and thunderstorm frequency. The cycles of solar flares display periods of 9 years, 2.25 years, and 3.3 months, which hints at the long-term and seasonal effects on weather.[304,340] These facts alone reveal that the solar linkage of FEM is responsible for solar-induced weather shifts.

Weather phenomena, in general, are correlated with the solar cycle. Particle flow ionizes the atmosphere, producing a partial vacuum that alters air pressure. One of the most widespread effects of solar activity is the alteration of atmospheric pressure worldwide.[141,265,266,301,457,522] Solar wind particles that enter the polar regions alter pressure, especially above 100 kilometers (62 miles), and the troposphere, varying markedly with the solar cycle.[306-309,374,392,583,619,628] However, the effect is much greater than anticipated, and while the effect is expected for only the circum-polar regions, it is instead global.

Because the particles released by FEM carry charges, they would flow along magnetic field lines, and therefore, should display relationships with the geomagnetic field (GMF). The average wind does follow the rather complicated contours of the GMF. GMF activity, always associated with solar activity, is correlated with the development of low-pressure systems (*troughs*).[510] Daily and semi-daily GMF activity, and weather fluctuations are associated with ionospheric gravity waves.[129,234] Atmospheric gravity waves are also generated during solar eclipses.[221] Likewise, longstanding planetary waves (in the troposphere) are triggered by solar events.[309,663] Observations such as these require a Field system interconnected with the geomagnetic field along which charged particles flow.

Solar relationships have been confirmed for many of FEM's processes, such as the magnetosphere[483,664], sea level[126,132,301], the upper atmosphere[317,583], and the physical processes of the atmosphere.[530] Atmospheric electricity[375-379,457,497-505], temperature[522,541,569], pressure[134,147,522], and circulation[654] are also correlated with solar activity. As could be predicted from an understanding of FEM, every geophysical aspect of weather-related phenomena is affected, even down to the ground and below the oceans (ridges, deep sea currents, etc.).[456-458]

15.2 Solar-FEM Linkages

The Interplanetary Magnetic Field (IMF) also has an influence on weather, revealing another aspect of the solar-FEM linkage. Low-pressure systems (*troughs*), or cyclones in the North Hemisphere, are at a minimum about one day after the IMF sector boundary crossing (SBC) is carried past the Earth. A sector boundary crossing (SBC) is when the IMF shifts from away from the Sun to towards the Sun, or vice versa.

Increased geomagnetic activity takes place around solar maximum. The size of storms, referred to as the Vorticity Area Index (VAI), increases during solar maximum, when the SBC sweeps by. The vertical electric field near the South Pole of the geomagnetic field (GMF) is at a minimum a few days after the SBC, but local winter shows a large effect (*amplitude*) with a maximum, for example, at Zugspitze in the Alps. Expected of an ionizing source, increases in isotopes (i.e., Be^7) at mountain peaks are also observed. When the IMF is moving away from the Sun, towards the Earth, it increases storms (*troughs*). FEM's solar linkage is responsible, as can be seen in this statement: "The physical mechanism may relate to the topological condition that an Interplanetary Magnetic Field line directed away from the Sun may merge with a geomagnetic field line going into the northern polar regions."[654]; p.762

The flux of electrons is greater when the IMF is moving away from the Sun (*a few hundred eV; an order of magnitude*). The SBC is correlated with lightning and thunderstorm frequencies, and the electric field variations conform to FEM, as they were noted simultaneously in the Arctic, Antarctic, and mid-latitude mountain tops. The formation of storms (*atmospheric vorticity*) takes place along with the SBC. Shortly after solar flares, atmospheric electricity responds with increasing electric fields and lightning frequency. Likewise, electrons increase in the upper atmosphere (*stratosphere*), during geomagnetic storms. For example, a high correlation exists between the 11-year solar cycle and thunderstorms in England with storm size (VAI) increasing one to four days after solar flares, and other solar eruptions.[27,375,376,437,654]

A particularly strong solar correlation exists for high latitude thunderstorms, but not equatorial thunderstorms. The Fields are located on the northern and southern extremes of the equatorial bulge, and point away from the equator, hence thunderstorms should display this relationship. Ionizing radiation from solar activity can cause large effects on atmospheric conductivity down to at least 15 kilometers (10 miles) at mid-latitudes (the effect is deeper, as will be discussed). Again, the Fields are along mid-latitudes with the exception of the polar Fields. Oceanic thunderstorms maximize in the northern latitudes during winter. Such an effect is due to the fact that the Fields are in the oceans and are more active in winter. Likewise, the North Pacific near the Japanese Current, and the North Atlantic by the Gulf Stream, the Japanese and North Atlantic Fields, are the largest sources of convective clouds that respond to solar activity. Moreover, the Brazilian Field (*South Atlantic Geomagnetic Anomaly; SAGA*) displays enhanced lightning frequency.[375,376,437] Depending on whether it is solar minimum or maximum, global thunderstorm activity increases by 50% to 70% four days after major flares.[260] The solar-FEM linkage and Field locations are conspicuous in these data.

Solar activity and thunderstorms are closely correlated, even in ancient Chinese records. A study of winter thunderstorms in China was undertaken for the period from 250 BC to AD 1900. The occurrence of winter thunderstorms indicates that there were global disturbances in the jet streams. The disturbed times are typical of history, as will be shown in *In Defense of Nature—The History Nobody Told You About*, and are AD 100-150, 300-500, 1100-1300, 1350-1400, 1450-1550, and 1600-1900.[147,642]

Another indication of this solar-FEM linkage is the effect produced by the Sun's movement around the center of the Solar System's mass; its spiral motion through space. The Sun moves along a very irregular helix around the elliptic line of motion of the center of mass, known as the barycenter. The distance from the center is known to effect solar activity with times of less distance between the center and the Sun being times of solar minimums.

Times of less distance include periods such as the Sporer and Maunder Minimums in solar activity. Just prior to these minimums there were solar maximums that produced climate and radiocarbon (C^{14}) fluctuations that, after a lag time, are noted during these minimums. The Maunder Minimum was a time of the Little Ice Age when glaciers most recently advanced down mountains around the world. The Little Ice Age was a time for low values in long-term temperatures for all seasons, enhanced variability of temperatures from spell to spell, and year to year, and an enlarged polar ice cap and frigid air over the Northern Hemisphere that was accompanied by jet streams that were weaker and further south than those of today.

Phases that are influential are typical of magnetic field interactions with 0° and 90° as potential peaks, and 180° and 270° as troughs in the swinging Sun. This motion around the barycenter is linked with solar activity, climate, earthquakes, and volcanic activity.[184,288,338,340,456-458,551,558] In contrast to conventional thought, but expected of FEM: "This previously unsuspected relationship tends to corroborate the reality of a solar motion-solar activity-terrestrial systems linkage."[551]

Impulses of torque in the Sun's motion around the barycenter are reflected in solar and climatic cycles. Both extremes are present in the (Gleissberg) solar cycle, and maximums are noted in sediment (varve) thickness from Lake Saki, Crimea back to the 7th Century. Also correlated is rainfall over central Europe, England, Wales, the eastern United States, and India, as well as temperatures in Europe for more than 130 years.[150,344] This relationship is supported by the historical record, as well.

Cold periods are noted at times when solar motion misses the barycenter. Two such times were the periods of 1632-33 and 1810-12. Climate in the 1630s and 1640s was comparable to the Little Ice Age, and on occasion colder. The period was marked by short growing seasons and glacial advance. The 1810-12 period was followed by a cold interval worldwide with cold summers and short growing seasons. A "year without a summer" (1816) was noted in the northeastern United States, and in western Europe it stopped Napoleon's advance into Russia. Conditions brought famine to Switzerland and the Ukraine. The years between 1812 and 1817 introduced three decades of economic pause, punctuated by recurring crises, distress, social upheaval, international migrations, political rebellion, and pandemic disease.[551] Again, a solar-FEM linkage and the effects of FEM are clearly discernable.

Planetary positions, which influence solar activity, are also found in weather. For example, severe winters in China are correlated with planetary alignments. Those times when the Earth is on one side of the Sun, and all other planets are grouped on the other side in a tight arc of 90°, or less, are times of severe winters. These times include 269 BC, 449 BC, AD 1126, AD 1304, AD 1483, AD 1665, AD 1844, and AD 1982. Cold periods, noted in both Chinese records and Greenland ice cores, were observed in the first half of the 12th Century, the 14th Century, the end of the 15th Century, the second half of the 17th Century, and the 19th Century, coinciding with the dates of some of the alignments. These planetary alignments are correlated with severe winters, and the spells can last 30 to 50 years.[277,228,230] Based on this understanding Chinese researchers, in 1982, predicted "cold winter and freezing disaster in the next two decades."[230] All of these periods coincide with or occur shortly after the historical cycles discussed in *In Defense of Nature—The History Nobody Told You About.*

The all-planet synod is linked to air temperature, air pressure, droughts, floods, and tree rings in China.[147,316] Long-term Chinese climatic history displays the all-planet synod of 178.7 years. The synod cycle can be traced to severe winters and abnormally cold periods, and unusually warm summers, extending back to 2453 BC. In addition, there is a general parallel of this record in Greenland ice cores.[506,509]

Other planetary alignments also influence climate. The Jupiter-Saturn Lap of 19.857 years appears to be responsible for the approximately 20-year cycle in climate.[506] However, other interpretations exist, such as a combination of solar and lunar cycles.[599]

Additional planetary influences are somewhat indirect. Jupiter's and Saturn's orbits, and Jupiter-Saturn conjunctions, are evident in sunspots, and changes in the Earth's rotation or the length of day.[226] Periods of the planets are apparent in temperature and rainfall in Helsinki for the interval between 1902 and 1971.[372] Planetary alignments were correlated to solar activity and climate change in other studies, as well.[224,226,228,349,416,557,685] Furthermore, long-term (*secular*) changes in the geomagnetic field, and the length of day are correlated with global temperatures on a decade scale.[184,341] Geomagnetic storms and planetary configurations are also correlated.[421] These planetary configurations and periods influence solar activity, which then interacts with FEM, creating shifts in weather.

15.3 11-Year and 22-Year Solar Cycles in Weather

This entire relationship with regard to weather is all the more obvious in studies on the effects of the 11-year solar cycle on climate. Numerous studies have been conducted that indicate the reality of a 11-year solar cycle in climate.[128,130-132,140,146,147,336,337,356,414,616,617,630,631,675] For example, the 11-year cycle is conspicuous in worldwide air temperature, air pressure, and ozone.[64,67,90,128,130,141,385,536]

The 11-year solar cycle has been correlated with air temperature, air pressure, droughts, floods, lake levels, snowfall, tree abundance, and tree-ring growth.[63,147,171,261,616,617] Rivers, such as the Nile, Ohio River, and Parana River (Buenos Aires, Argentina), rise and fall in accord with solar activity.[113,189,261] Numerous examples exist that reveal an 11-year solar cycle influence on weather-related phenomena.

Consider some of the observed effects of the 11-year solar cycle on weather. Droughts in Germany show double waves of the 11-year cycle covering a 124-year period, which had 23 very severe droughts. These droughts took place during the minimum and maximum of the solar cycle (*Wolf numbers*). Likewise, annual records of precipitation at Klagenfurt, Austria, covering a period from 1854 to 1964, displayed dry summers at times of both the minimum and the maximum of the solar cycle.[418] Droughts and accompanying famines in India are correlated with the solar cycle, as well.[156] Rainfall and temperature records for West Germany are completely parallel to the sunspot curve (*Wolf numbers*). This study was comprised of monthly figures from the period of 1950 to 1962, and revealed a drop during minimum, and a rise during maximum, solar activity.[261]

Likewise, pressure and rainfall in other parts of the world are correlated with the 11-year cycle. For example, North African, Middle Eastern and northwestern Indian records conform very well with solar activity.[666] Abnormal rainfall in Kew, England reveals that the driest winters take place at the time of solar minimums, or the year following.

Dry springs, summers and autumns came at times of increased solar activity. In addition, rainfall in spring occurs at the extremes of the solar cycle with the driest at times of maximums, and the wettest around minimums. Similarly, the most probable times for severe winter in the northeastern United States, and north-central Europe are times of minimum and maximum solar activity.[261] In North America, droughts in the high plains, and rainfall and drought fluctuate with the solar cycle.[509,669] One effect was the mid-1980s drought in the Southeast, which was linked to solar activity changing the jet stream and pressure.[417] Atmospheric circulation and wind in the former Soviet Union displays solar cycle influences, as well.[66,242,433]

A half cycle of 5 to 6 years is also apparent in the data. Precipitation cycles of this length were noted for England and Russia, and droughts in Kazakhstan.[53,54,471] The duration of north and east winds in Greenwich, and air pressure in London of this length are also evident.[54] This shorter cycle, and the major 11-year cycle, comprise other longer cycles.

Such is the case with the 22-year solar cycle, also referred to as the double sunspot cycle, when the Sun's magnetic field reverses. The solar-FEM linkage is immediately apparent in the fact that the 22-year solar cycle is noted in geomagnetic activity.[103] A 22-year cycle is also noted in tree-ring growth, as noted in fossilized tree wood.[164,180] In fact, the 22-year cycle of fossilized tree wood suggests that there are "cyclic variations in air or water temperature that could be driven by an unspecified solar variable."[180]; p.131

Sediment runoff is a measure of rainfall and the 22-year solar cycle is seen in the thickness of annual sediment layers (*i.e., liminological studies*).[180,192] Similarly, the 22-year solar cycle is clearly shown in droughts in the Great Plains and western United States, and also mean surface temperatures for the Northern Hemisphere, and for planetary waves.[32,261] Severe winters and droughts are noted during the minimum of the 22-year cycle.[376] As could be predicted of FEM and life's role, the double solar-cycle is even more apparent in ancient climates prior to a diversified and complex life system on Earth (*i.e., the Precambrian*).[33,429,660,682]

15.4 Three- to Four-Day Phenomena

Aside from these effects there is also a three- to four-day period in a number of observations. The influx of stratospheric air into the troposphere takes place three to four days after solar flares.[501-503,542] Zugspitze, a mountain peak in the Alps, undergoes an increase in isotopes (Be^7) about three to four days after the passage of the IMF (SBC).[497-505] Global thunderstorms increase by 50% in solar minimum years, and increase by 70% in solar maximum years, four days after major solar activity.[261]

The electrical characteristics of the atmosphere and the Earth (*atmospheric potential gradient and air-Earth current density*) increase shortly after a flare, and peak in three to four days.[497] Wind formation displays the same periods with high altitude winds three days, and low altitude winds four days, after flares. Likewise, three days after a geomagnetic disturbance the winter side of the Earth shows pressure changes at sea level.[261,277] Thunderstorm occurrence east of the central United States (i.e., the half closer to the North Atlantic Field) is related to lunar positions, with Full Moon the most influential, two to three days after which thunderstorms take place.[354,355] These observations are in addition to other geophysical phenomena that display a three- to four-day period.

15.5 Isotope Fluctuations

Fluctuations in isotopes indicate that electrostatic forces, produced by particle flow and ionizing radiation, are responsible for climatic shifts. Isotopes, such as radiocarbon (C^{14}), were shown to be mediated by solar activity, as revealed by tree rings, extending back to 5300 BC, or more than 7,250 years.[147,178,340] Levels of radiocarbon are correlated with the (*heliomagnetic*) cosmic-ray flux, as supported by historical observations of aurora and naked-eye sunspot sightings. Likewise, the 22-year cycle is observed in fossilized tree wood, as tree-ring growth and isotope ratios (*D/H isotope analysis*).[179] An increase in beryllium isotopes (Be^7) occurs about three to four days after the Sector Boundary Crossing (SBC) of the IMF sweeps past the Earth.[504] Likewise, other beryllium isotopes (Be^{10}) in polar ice have been used to trace the 11-year cycle.[56] Isotopes of oxygen in tropical tree species from India[491], Antarctic ice[56,58], and marine sediments[517], and nitrogen in fossil bones[257,412] have been used in climate studies, as well.

15.6 Lunar Effects

A number of facts indicate that the Moon affects geophysical phenomena far beyond what would be expected from its gravitational pull. According to statistical analysis there is a lunar effect on a variety of geophysical and meteorological activity when the Moon is within 4° of the (*ecliptic*) plane of the Earth's orbit around the Sun. For example, between Full Moon and Last Quarter (on the morning side of the Earth), there are geomagnetic storms. The effect is known to be due to energetic particles that are precipitated into the upper atmosphere, producing electrical currents that perturb the magnetic field (i.e., electromagnetic induction).[261]

This effect also involves the entire field system of FEM. The geomagnetic field is also affected by eclipses, which increase the conductivity of the atmosphere (*E-region*).[70] Lunar effects are more dynamic than previously realized because the effects are due to interaction with the Fields and particle flow (*electrostatic repulsion, plasma torus, bow shock, etc.*), not gravity.

Extremes in the Moon's orbit are referred to as the lunar nodal cycle, an 18.6-year lunar cycle, which influences weather and other geophysical phenomena. This lunar cycle is apparent in atmospheric pressure, sea level, precipitation, sea-ice conditions, tidal currents, currents in submarine canyons, sea-surface temperatures, geyser eruptions, volcanic eruptions, earthquakes, thunderstorms, auroral frequency, and biological growth series, including tree-ring widths. The lunar influence is more pronounced in mid-latitudes, due to the mid-latitude Fields, and is more clearly represented in records during solar minimums when the effect is not obscured by solar activity. These correlations immediately suggest that FEM is responsible. Meanwhile, present-day, conventional theory is at a loss for an explanation: "The lunar linkage mechanism has not been established and evidently needs research."[193; p.466].

Chinese work suggested solar forcing in certain weather phenomena, but a lunar tidal cycle reinforced by solar is a more substantial interpretation of the data.[147,193,567] Floods in China's Szechwan Basin, in July 1981, were due to the positions of the Sun, Moon and Earth. Likewise, lunar phase, and solar or lunar eclipses were correlated with floods that took place from 1951 to 1980.[295] Other solar-lunar effects in Szechwan climate have also been noted.[102]

Lunar effects on precipitation are more evident seasonally for certain geographic regions (i.e., FEM's Field contours in relation to the GMF-IMF linkage).[246] Both drought and flood in China exhibit the 18.6-year lunar cycle. Also correlated were air temperature, sea surface temperature, air pressure, and tree-ring growth.[147,362] Beijing, China precipitation records for the period from 1470 to 1974 also disclose a lunar cycle influence.[240] The 18.6-year lunar cycle is also apparent in prolonged and widespread droughts in the western United States, as indicated by 300 years of tree rings.[32,138,140]

The 20-year drought cycle occurs near the end of the lunar cycle when solar activity is low (after maximum), but did not consistently follow this pattern from 1800 to 1900. It was after 1900, when the Industrial Revolution began a sweeping impact on the biosphere, that the cycle is first noted. This drought cycle can also be explained as a combination of the 22-year solar and 18.6-year lunar cycles.[599]

Sea level, sea temperature, air temperature, and air pressure are known to fluctuate in an 18.6-year cycle.[92,93,133,136,138-143,356,359] The lunar cycle is also noted in the formation of hurricanes and tropical storms.[95] In addition, the lunar nodal cycle controls the latitude of subtropical high-pressure belts in both hemispheres, and drought-flood cycles in South Africa and Argentina.[492-494] Expected of life's interaction with FEM, lunar cycles were even more apparent in ancient climates prior to a fully developed life system on Earth (*i.e., the Precambrian*).[33,429, 660,682] Additional lunar effects confirm this lunar linkage of FEM.

Declination and position of the furthest (apogee) and closest (perigee) approaches of the Moon display a lunar phase influence. As is typical, conventional gravitational models do not explain these effects. For example, the heaviest rainfalls of the month at stations in New Zealand, and the Spanish peninsula were correlated with lunar effects, while the magnitude rules out gravitational forces as the primary influence.[5] Yet, the West Australian Field is near New Zealand, and the Mediterranean Field is near the Spanish peninsula. An electrostatic trigger is responsible for these lunar effects, which involves the Moon causing particle cascades along Field lines (*electrostatic repulsion, bow shock, plasma torus, etc.*) that ionize the atmosphere. In accord with FEM, the forcing mechanism is a time-varying balance between the Coriolis force and the tractive force of the 18.6-year lunar cycle, which reaches a maximum at the 35° latitudes.[147]

As a result, a number of studies demonstrate that lunar phase and lunar tidal forces trigger individual events. In the Alps and Po Valley there are lunar-solar "tidal" effects, especially for rainfall.[99] Indian Ocean depressions, which are responsible for heavy rainfall, are correlated with lunar phase.[632] Lunar effects are observed in fluctuations of precipitation, particularly rainfall, in many regions the world over.[74,75]

Likewise, the phases of the Moon are known to effect widespread, heavy rainfall in the United States.[72] According to the records of 108

stations, thunderstorm occurrence east of the central United States—
the half closer to the North Atlantic Field—is related to lunar positions
for the years 1930 to 1933, and 1942 to 1965. Full Moon is the most
influential, two to three days after which increased thunderstorms take
place.[354,355] Even the degree of cloudiness or sunshine is related to the
lunar month (*synodic cycle*).[364]

Daily data for the period 1900 to 1980, in the United States, revealed
a lunar influence (*phase progression*) on variations in precipitation. The
Moon's revolution around the Earth, or lunar month, known as the
lunar synodic period (29.531 days), and also half that period (14.765
days) were detectable. Again, the effect is not explainable by gravitational models, especially with regard to geographic region and season.[246] The geographic effects are due to Field location and contours,
and the seasonal shifts are the result of the solar-FEM linkage.

The impact of these influences can be understood by this statement
offered by two climatologists: "It is observed that when the maximum
lunar tidal epoch is in phase with the maximum solar activity epoch,
climate and economic impacts are amplified."[147] History shows us that
this is the case, and could be predicted with an understanding of the
solar-lunar-FEM linkage.

15.7 FEM and the Geographic Distribution of Weather Phenomena

The new model of the Earth is even more obvious, because weather-
solar connections are "manifested differently in different geographical
regions and depends on longitude, as well as latitude. This has been evident in studies of 11-year periodicities in pressure, and also shows up in
short-term correlations with solar flares and magnetic activities."[261];
p.125. Others have also commented on this relationship.[94,212,457,458]

The Fields are in the oceans, and as a result, the solar cycle is noted in
sea level, as well as air temperature and pressure over the
oceans.[126,133,135,141] Ocean circulation and shelf-area currents are at a
maximum off the North and South Coasts in high to mid-latitudes and

equatorial latitudes. Meanwhile, in accord with FEM, certain types of ocean circulations are near zero values around the 30° to 40° latitudes, the latitudes of the Fields.[188]

Other weather observations disclose the latitude restrictions of the Fields. The Fields are situated on the northern and southern extremes of the equatorial bulge, and point away from the equator. Hence, a high correlation exists between the sunspot cycle and high-latitude thunderstorms, while it is low or negative for equatorial thunderstorms.

Around solar maximum the annual mean pressure is shown to be lower than normal in equatorial regions, and higher than normal in the temperate regions, where the Fields produce the greatest effects.[261] Ionization causes a time-varying (periodic) partial vacuum in the Fields. Afterward, equatorial air flows into the partial vacuum produced by the Fields, thereby creating high pressure in temperate regions (damming) and low pressure in equatorial regions.

Observations indicate that during solar maximum there is more snow, and two to three times more icebergs in high latitudes, and lower temperatures between 45° North and South Latitudes. An exception to this is that the arid subtropics have higher temperatures due to the formation of a high pressure system by the Field (this will be included in a discussion on deserts). In addition, there is 10% to 20% more precipitation between the Azores and Bermuda Highs (North Atlantic and Mediterranean Fields), mid-Australia (between the East and West Australian Fields), South Africa (South African Field), Southeast Asia (Japanese Field), India (near the Persian Gulf Field), and Brazil (Brazilian Field). Three times more hurricanes develop in Southeast Asia, 65% more hurricanes form in the Indian Ocean, violent hurricanes are six to eight per year (during solar minimum there are one to two), and there are 40% more storms on the storm tracks.[73,261,573]

Mean annual temperature curves for the tropics are correlated with sunspot curves. This effect is discernible in monthly air temperature data for the Northern Hemisphere, particularly in winter.[434,481] The

Fields are more active in winter, especially in the Northern Hemisphere due to the solar linkage in relation to orbit.

FEM is an integrated system of Fields that are structurally related to the geomagnetic field (GMF), and therefore, geomagnetic changes should be reflected in weather. A twelve-month wave is noted in geomagnetic activity with maximums at the vernal (March) and autumnal (September) equinoxes due to interaction with the Interplanetary Magnetic Field (IMF).[608,609] As the IMF crosses the Earth (Sector Boundary Crossing; SBC) there are changes in the magnetosphere, ionosphere and atmosphere. What follows are geomagnetic disturbances, changes in storm size (Vorticity Area Index; VAI) and wind direction, but as is usual for present theories, "the mechanism is unknown."[350]

A number of observations show weather shifts that correlate with changes in geomagnetic activity. Because atmospheric particles carry charges, the average wind follows the rather complicated contours of the GMF, with a particularly strong relationship with the "twin north magnetic poles," the south poles, and the Brazilian Field (*South Atlantic Geomagnetic Anomaly; SAGA*).[306-309,460] This can also be seen in that both the Azores High and the Icelandic Low gradually migrated northward between 1889 and 1940, and then began moving southward and westward.[21] This "climatic drift" is also noted in the drift of the geomagnetic field. Another correlation is between geomagnetic storms and winter storm development in the northern Pacific and North America.[511] Clearly relating to FEM and its solar linkage, a correlation exists for a 120-day oscillation in geomagnetic activity, the Earth's rotation, atmospheric zonal circulation, the IMF, and solar activity.[169]

Atmospheric vorticity over most of the Northern Hemisphere is related to geomagnetic activity. Sharp rises in geomagnetic activity, and the SBC of the IMF are followed by significant changes in storm size (VAI). During the period between 1950 and 1971, a 27-day solar period was linked with (*500-mb*) pressure and geomagnetic activity. This pressure change was more prominent in the colder part of the year (October to March; basically from equinox to equinox), and the (*500-mb*) pressure level at all latitudes from 20° North Latitude to the Pole.[683]

During geomagnetic storms, circulation increases greatly, extending to the equator and lower altitudes, while wind speeds increase. Normally, circulation is between the 20° latitudes and at high altitudes. Substantial zonal flows take place around the 30° to 40° latitudes at the equinoxes and solstices, as well.[512]

Geomagnetic disturbances produce monthly variations in the air-earth current near the North Pacific Field (*Mauna Loa, Hawaii correlated with Bartel's Cp Index*). Surface pressure modifications, increases in anticyclones and decreases in cyclones occur after geomagnetic storms. Geomagnetic activity (*aa-Index*) and mean monthly temperatures at 32 stations in the United States are correlated. In fact, long-term (*secular*) changes in the GMF and the length of day are correlated with global temperatures on a decade scale.[184,341,342] Larger (VAI) low pressure troughs or cyclones develop during or following geomagnetic storms.[653-655] FEM is particularly revealed in the fact that geomagnetic drift and shifts in weather centers are uniform.[21] The solar linkage is seen in the fact that geomagnetic drift and geomagnetic storms are triggered by solar activity.

Local winter above 40° latitudes exhibits a minimum in pressure for both hemispheres, during solar maximum. Low pressure troughs develop more often in those years with high solar activity. Also, greater than average rainfall develops in equatorial regions (between the 20° latitudes).[261] All of these observations can be readily explained by increased ionization in the Fields and particle flow along Field lines producing a partial vacuum, leading to low pressure with the colder, upper atmosphere flowing in, thereby producing storms, reducing temperatures, and increasing precipitation in certain locations, especially along mid-latitudes and the poles.

Short-term climate changes occur along with changes in geomagnetic intensity, and again, the effects of FEM are apparent. When there is decreased GMF intensity, there are increased temperatures in Canada (North Polar Field), Mexico and the United States (North Atlantic Field), Brazil (Brazilian Field), South Africa (South African Field), and New Zealand and Samoa (East Australian Field). When increased GMF intensity occurs, there are decreased temperatures in Greenland, Norway and Sweden (North Polar Field), the former Soviet Union (North Polar, and Persian Gulf Fields), Egypt and Germany (Persian Gulf Field), and Scotland (Mediterranean Field). Abrupt changes in GMF intensity are correlated with abrupt changes in temperature.[667] Likewise, radiocarbon (C[14]) levels are affected by geomagnetic activity, with deviations extending more than 7,250 years, as revealed by tree rings.[147,178,340,562]

Abrupt changes in the GMF can also cause temperature changes 3 to 7 years later. This is in addition to the 27-day recurrence in temperature distribution and geomagnetic activity, though geomagnetic activity is not generally accepted as an influence on weather. However, first there is geomagnetic activity then weather changes, and this is especially true in winter.

A 3-year lag in temperature changes behind geomagnetic fluctuations at mid-latitudes was revealed by comparing the Kakioka Magnetic Observatory (36°) with Tokyo (35°), and the Tucson Magnetic Observatory (32°) with Phoenix (33° Latitudes). Meanwhile, there is a 7-year lag at low latitudes, as revealed by comparing the San Juan Magnetic Observatory (18°) with San Juan weather.[667] Each Field area is represented in the magnetic observatories' data, while weather satellites show even more phenomena relates to FEM. Again, present theory is lacking:

> "The influence of geomagnetic activity on the weather is not used for forecasting purposes, and a widely accepted physical mechanism for any connection between geomagnetism and weather has not emerged. Nevertheless, a few common features

appear so widely in the otherwise disparate literature that they suggest the existence of a valid basis for the conclusion that they are related."[667]; p.242.

The Earth's wobble on its rotational axis is also correlated with weather. The Chandler Wobble produces an oceanic tide that is several times larger than the 18.6-year lunar nodal cycle.[129] A "strong signal beat" period of 6.25 years occurs between the Chandler Wobble and seasonal fluctuations.[125] Atmospheric tide and the 14-month Chandler period play a role in climate, particularly at high latitudes.[390,391] This Chandler Wobble-climate effect is easily understood as the wobble of the Field-system of FEM due to its solar linkage.

This is why standard models do not explain much of the evidence. Long-term oscillations of the upper atmosphere (*thermosphere*) are triggered from *below* on special occasions in the winter. This winter effect, with the Earth tilted away from the Sun and being triggered from below, is not what could be predicted by present models, which instead predict effects from above in the summer. The consensus is clear: "The evidence shown for an intense coupling between the basic states in the strato-, meso- and thermospheres during late winter is physically not yet explainable. But why are the internal atmospheric processes so well reflected in the upper region during winter?"[564]; p.302. As stated previously, the Fields are more active in winter, due to the solar linkage of the IMF.

Solar flares can increase electrical conductivity and produce a transient reversal of the vertical electric field within the middle atmosphere. The origin of large vertical electric fields near the 65-kilometer (40-mile) altitude is "unknown," and is therefore controversial.[119,267,636,637] These electric fields are the result of particle flow along the Field lines of FEM activated by solar wind particles.

The interconnectedness of the Field system of FEM was demonstrated in a study comparing drought in the Sahel with intense hurricanes that hit the United States and the Carribean. Sahel is a semiarid region south of the Sahara, stretching from the Atlantic Coast through to the Sudan, in Africa. During years of drought in the Sahel fewer intense hurricanes develop in the North Atlantic. When there are wet years in the Sahel, there are a greater number of intense hurricanes that hit the United States and the Carribean. The meteorologist that conducted this highly significant study predicted that more intense hurricanes would form in the 1990s, because the Sahel was experiencing a wet period.[222]

This has already proven to be true, as for example, with Hurricane Andrew, the costliest disaster in United States history, with damages at more than $30 billion. There were also more tropical storms and hurricanes in the Atlantic than any previous time in history, with at least 19 developing in 1995 alone. In the 1990s, there were hurricanes Bob, Andrew, Marilyn, Opal, and Georges, just to mention a few, that made landfall in either or both the United States and the Caribbean, causing extensive damage. Here is the evidence of greater ionization of the atmosphere in both the Mediterranean Field (just southeast of the Sahel) and the North Atlantic Field at the same time.

Throughout the world, deserts are mostly situated along the 30° to 40° latitudes. The distance from any given Field longitudinally is governed by the extent of life's destruction through time. As the Field(s) weakens it creates a convex lens of high pressure, which increases temperatures and prevents rainfall (see Figure 5.3). The influence of civilization is confirmed by archeological evidence of civilizations lost beneath the desert sands, petroglyphs showing the Sahara as an open grassland, roads in the deserts that are no longer open to travel, historical records, and myths and legends (see *In Defense of Nature—The History Nobody Told You About*).

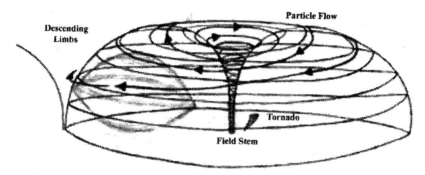

Figure 5.3. Field Contours in Weather. The Field looks very basically like this drawing. The arrows represent particle flow. The gray lense-like structure is a high pressure system caused by Field contours and particle flow. If this were the North Atlantic Field, the stem of the Field (near the tornado) would be east of the Southeastern United States, a descending limb (far left) would be along the West Coast, and high pressure would be over the southwestern desert, known as the Sonaran and Mojave Deserts. This scenario changes according to season, and solar and lunar influences.

For example, the Mediterranean region and the Near East have been the seat of the oldest civilizations: Mesopotamia, Babylonia, Indus, ancient Egypt, ancient Greece (Minoans, Myceneans, etc.), Etruscans, the Roman Empire, and others. Due to the life-destructive characteristics of these civilizations, the Mediterranean and Persian Gulf Fields have weakened creating a strong high-pressure system over northern Africa and southwestern Asia, bringing into being the Sahara Desert, the largest desert in the world, as well as the Syrian Desert, Dashte Kavir (Iran), and the Great Indian Desert (Thar).

As a result of the ancient Chinese and other civilizations in Southeast Asia the formation of the Gobi Desert, Eastern Turkestan Dessert (Takla Makan), and Ordos Desert (Mongolia) developed. In fact, Genghis

Khan once used a lush hunting ground that is now the Gobi Desert. These deserts are the result of the weakening of the Japanese Field.

Comparatively, the South African Field has not produced widespread deserts. This is because civilization has not been longstanding in the region. However, there is the relatively smaller deserts, known as the Kalahari and Namib Deserts, located in Southern Africa.

Australia has deserts between the East and West Australian Fields, and are called the Great Sandy, Gibson, Great Victoria, and Simpson Deserts, which are actually a single extensive desert sectioned into regions of different names. This desert area is shifted to the west towards the West Australian Field, which is (*magnetohydro-*) dynamically affected by the Persian Gulf and Japanese Fields. Australia has not been throughly investigated archeologically, and therefore, relationships with civilization cannot be made.

The Americas also display this Field-desert relationship. In South America deserts are few with one narrow desert that stretches nearly the length of Argentina. South America has had few ancient civilizations, and correspondingly few deserts. North America has the Mojave and Sonoran Deserts, which are fairly extensive deserts due to the ancient civilizations in the southern regions (Mesoamericans: Olmecs, Toltecs, Mayans, Aztecs, etc.).

One Field that clearly influences weather is the North Atlantic Field. The North Atlantic Field is off the Southeast Coast of North America around the 35° latitude. When considering the Field's descending limb in the United States, it is along the West Coast. Both the 11-year solar and 18.6-year lunar cycles are apparent in temperature records, but only east of the Rockies along the West Coast, and north of the 35° latitude, which also corresponds to the storm track.[130,138,147, 229] The flow of air down the Mississippi Valley into the Field area also creates some of the greatest air temperature fluctuations to occur in the Great Lakes.[128.130,135,245] An 11-year period is apparent in drought records for western North America, near the descending limb of the Field.[138.147]

More rainfall and lower temperatures occur between the stem and east of the descending limb, while west of the limb drought occurs, during solar maximums.

Also evident in these western droughts is the 18.6-year lunar nodal cycle, and the entire Northern Hemisphere most often displays solar and/or lunar periods.[32] The phases of the Moon are known to effect widespread heavy rainfall in the United States, as well.[72] The 18.6-year and 22-year cycles combine in producing other effects, with the 18.6-year cycle having a maximum influence during dry periods.[599]

These weather phenomena are also reflected in the United States agricultural and livestock production. This includes corn, wheat, oats, barley, rye, buckwheat, flaxseed, sugar beets, potatoes, sweet potatoes, and sugar cane, all of which display the 18.6-year lunar cycle.[145,146] Also, the livestock production of horses, hogs, chickens (including eggs), and other livestock are in accord with the lunar cycle.[144,145]

Winter temperatures for the East fluctuate nearly in phase with sunspot cycles.[336] Likewise, thunderstorm occurrence east of the central United States, according to the records of 108 stations, is related to lunar positions. Full Moon is the most influential, two to three days after which there are increased thunderstorms.[354] Moreover, it is the North Atlantic by the Gulf Stream that is one of the largest sources of convective clouds and enhanced lightning activity that is influenced by solar activity.[376] Particle flow along the contours of the North Atlantic Field are quite obvious in these observations.

A study of the latitude restrictions of lightning flashes disclosed that there is a much greater frequency in the Southeast than in the Northeast, or elsewhere. There is 30% more volume to the electrical charge, and a 20% increase in altitude of lightning-channel, cloud tops in Florida than in New England.[447] As can be plainly seen the East, and particularly the Southeast, are affected more. These areas are closer to the stem of the Field where the atmosphere is first ionized by the release of particles in the North Atlantic Field. This is also why the positively charged superbolts strike areas of the southeastern United States. In contrast, thunderstorms are a relatively rare phenomena in West Coast weather.

The southeastern drought and heat wave of the mid-1980s reveals the influence of the North Atlantic Field, as well. First there was an unusual jet stream in winter and spring, which resulted from the winter peak and vernal equinox effects on weather. Second, there was an abnormal trough of low pressure parallel to the East Coast, and from the west of Bermuda to Cuba, in spring. This low pressure was due to a partial vacuum caused by ionization from particle flow in the Field's stem. Third, in June the Bermuda High shifted to an unusual position with a wedge covering the Southeast. This was created by the IMF-GMF interrelationships around the June solstice, and the flow of air masses into the partial vacuum, which eventually produces high pressure (damming). These interrelationships are reflected in the fact that there is a 30- to 33-year cycle of droughts in the Southeast that are linked to solar activity.[417]

The so-called "Mystery Booms" off the East Coast of the United States are another phenomena associated with the North Atlantic Field. Window shattering, house rocking booms were experienced some 594 times during the winter and spring of 1978-1979. Only 413 of the events could be associated with supersonic aircraft, leaving at least 181 events for which no known cause could be ascribed.[239] The cause is the displacement of air due to ionization creating a vacuum into which air masses flowed (damming), thereby clapping together in booms. This is very much like the formation of thunder as lightning cuts the air, causing a vacuum and the colliding air masses producing thunder. As is typical of Field phenomena, the events occurred mostly in winter, and also spring, when there are more ionized particles released by the North Atlantic Field.

As discussed in Tome One, the North Atlantic Field is situated where the Bermuda High forms, and is an area of hurricane development. The southeastern United States has a higher occurrence of tornadoes and hurricanes. As will be discussed, unusual winter-spring snow storms, known as weather "bombs," and other weather phenomena that are unique.

The Japanese Field is also one of the most active Fields. Both the 18.6-year lunar and the 11-year solar cycles are noted for pressure changes in Japan.[229] Flood-drought cycles in northeastern China display lunar and solar cycles from at least 1470 onwards.[147,240] In Taiwan, tree rings of the Formosa grow in phase with the 11-year solar cycle, reflecting cycles in rainfall.[147,414] Both direct and indirect evidence indicate winter temperatures in Tokyo, spanning the years 1443 to 1970, display a solar-cycle influence.[336] The North Pacific, near the Japanese Current, is one of the largest sources of convective clouds and enhanced lightning frequency that respond to solar activity.[376]

Global anomalies of atmospheric circulation accompany droughts, such as occurred in 1988. At times of drought there are heating anomalies in the Western Pacific, south of Japan (the Japanese Field). In fact, the presence of unusual negative anomalies in sea surface temperatures took place in the Pacific Fields (Japanese, North Pacific and South Pacific Field areas). This pattern was also linked to previous droughts.[605]

Also in the Japanese Field are winter thunderstorms that produce superbolts, which are 89% positively charged (in summer all are negative).[590,592,615] As usual, present models are inadequate: "the reason for the occurrence of dominantly positive return stroke flashes in these winter storms remains unclear."[591]; p.2385. A source of ionization is required for the positive charge.

Negative charges are suppressed and the superbolts can be 100 times stronger than ordinary lightning. This is a scenario that cannot be explained by present theories. The Tsushima current flowing parallel to the coast is 10° to 12° Celsius (50° to 54° F) warmer than the surrounding gulf. These so-called superbolts take place mostly over Japan and the Northwestern Pacific Ocean in the region of the Japanese Field. Most are observed along the 30° to 40° North Latitudes from mid-December to mid-April, or in winter, with a peak in March and April, around the equinox.[28-30,475,590,591,615] These superbolts are notorious for striking artificial structures, especially electric power stations and transformers. As discussed in Tome One, the Japanese Field is an area where typhoons develop, and high and low pressure systems form.

The South African Field is another of the weather centers. The mechanism may have been observed when the so-called "Atomic Blast" of 22 September 1979 was recorded by the Vela Satellite. It was a bright flash aloft between South Africa and Antarctica, which matched the signature of a nuclear blast (i.e., hydrogen fusion). A government committee suggested that a nuclear bomb had been detonated, others, such as the South African government, said it was a natural phenomena, because they had no nuclear weapons testing at the time.[380]

However, about 18 years later, a government official said that they did detonate a nuclear weapon. Was the South African government lying about the detonation for political advantage? They would become more a part of the global political scene after saying that they had detonated a weapon. For example, they were then allowed to vote on a global ban on nuclear weapons testing. Notwithstanding, no atomic blast was reported to have been recorded on seismographs. Or, had there been the release of particles (the by-products of hydrogen fusion) through the South African Field? It is difficult to discern when the South African government was lying.

Nevertheless, a drought-flood cycle for South Africa matches the lunar nodal cycle of 18.6 years. Flood-drought cycles in South Africa also display the 11-year solar cycle.[147,492-494,616,617] The South African Field is noted for high and low pressure systems, and the development of cyclones.

The Persian Gulf Field is surrounded by land, and as a result, the formation of certain types of weather phenomena occur more displaced from the area. Furthermore, little has been done in the way of research for solar or lunar effects on climate in most of those countries surrounding the Persian Gulf (i.e., Iran, Iraq, Kuwait, Arabia, etc.). However, the nearby India Ocean has been studied somewhat, and research shows that Indian Ocean depressions are related to lunar phase, which is also noted in heavy rainfall.[632] The 18.6-year lunar nodal cycle and the 11-year solar cycle are evident in India's drought-flood cycle.[92,93,139,147] The same is true of nearby northeast Africa drained by the Nile, since 650.[113,147,238,239] The altitude of a (*500-mb*) pressure level at nearby Beirut is highly correlated with solar activity.[460] The Persian Gulf region is a weather center, as discussed in Tome One.

The Mediterranean Field is also represented in climatic data. In the nearby Alps and Po Valley there are what has been called lunar-solar "tidal" effects, especially for rainfall.[99] Zugspitze, in the Alps, displays an increase in isotopes (Be[7]) about three to four days after the passage of the IMF (SBC).[497-505] Rainfall feeds sea level, which is evident in a European 18.6-year lunar nodal cycle in sea level records.[233] Tornadoes and lightning in Britain are also correlated with solar activity.[454,577] Central England's monthly mean temperatures for the period 1659 to 1973 display a nearly in-phase relationship with solar activity, as well.[384]

Lunar effects were first noted for the very close-by Spanish Peninsula with the declination and position of furthest and closest approaches (*apogee and perigee*) displaying an influence. As is typical, conventional gravitational models do not work: "However, a calculation of the magnitude of such an effect shows that it is unlikely that gravitational forces alone could produce a variation in rainfall as large as that shown by the accompanying curves. It may therefore be necessary to look for some other explanation for the phenomena."[5] The Mediterranean Field, just south-southwest of Spain, is another area affected by high and low pressure systems, such as the Azores High, and storm development.

The Brazilian Field also displays an impact on weather. In fact, the 18.6-year lunar nodal cycle is apparent in flood-drought cycles of South America extending back at least as far as 1600.[137,147] The 18.6-year lunar nodal cycle and the 11-year solar cycle are noted in the drought-flood cycle of Argentina.[137,147,492-494] A solar influenced lightning enhancement occurs in the Brazilian Field (SAGA), as well.[376] This Field region generates pressure and storm systems.

The East and West Australian Fields are also represented in weather data. The heaviest rainfall of the month at 50 stations in New Zealand, near the East Australian Field, are correlated to the closest and farthest approaches of the Moon. Like the effect on the Spanish Peninsula, the magnitude of such an effect rules out gravitational forces as a sole influence.[53]

The Field's lunar trigger is clearly discernible in a correlation between radio signals (the signature of particle flow along Field lines) and lunar phase for New Zealand. The variations took place with long distance transmission at short wavelengths. The phase of the Moon was noted in that two to three days before and after Full Moon a minimum of background noise was observed at high signal strength. This effect changed to maximum with poor signals around the New Moon.[23]

Other studies show that Australia and those places nearby have the 11-year solar and 18.6-year lunar cycles in rainfall and tree-ring growth. Both the East and West Australian Fields are centers of cyclone development, pressure systems and weather front formation. Australia and the surrounding environs have only been the subject of a few solar and lunar cycle studies.

The North and South Polar Fields are more easily affected by solar fluctuations because the magnetosphere is open in these regions, and so the solar wind enters with little resistance. The solar cycle correlation with weather is the strongest over the North Pole, controlling 64% of the winter temperatures and air pressure there. Yet, the effect is still much greater than anticipated, as can be seen by this statement: "And no one can yet explain the mechanism of the correlation. They wonder how a small oscillation in several solar properties can exert such a drastic influence on Earthly weather."[410] Helsinki, Finland is easily affected by weather systems that form near the North Pole, and displays the lunar cycle in temperature and rainfall in about 75 years of records.[372] Likewise, the solar cycle is apparent in Swedish weather according to 120 years of records.[586]

When the IMF sector boundary crosses the polar region the size of storms (*Vorticity Area Index; VAI*) increases with the greatest effect in about four days.[510,511,653-655] This effect is more pronounced when the IMF is moving away from the Sun, towards Earth, and during solar maximums.[173,253,504,653-655,679] When the IMF is moving toward the Sun (away from Earth) the average surfaces pressure near the North Pole is higher than at times when it is moving away (toward Earth). Also, when the IMF is moving away from the Sun it produces counter-clockwise vortices in storms, and when toward the IMF produces clockwise storm vortices. The reverse is true of the South Pole.[374,652] The degree of disturbance in the geomagnetic field at high latitudes exhibits the same relationship with the IMF as does atmospheric pressure.[652-655]

As the Northern Hemisphere was in winter the vertical electrical field was greater in Zugspitze, in the Alps, than at Vostok, in Antarctica.[453,654] Unlike the new model, FEM, present models cannot explain this: "It is not clear why a minimum was found at Vostok and a maximum at Zugspitze."[654]; p.758. Notwithstanding, the mid-latitude Fields (Zugspitze is near the Mediterranean Field) are more active in winter than the polar regions (Vostok is near the South Polar Field), and this is an observation of that effect in the vertical electric field.

Solar flares and solar maximums are recorded in the Antarctic ice.[575] Following geomagnetically disturbed days pressure rose in the Greenland-Iceland region.[306,308] Ice along the coast of Iceland, during the period 1791-1906, was shown to last longer (2.4 months) near sunspot maximum than near minimum (2.0 months), indicating colder conditions.[113] The Gulf of Alaska and Iceland, in the vicinity of semi-permanent, low-pressure centers, show greater rainfall during solar maximums.[261]

Both polar regions show ground-level pressure changes at times of recurrent geomagnetic storms, caused by solar activity. The effect is pronounced in local winter, and is almost non-existent in summer. An increase in magnetic disturbance is accompanied by a pressure drop (as ionizing radiation is released, creating a partial vacuum). Afterward there is a decrease in magnetic disturbance that is accompanied by an increase in pressure.[418] This effect is due to the time-varying characteristics of the Field, which when no longer active allow the air masses to flow back into the previously created low pressure. This mass flow of air then produces high pressure (i.e., damming). Upper atmospheric pressures in the North Pole change more drastically than the South Pole, because the North Pole receives 30% more aurora.[307,308]

Sudden commencements of geomagnetic storms are related to a highly localized sudden warming in the polar region. Of 66 magnetic storms during the winters of 1978-79 and 1979-80, 61 were accompanied by a sudden warming.[340] Just following the equinox, and continuing in spring and summer, circulation zones approach the poles. Geomagnetic storms also shift air masses toward higher latitudes at high altitudes.[481] Fluctuations in the geomagnetic field, atmospheric disturbances, and distinctive cloud organization occur *before and after* solar flares, displaying FEM's intimate solar linkage.[460]

High pressure occurs at high latitudes in both the Northern and Southern Hemispheres during solar maximum years, indicating a mass displacement of air towards the poles (damming).[261] Likewise, pressure at the center of the Aleutian low, around the North Pole, and its position displays a pronounced 11-year variation near solar maximum, when the low fills and weakens (*anticyclonic processes increase*).[390] The Polar Fields are also regions of storm formation and pressure systems.

Every summer since 1981, about 90 kilometers above the North and South Poles, radars have picked up "unexplained" high-frequency echoes. Over the same period, scientists have noted noctilucent clouds.

These clouds are the highest clouds on Earth, and they streak the polar mesosphere just after dark on summer nights.

Noctilucent clouds have been steadily growing brighter and more common. Scientists know that the echoes and the clouds are the result of electrons interacting with dense layers of particles. What they do not know is that it is particle flow along the Field lines of FEM. Some researchers believe that these effects are due to human activities. How right they are, because it is the weakening of the Polar Fields due to human activities destroying life that produce both phenomena.

15.8 FEM and Ozone Depletion

Another Polar Field weather phenomena is the ozone holes that development over each pole. The seasonal effect is evident in the fact that ozone falls considerably in spring, after the equinox, over Antarctica.[194] A solar influence on ozone levels has been known for a long time.[123,254,255] The loss is confined to altitudes between 12 and 20 kilometers (7.5 and 12.5 miles), when solar effects should be above 20 kilometers, according to present models. Like other weather this appears to be due to a source of ionization that comes from below, rather than above. Moreover, nitrogen dioxide, which is destructive to ozone, and believed to be the result of the observed solar influence, is the lowest in the world within the hole.[300]

Yet, in spite of the contradictions the solar influence persists, as a scientist that demonstrated the solar effect comments: "We still see many pieces of evidence that tend to confirm the solar cycle hypothesis."[300] A researcher at the Goddard Space Flight Center practically asks for the Field-dynamical Earth Model when he says: "The year-to-year deepening of the hole definitely has some dynamical component."[300]

Both ozone and nitrogen dioxide from 1979 to 1984 display springtime minimums above Antarctica. In accord with FEM, mid-latitude ozone (particularly within the 30° to 40° latitudes) decreases in spring, around the equinox, and in winter. Strong ozone gradients are associated with the jet streams, which are produced by air ionization. No increases of nitrogen dioxide occurred along with the solar maximum, but odd nitrogen was present at high altitudes during the ozone minimum. Furthermore, "strong transients" (i.e., time-varying effects) in the polar vortex and *(quasi-stationary)* planetary waves were observed.[89,342]

Chlorine mono-oxide, active in ozone destruction, is missing when the hole develops over Antarctica, and this is unaccounted for by theory. Still, an unexplained relationship remains between total ozone and the long-term solar cycle.[496] Typical of FEM's solar linkage, there are predictable cycles in geomagnetic activity that fluctuate along with the ozone levels.[342]

The ozone loss is greater than just the ozone hole itself. A ring extends 5° to 10° in latitude beyond the hole, and is not over any populated areas emitting CFCs (*chlorofluoro-carbons*) that destroy ozone. This is evidence of a dynamical component that channels the CFCs and other compounds, which along with air ionization, create the hole. For the same reason, the ozone is lower in the region near, but outside, the polar vortex over Antarctica.

The conclusion of researchers asks for FEM: "Scientists will have to determine the mechanism causing the ozone destruction. They must also explain the timing of the loss."[411] Both poles show ozone depletion with the same seasonal and solar effects, and energetic electrons were noted to be partly responsible for producing the holes, as expected of FEM.

Ozone concentrations, like so many other phenomena, fluctuate with the center of mass of the Solar System. Long-term (*secular and super-secular*) cycles of solar activity are related to climate, particularly in the Northern Hemisphere. Also, a cycle in the angular relationship between Venus, Earth and Jupiter triggers strong flares (energetic x-ray bursts), which undoubtedly induces weather changes, including ozone depletion.[339-342]

15.9 Lightning

Thunderstorm and lightning frequency, and locations also exhibit what could be predicted of FEM, including its solar linkage. Lightning is ten times more frequent over land than over the sea. This is not explained by present theory, because moisture separates charges, which should increase the likelihood of lightning over the oceans instead.[31]

The reason for the greater rate of lightning over land confirms the role of electrostatic forces in the dynamics of FEM. Land attracts lightning because the net charge of the land has been offset by life's destruction. Also, the building of structures using insulating materials, which prevent the grounding of charge, offset the weather field of the area. FEM is evident in that oceanic thunderstorms maximize in the northern latitudes in winter, because the Fields are in the oceans and activate more in winter. The north Pacific near the Japanese Current, the north Atlantic by the Gulf Stream, and the south Atlantic near Brazil (Japanese, North Atlantic and Brazilian Fields) are the largest sources of convective clouds, including thunderstorms, that respond to solar activity.[376,640]

The Fields and the role of solar activity are obvious in the observations. The Japanese Field is responsible for the superbolts that occur in winter storms over Japan and the Northwest Pacific.[28,590,591,615] Likewise, superbolts have been noted in the southeastern United States as a result of the North Atlantic Field.[29,30] Winter thunderstorms in ancient China are correlated to times of solar maximums, as well.[147,642] All of these phenomena are not explained by present models of the Earth, but are clearly predictable of FEM.

Auroras in the upper atmosphere are followed by thunderstorms in the lower atmosphere. However, the energies of the auroras are many times less than the thunderstorms, and any connection cannot be accounted for by present models of the Earth.[566] Even, ball lightning is correlated with solar activity.[35,200] Lightning and thunderstorms increase after solar flares and IMF sector boundary crossings, as well.[69,261,348,355,376,377,497,500] Lightning and thunderstorms occur mostly around the 30° to 40° latitudes, and along the longitudes of the Fields. The role of the Fields can also be noted in the numerous examples of luminous phenomena that issue from above the cloud tops into the upper atmosphere, ionosphere and magnetosphere, as well as the other way around. The only model that can explain all of these observations completely is FEM.

15.10 FEM's Solar Linkage

Seasonal effects or shifts that accompany the equinoxes and solstices are clearly discernible in weather, displaying FEM's solar linkage. Geomagnetic activity and season strongly influence the circulation and composition of the thermosphere (season meaning those transitions that occur around the solstices and equinoxes).[512] More vertical motions in the atmosphere take place in winter from solstice to equinox than at other times.[99] Oceanic thunderstorms also maximize in the northern latitudes during winter.[375,376,437] Expected of FEM, a correlation exists between geomagnetic storms and winter (300-mb) trough development in the northern Pacific and North America.[510,511] An intense coupling between the basic states of the layers in the upper atmosphere (strato-, meso- and thermospheres) occurs during late winter, when the Fields are more active.[418,564] Mid-latitude (particularly 30° to 40° latitudes) ozone decreases in spring, around the equinox, and during winter, again demonstrating FEM's solar linkage.[89] Likewise, superbolts occur in winter storms over Japan and the Pacific Northwest, particularly around the equinox.[590,591,615] In ancient China more thunderstorms also took place during the winter.[147,642] This winter effect is due to a solar linkage that activates the release of particles in the hemisphere that is pointing away from the Sun (as discussed in section 9.8 in Volume One).

Air masses display shifts in relation to the equinoxes, as well. These effects include air masses moving towards the equator during the autumn and winter, and towards the poles in spring and summer, at times of solar maximums.[481] Gravity waves are also triggered by the solar linkage, and were detected around the June solstice and March equinox, and are correlated with geomagnetic disturbances and solar activity.[24,645]

Very, low-frequency (VLF) waves were noted around the June solstice in 1962, at Salisbury, South Australia, and Lower Hutt, New Zealand, which is a region surrounding the East Australian Field. In 1963, other anomalies took place around the vernal (March) equinox, as observed in New Zealand (East Australian Field) and New Jersey (North Atlantic Field). In 1965 the fluctuations occurred around the December solstice.[17,495] VLF waves are typically produced by particle flow along field lines (*plasma, whistlers, etc.*), and these observations show the solar linkage.

Many phenomena that take place around the equinox seem to be unrelated to the changes that occur at that time, but again show the solar linkage. For example, the largest amplitudes of the lunar tide in geomagnetic declination occur in the morning, during equinoctial months.[571] The largest known outbreak of tornadoes within 24 hours, claiming 315 lives, occurred on 3-4 April 1974, hitting 14 states with 148 tornadoes. This event occurred shortly after the vernal equinox at mid-phase prior to the Full Moon. Many of the most powerful flares and coronal mass ejections occur around the equinoxes, especially the vernal equinox.

Reports of visually observed lightning discharges from thunderstorm cloud tops into the clear air *above* also cluster around the equinoxes and solstices. At times these lightning strokes are greater than the cloud-to-ground strokes. They involve strong electric fields arising from "large local accumulations of charge" (i.e., masses of particles). One of the more recent observations occurred on 22-23 September

1989, around the equinox, and were associated with hurricane Hugo.[202] This discloses how the atmosphere was highly ionized around the time of the development of a devastating hurricane, and the source of the particles came from *below*.

Other incidents that were recorded were noted within less than a month of either the equinox or solstice, with the March equinox represented the most. For example, on 11 April 1965 there was a cloud-top stroke noted, at which point 47 tornadoes touched down in Illinois, Indiana, Iowa, Michigan, Ohio and Wisconsin. The mechanism responsible is, conventionally speaking, unknown. However, it is known that it involves the acceleration of electrons (and other particles) upward in a direct process from *below* that penetrates the upper atmosphere (*ionosphere*) and magnetosphere.[201,368,483,627,639] The observations are totally supportive of what would be expected of FEM.

15.11 The QBO and FEM

The Quasi-Biennial-Oscillation (QBO) is a less than perfect periodic reversal of stratospheric winds over the equator (every 13 months or so). QBO's effects outside the tropics fluctuate with the 11-year solar cycle. However, the effects are masked by the interaction with an internal control on the atmosphere (i.e., Field dynamics).

The QBO displays certain features. Stratospheric winds over the equator change with solar maximum, especially temperature. Correlations with sea level, atmospheric pressure, and surface air temperature are noted during the summer and winter. In the winter, when the mid-latitude Fields become more active, the effect is particularly over the oceans. The East Phase is in the Southern Hemisphere, especially over Antarctica, and the West Phase is in the Northern Hemisphere. There is no known physical link, but the observations have passed many stringent tests, and the correlation with solar activity is highly significant and compelling.

The relationship with temperature is apparent from the northern polar stratosphere into the troposphere, and to the surface in large regions over the Northern Hemisphere.[301] Furthermore, the solar-ocean correlation holds true from at least 1925 to the present. Prior to 1925 the sea surface temperature data is poor and only a weak solar variability is observed.[49] After 1925 the global economic system began a new era of industrialization. The world economy had just emerged from World War I, a worldwide depression, and the stock market crash into a new era that brought a devastating impact on the biosphere, off-setting FEM's dynamics and producing these highly significant QBO observations.

Winter storms are 2.5° to 6.0° farther south during solar maximum than minimum. The data urge that the inevitable be asked: "How could the feeble variations in solar output over a cycle be amplified to give the apparent changes in the weather, a required amplification of over a million?"[301] Those who uncovered the best statistical representation of the effect discuss the pervasive mind-set: "'Serious' meteorologists still prefer to dismiss any claim that there is a noticeable relationship between the activity of the Sun and events on Earth. And yet, to our surprise we have found a highly significant correlation between the state of the atmosphere and solar activity, from the ground to the top of the stratosphere."[624]

The new model is clearly responsible for the effects, especially since the effects reach ground level. During the most active time of the Sun, the West Phase creates pressure over North America with lower pressure than usual over the water on either side of the continent. That is, high pressure forms between the North Atlantic Field's stem and its descending limb, while low pressure occurs within the stem and west of the descending limb (particle flow along Field lines produces the effects). Furthermore, the winds produce a vortex or cyclone in the region of the Field, and there are lower temperatures in winter along the East Coast. This effect is also due to particle flow in the Field producing a partial vacuum, leading to cyclones and lower temperatures as the upper, colder atmosphere flows in. In this way more storms are developed at mid-latitudes where the Fields are located. During solar minimums, as could be predicted, the opposite is true with low pressure over the continent, high pressure over the oceans, and the pattern of air flow is reversed.[331,624-626]

In contrast to what would be suspected, a 2% decrease in the blue end of the spectrum and a similar increase in the red end of the spectrum occurs during solar maximums. This spectrum shift should cause a slight warming of the Earth, not the opposite as is observed, indicating the need for some unknown mechanism(s).[270] The data force us to conceive of a new model of the Earth, and FEM is ideal.

15.12 Weather "Bombs"

Another phenomenon that relates to the new model is the paralyzing winter-spring storms that explode without warning along the East Coast, typically from the Carolinas northward. Cyclones are known to develop in less than a day. By the following day, twenty cities can get more than 50 centimeters (20 inches) of snow. The storms can affect 30 million people, and cause more than a billion in damages, on the average, per year. The lightning striking the ground and the ocean surface from these storms is positively charged, much like the superbolts. This should not occur over the oceans in the usual meteorological sense, because of the moisture and energy present, which separates charges. Yet, it does, again forcing us to consider new models of the Earth.

Examples of these "weather bombs" have been plentiful. One was the 18-19 February 1979 snowstorm that deposited 60 centimeters (23.5 inches) on the middle Atlantic States. Another is the 6-7 April 1982 snowstorm and windstorm in which more than 50 people were lost. The 11-12 February 1983 blizzard with record-breaking snowfall and freezing rain paralyzed the Northeast and took 70 lives.[167,646] The most recent "white hurricane," considered the blizzard of the century, deposited as much as 140 centimeters (55 inches) on 12-13 March 1993. It battered Florida with tornadoes and gale-force winds within hours of its formation, while it burned cities from Alabama to New England with record snowfall, claiming 200 lives. These storms led to the Genesis of Atlantic Lows Experiment; abbreviated GALE.

GALE has revealed that these storms are very much like winter hurricanes, an unforeseen effect, because hurricanes are thought to require very warm waters to develop. There remains many gaps in the scientific basis of how these storms develop. The effects of the North Atlantic Field are, however, quite detectable in the observations.

Ionization of the air, creating a vacuum causing upper, cold-air masses to move in, is obvious in the observations, and is referred to as damming. This is especially evident in the fact that the lightning discharges from the clouds to the ocean surface are often positively charged. Research flights have also found indications that the cloud tops are negatively charged, indicating a reversed polarity; again, like the superbolts. As could be predicted if sudden ionization caused the colder, upper atmosphere to flow in, air-mass transformations during cold-air outbreaks are *very* rapid with the cloud bases descending and the cloud tops ascending, and masses of air smashing together (damming).[167]

Because the North Atlantic Field is in the ocean there should be observable effects in the ocean waters. Observations show that the continental shelf waters are particularly responsive to these storms.[37] Temperature drops are very rapid with pronounced cold-air damming, especially with "bomb" category storms. Differences were greater than 24° Celsius (70° F) between the sea surface and cold air flowing over the Gulf Stream. The heat transfer is an important contributor to the storm's formation and intensity. As a result, rapid, intense coastal storms have a warm-core structure more typical of tropical hurricanes. Large amounts of heat are extracted from the Gulf Stream through air-sea interaction during cold-air outbreaks.[167] These observations are exactly the type of effect one would expect from sudden ionization.

Other observations support this conclusion. Maximum heat flow occurs during the times of cold outbreaks, and sharp contrasts in sea surface temperatures are reflected in the overlying air.[65] Changes in wind direction and/or speed frequently occur together with the heat transfer and flow of moisture. Upper atmospheric (stratospheric) air descended to lower levels (*below the 500-mb level*) and several storms produced slantwise flows. Migrating rain-bands, intense and well-organized, occur frequently, as could be expected of "bundles" of particles undergoing time-varying acceleration. In fact, rapidly formed and bomb category storms involve cold-air in mass flows smashing together (damming), and changes in jet-stream position.[167]

The largest gradients in wind stress are associated with times immediately *preceding* the outbreaks of cold air over the continental shelf. The along-shelf stress changes abruptly to across-shelf stress along the coast. These events trigger oceanic fronts that accelerate winds connected to a low pressure system off the Southeast Coast, where the North Atlantic Field is situated.[65] The pressure-wind relationship was not typical of (*linear*) gravity waves, which are horizontal, because it also had a vertical component; typical effects of a funnel-shaped field accelerating particle bundles in the form of a helix. The precipitation was associated with a wave that occurred ahead of the disturbance.[154] Vorticity in the ocean waters occurs along with the formation of the fronts. The storm's vorticity exceeds what could be expected of the effects of the Earth's rotation on air masses, and coincides with a maximum in heat flux.[65] The observations are clearly supportive of FEM.

The fact that these storms form fairly close to the vernal equinox, and are generated during lunar phases or mid-phase, are rarely, if ever, considered by meteorologists. This is because solar influences are thought to be negligible, and the gravitational effects of the Moon are minor. However, solar wind and gravity are not the only triggering mechanism, but FEM's solar linkage and the electrostatic effects of particle flow, partially triggered by the Moon's effect on particle flow, are much more at work.

15.13 El Nino

Another phenomena that is poorly understood using present models, but confirms the role of FEM, is El Nino (-Southern Oscillation). The El Nino event is now recognized as part of the variability of the coupled ocean-atmosphere system in the Pacific Ocean. The event is associated with an unusual warming of the surface waters of the East and Central Tropical Pacific. Many studies suggest that the northern winter conditions, when the Fields are more active, are linked with this warming and the unusual weather conditions that accompany it.[450]

One of the effects of El Nino is the development of droughts and heavy rainfalls in the United States, and elsewhere. The most extreme, extensive and severe droughts include the 1930s Dust Bowl, and the droughts of the 1950s and 1980s. Heavy rains and pounding surf hit the west coasts of North and South America, and weather around the globe is altered. Large-scale atmospheric circulation shifts are seen as the primary cause.[605] As discussed, droughts and rainfall are correlated with solar and lunar cycles, and data indicate El Nino is also.

Many of the typical weather-altering factors of FEM are noted in studies of El Nino. During El Nino the sea surface temperatures are above normal along the equatorial region from the Central Pacific to South America. This was associated with large displacements of major rain-producing zones in the tropics and shifts in atmospheric circulation.

Observations disclose that the Field areas undergo changes as El Nino develops. There were unusual temperatures in the region of the North Pacific Field, which also had high pressure. In addition, a low pressure system developed in the North Atlantic Field (Bermuda Low) and the North Polar Field (Aleutian Low), and enhanced wind patterns took place near the Japanese Field (Western Pacific south of Japan).[605]

The annual cycling of water between the equatorial Pacific and a band in the northern tropics, which is driven by a seasonal wind shift, does not occur during El Nino. The amount of water between 20° North and South Latitudes remained unchanged.[302]

Both the equinoxes and the solstices are times of transition. El Nino evolved shortly after the December solstice. In fact, this is how it got its name, as it occurs around the time of Christmas, and hence, the name El Nino, which means "little boy child," or little Christ child. Alterations in atmospheric circulation were generated around the equinox, March through April, when changes in the Earth's rotation also took place.[95,605]

Likewise, the most prominent variations of methane occurred during El Nino, when concentrations fell far below normal. The variations display seasonal shifts around the solstices and the equinoxes.[303] When methane is bombarded by ionizing radiation it is altered into complex hydrocarbons, hence particle flow would lower methane.[237]

A 2.2-year (26-month) cycle is noted in El Nino, and this is similar to a period in solar particle flow (*i.e., neutrino flux and solar ultraviolet*). Between 1950 and 1978 the pressure profile at about 9° North Latitude (*through the 100-mb to 7-mb levels; 16- to 32-kilometer or 10- to 20-mile levels*) disclosed a high velocity core in the jet stream that oscillated in height with the 22-year solar cycle. White light flares, which have very high energy, preceded El Ninos, such as those in 1891, 1926, 1938, 1957-58, and 1963. The Chandler Wobble (*a beat frequency of 6.2 years*) is noted in about 52% of El Ninos. A similar period is found in tree rings and lake sediment layers (*varves*). Between 1525 and 1987, 60% of El Ninos were correlated with the 18.6-year lunar nodal cycle. Of 107 El Ninos prior to the 1997-98 El Nino, 43 do not match the lunar cycle, but of these 66% match the Chandler Wobble. Furthermore, of these 107 events, 53 coincided with the QBO. Changes in the length of day (LOD), or the Earth's rotation, were the most extreme in 40 years just *prior* to the 1986-87 El Nino.[97] NASA also noted a similar LOD for the 1997-98 El Nino. The entire spectrum of cycles, both geomagnetic and solar, appears in El Ninos' timing, including the 179-year solar cycle.[188] El Nino has been responsible for many devastating effects on weather, has been correlated with earthquakes, and has occurred more often in the recent past than previously, as life-systems are devastated and solar activity is at an all-time high.

Other El Nino-like phenomena have been noted. One took place in the Atlantic where surface waters were unusually warm and winds were very light with a band of clouds over the equator. This cloud band usually migrates 15° in latitude to the north in March, around the equinox, but did not move until June, around the solstice.[472] Another El Nino-like pattern was noted in the Indian Ocean by Columbia University scientists, it began just south of the Persian Gulf Field and migrated to near the West Australian Field.**** A latitude restriction and global effect, and solar linkages, typical of FEM, are observed.

The conditions surrounding El Nino have been responsible for droughts in the United States. The most extensive drought in many years for the West Coast and northwestern United States, during 1986-88, developed in association with the 1986-87 El Nino. Record low rainfall characterized the April to June period of 1988; a period from equinox to solstice. Strong anticyclonic conditions and a northward displacement of the jet stream took place in the upper atmosphere over North America throughout the period. A strong West Coast ridge of high pressure prevented rain from reaching the area. This high pressure also stopped the development of the normal low-pressure systems and cold fronts from reaching the western states, and subsequently, the Northern Plains. This circulation pattern also took place from September 1987 to March 1988, between the equinoxes. By July 1988, 43% of the contiguous United States was in a severe drought that was only exceeded in 1934, 1936, 1954 and 1956, which had similar conditions.

****   From www.columbia.edu/cu/record/record2012.17.html; accessed 8/1/98

The effects of the North Atlantic Field are evident, as pressures were high along, and low off, the West Coast where the Field's descending limb guides particle flow. Likewise, in October 1986, just prior to the drought's development, soil moisture content was above average in much of the United States, except in the Southeast, near the Field's stem in the North Atlantic. Similar circulation patterns were evident in earlier droughts and were in association with other Field areas.[450,605] Droughts in the United States are linked with the 22-year solar and 18.6-year lunar cycles, as well.

15.14 Glacial Advance

Glacial advance is also understandable with FEM and its solar linkage. Mid-latitude glacial advance takes place during solar minimums, but variations in solar output "do not offer an explanation."[340] This timing is actually due to the solar maximum that precedes the minimum. The solar maximum supplies more solar plasma, and after a lag period ionizes the atmosphere through the Fields, producing glacial advance. Cold upper atmospheric air flows into low levels, and thereby, increases snowfall and decreases temperatures, causing glaciers to advance. Worldwide glacial advances, as a result, are correlated with times of solar minimums since the end of the last ice age.[340,650]

The connection between solar activity and glacial advance involves studies of isotopes (the products of ionization), geomagnetic reversals, orbital eccentricities, and the precession of the equinoxes.[283] Cold epochs correspond to lower sea level, and this is strikingly true for the Han Dynasty, Chinese records (206 BC to AD 8). In contrast, sea level rose in the Chin Dynasty (AD 265 to AD 420), as glaciers retreated, wells dried, and blown sand was reduced.[147]

FEM is clearly involved simply by noting the fact that glacial advance in the Northern Hemisphere, during the last 100 years, shows a lack of synchronicity with that of the Southern Hemisphere. That is, the solar linkage activates one hemisphere in one period, and the other in another period, depending on which hemisphere is pointing away from the Sun, and the polarity of the Interplanetary Magnetic Field (IMF) at the time of peak solar activity. Each glaciation can be matched with acidity levels recorded in polar ice cores, and typical of FEM's interconnected phenomena, the frequency of volcanic eruptions.[485] The 18.6-year lunar and 11-year solar cycles, as well as long-term solar variations, are also observed in glacial advance and ice ages.

The last glacial stage offers a good example of how FEM produces glacial advance and ice ages. Aerosol content of the eastern Antarctic ice indicates that there where large marine and continental inputs at the end of the last glacial stage. Aerosols are airborne droplets of water, and indicate the intensity of wind speeds at the time. Glacial age climate had stronger atmospheric circulation, enhanced aridity, and more aerosol production, as noted in the Antarctic ice. In spite of the greater extent of sea ice, sea salt content in aerosols over central Antarctica was higher, which is due to more aerosol production driven by greater wind speeds over the oceans' surface. The greater salt content indicates greater storminess.[470]

The continents were drier, and active deserts between the 30° latitudes were five times larger than those of today. Stronger ground level winds left behind sand dunes and wind-erosion features (ventifacts, eolian deposits, and loess). About 13,500 to 12,000 years ago, there was an end to the aridity maximum in Africa, the stabilizing of sand dunes in the Australian desert, and a relaxation of the once vigorous circulation.[470]

Just *prior* to this glaciation there was a burst of solar activity so huge that it pitted the rocks on the Moon![687,688] This solar burst is the triggering mechanism for the sudden glacial advance. A solar-FEM linkage was clearly an aspect of producing this ice age.

The nearly instantaneous onset of glacial advance of the ice age cannot be explained by anything other than FEM. Mammoths that have been found illustrate just how fast the advance was. Whole mammoths, without a single claw or tooth mark of a predator or scavenger, and food still undigested in their stomachs and stuck between their teeth, were uncovered a number of times. Often they were found still standing frozen in the ice. Had the snow fallen slowly or even quickly by modern standards, they would have undoubtedly finished digesting the meal in their stomachs, licked the last morsels between their teeth, and fallen on their sides to be consumed by some predator or scavenger. Instead, the snow fell so quickly that when one body was uncovered in Siberia, sled dogs ate is without the slightest ill effects, as if it were freshly thawed beef.

When reindeer or caribou die in the Arctic, they have been noted to quickly decompose due to remnant body heat and bacteria. This, however, did not happen to the mammoths. Moreover, the undigested vegetable meals in their stomachs should have also quickly decomposed. In fact, the very delicate buttercup found in the stomach of a mammoth should have decomposed within 10 hours. The air temperature would have to be 150º F (62º C) below zero in order to cool the stomach from 74º to 40º F (14º to -7º C) within ten hours, and thereby, preserve the buttercup. Yet, the initial body temperature was closer to 100º F (30º C), which would require a temperature drop to about 200º F (93º C) below zero.[117]

Moreover, on examining mammoth blood it was discovered that red and white blood cells were still whole, and had not burst.[220] Slow freezing, such as is typical even of a blizzard, would have expanded the cells, causing them to burst. The only known method of preventing blood cells from bursting upon freezing is to use the extremely rapid freezes of cryogenic methods. Likewise, fatty tissue under the skin had survived even after the exposed skin began to decay.[252] Again, had the freezing been slow, or had the mammoths died slowly, the fatty tissue would have been gone long ago. There is no way that conventional theory can explain such rapid glacial development, and while this has been reported in the literature with astonishment, it has gone unexplained, and hence, ignored and even denied. Yet, with FEM there can be the sudden ionization of the atmosphere, triggered by a burst of solar activity, which could cause such a rapid glacial advance.

15.15 The Little Ice Age

Another good example of the new model at work in glacial advance is the Little Ice Age. Some scientists have considered the beginning of the Little Ice Age to be about 1430, and lasting until 1850. Its more definite bounds encompasses the Maunder Minimum, or about 1645 to 1715. However, the real glacial advance in the Alps, Scandinavia and Iceland began in the 1570s to early 1600s.[353] The connection with solar activity is apparent in that there was a remarkably high maximum in 1433, and a series of maximums from 1525 to 1630 (very strong in 1558-1572).[540] This was a time for low values in long-term temperatures in all seasons, enhanced variability of temperature from spell to spell and year to year, and an enhanced polar cap of frigid air over the Northern Hemisphere that was accompanied by jet streams that were weaker and further south than those of today.

At the time of the Little Ice Age, the Colorado Plateau had a number of serious droughts that also affected southern California and Oregon. An era of bitter cold and heavy snowfall brought glacial advance as the 17th Century progressed. At that time, Alpine glaciers moved into valleys overrunning settlements, pastures, farms and more. Evidence of glacial advance is noted for Iceland, Greenland and North America, as well. Sir Francis Drake's circumnavigation of the world, and maps of the 15th and 16th Centuries indicate that the ice of Antarctica was much less than during the Little Ice Age that followed, or even today.

An increase in solar activity had activated the release of ionizing radiation, which created a vacuum bringing the much colder upper atmosphere to lower levels. This caused heavy rainfalls and snowfalls that brought floods in some places and droughts in others, as well as glacial advance. The glacial advance of the Little Ice Age was a product of these phenomena and the solar linkage of FEM.

Solar activity indicates that the period was preceded by the Late Medieval Maximum, and other maximums (very strong in 1450 and 1525-1580) that ended in about 1630. Contrary to popular opinion in science, it is the increased solar activity that brings the Little Ice Age, not the Maunder Minimum that follows. Support for this understanding is found in tree ring research, which indicates that an increase in radiocarbon (C^{14}) accompanied the Maunder Minimum, an expected result of the atmosphere being ionized.[175,178, 179,338]

15.16 Life, FEM and Climate

The importance of life in FEM's dynamics is clearly demonstrated in an ancient climate. Lunar and solar cycles were observed in the Elatina formation in southern Australia. The cycles were 20.3 and 10.8 years, because the Earth's rotation was faster, producing a day that was 21 hours long. The correlation was very strong in contrast to today when the observed correlation is comparatively weak. Life's stabilizing effect can be seen in the fact that the impact is greatest in an ancient climate before an advanced and diversified life system developed on Earth (i.e., the Precambrian).[33,429,660,682]

The scientist who made this discovery tells us about this enigma: "It is puzzling, to say the least, that something as indistinct in the modern climate as the sunspot cycle should apparently dominate a local climate 680 [million years] ago, but should have left no other comparable records yet discovered."[682; p.282]. What is even more surprising is that the record turned out to be mostly lunar, for which the correlation is even weaker today than that of the solar correlation.[33] Again, FEM is supported very well by these facts, including the importance of life in the stabilizing of Field dynamics.

CHAPTER 16

Solar and Lunar Triggers on Earthquakes and Volcanic Eruptions *(The Solar-Lunar-FEM Linkage in Plate Tectonics and Volcanism)*

In Tome Three, Volume One, the subject of plate tectonics, more commonly known as continental drift, was discussed to some extent. That discussion brought forth the understanding that electrostatic forces are at work in producing the tectonic plates' dynamics. These electrostatic forces result from particle flow along Field lines with the descending limbs interacting with plate boundaries. Further confirmation of this understanding can be found in studies that correlate solar activity, and lunar phases and cycles with the triggering of earthquakes and volcanic eruptions.

16.1 Solar Earthquake Triggers

One fact that calls for new perspectives on plate tectonics is that earthquakes are correlated with solar activity.[46,560] Solar activity, as indicated by sunspots, solar flares, radio noise, and solar-induced geomagnetic activity, plays a significant role. A maximum of earthquakes occurs at times of moderately high and fluctuating solar activity, particularly solar flares.[553] Strong earthquakes take place when the Earth crosses the central meridian of the Sun, and this understanding has been used to successfully predict some quakes.[607] Moreover, the Sun's retrograde motion is linked to earthquakes and other geophysical phenomena, including climate and volcanic activity.[288,551] These facts alone indicate that the solar-FEM linkage and electrostatic forces are involved in earthquake occurrence.

However, there is much more, as for example in observations of California earthquakes. There is a 22-year solar cycle in San Andreas Fault earthquakes, and an 11-year solar cycle in large earthquakes in southern California.[320,355] Recent observations of earthquakes in the region disclosed that they have doubled between 1986 and 1989. Furthermore, from August 1988 to February 1989, they were 3.2 times more frequent than the historic rate. Quakes of magnitude 4.5 and greater went from one every four years to one every two months.[293] This increased earthquake activity transpired as a peak in solar activity was taking place.

One half of a solar cycle, 5.5 years, was noted for quakes in the Far East, with the years 1947, 1958 and 1969, being both earthquake and solar peaks.[431] Planetary influences on solar activity have been discussed, and the relative positions of Uranus and Neptune were correlated to peaks of energy released by earthquakes.[40] Variations in gravity, earthquake energy, and solar activity were correlated in another study.[560] Solar activity and quakes have been linked by many researchers, and they all agree that gravitational models do not work.[231,232,278,284,396-398,435,451,456]

Also revealing the need for new models is the fact that a maximum in earthquakes in some regions occurs shortly after an epoch of very low sunspot activity.[148,318,320] A maximum in quakes that occur shortly after such an epoch, as in 1763-65, reveals the global network:

"It is interesting to notice that the same periodicity holds in widely separated regions. Taking the three intensities together, the maximum occurs in 1764 in Europe, in 1763-64 in Asia, in 1764 in Italy, in 1764-65 in China, and in 1763-64 in the island groups of the Western Pacific. Even in the slight earthquakes of Great Britain, the same period is present with its maximum in 1763-64"[148]

These earthquake maximums occur shortly after the Maunder Minimum (1645-1715) when solar activity was shifting from an all-time low to increased activity (from 1718 to 1788 there was moderate to very strong solar activity).[540]

Also demonstrating the global system, plate motions have been observed to follow solar activity at 71 stations around the world. The plates more back and forth as the 11-year solar cycle goes up and down.[39] This is why some regions experience quakes during epochs of low solar activity, while in other regions it is increased solar activity. This indicates that there is a global system, regulated by the Fields, and also involves the solar linkage of FEM and electrostatic forces.

Illustrating the effects of this solar linkage are the correlations between earthquake activity and the Sun's Interplanetary Magnetic Field (IMF). The IMF sector boundary crossings (SBC) cause changes in the Earth's magnetosphere, ionosphere and atmosphere. Enhanced precipitation of energetic electrons take place as the Earth's magnetic field is disturbed, which is accompanied by changes in wind direction and the size of storms (Vorticity Area Index; VAI), about four days before and after the SBC, with the greatest effect in winter.[350] The SBC are also correlated with lightning and thunderstorms, which display a maximum in winter, as well. Also, large changes in conductivity and electric field variations are global.[376] The IMF and the geomagnetic field (GMF) interact, producing a twelve-month wave with a maximum at the vernal equinox, and are the "result of a common cause."[608] These observations reveal the dynamics of FEM's electrostatic forces, which trigger earthquakes.

Other studies confirm the electrostatic forces along Field lines play their role. Geomagnetic disturbances influence monthly variations in the air-earth current, mean temperatures, and alter surface atmospheric pressure and the development of storms.[341] Numerous studies show that weather, unexpectedly, displays more deep-seated effects in winter.[608,609,624-626,668] This includes the positively charged superbolts, which occur mostly in winter with a peak around the vernal equinox.[475,590,591,615] As will be shown, these phenomena are indicative of the Field dynamics that also trigger earthquakes and volcanic eruptions.

16.2 Solar-linkage Triggers

The Sun's motion around the barycenter, or the center of mass of the Solar System, which is determined when Jupiter is in conjunction with another of the large planets, has a triggering effect on earthquakes. This and other correlations have led scientists, including the American Geophysical Union, to state that a solar-terrestrial linkage exists.[184,341] The center of mass also has an influence on solar activity, and is likely to be behind this correlation.

The solar linkage is noted in the fact that solar flares abruptly change the Earth's rotation or length of day (LOD).[100] This altered rotation has been shown to trigger earthquakes.[223] A 120-day oscillation in the LOD, atmospheric zonal circulation, solar activity, the IMF and the GMF are also known.[169]

A global system and its solar linkage have been observed in a number of ways. Different earthquake belts have common active periods, indicating that they are strongly coupled on a global scale.[1,2,25,59,105,297,373,409,420,430,476,486] Isotopes, such as radon, are noted at times of earthquakes, and indicate an electrostatic trigger, as well.[280] A global network of earthquakes reveals the need for a new model of the Earth that includes electrostatic forces and global field dynamics (see Chapter 29 of Volume One for other discussions).

The seasonal occurrence of earthquakes also indicates a solar-FEM linkage. A study of earthquakes along the San Andreas Fault before the 18 April 1906 San Francisco earthquake shows that the majority took place around the vernal equinox in spring, and a secondary peak in winter.[393] This study supports the solar linkage and its relationship to earthquake occurrence.

One report disclosed an example of both the seasonal effect and the associated weather phenomena of particle flow. The Parkfield, California section of the San Andreas fault shifted suddenly in March in 1991. This surface shift took place immediately following heavy rains (caused by atmospheric ionization along Field lines). On March 19 the United States Geological Survey issued an earthquake warning after two instruments detected a substantial acceleration in the normally slow surface creep. This effect was noted previously in the Parkfield area before earthquakes, and the region was expected to have a major quake in 1992 or 1993.[513] While Parkfield did not experience a major tremor, a moderate one hit in early April 1993. Also revealing the solar linkage is the fact that these quakes occurred around the vernal equinox.

Another study displayed a daily peak in earthquake activity in some areas. A nighttime maximum in quakes displayed a peak around midnight. In Japan, Italy and other countries there is also a noon maximum. Annually, the noon maximum is identified with a summer maximum, and the midnight maximum with a winter maximum. Noon and summer quakes are associated with an elevation, and midnight and winter are associated with a depression, of the Earth's crust and atmosphere. Noon and summer (elevation) appear to generate the most destructive shocks, and midnight and winter (depression) generate slight to moderate quakes.[149] Another analysis of 15,325 events shows a higher occurrence at night and in summer.[547] Seasonal, and daily peaks of noon and midnight, are beyond the scope of gravitational theory and plate tectonics as it is presently described, and requires a solar linkage and electrostatic forces.

16.3 Three- to Four-Day Phenomena

Three-to four-day delays in the occurrence of earthquakes, and other electrostatically induced phenomena have also been observed. A maximum in thunderstorms occurs three days after solar events, which reveals another aspect of the electrostatic trigger.[354] The electrical potential of the troposphere and highly ionized isotopes (*radionuclides*) show the greatest fluctuations occur three to four days after solar eruptions (*especially hydrogen-alpha flares*).[497-505] Likewise, geomagnetic storms bring changes in four to eight days.[510] Together these weather phenomena indicate the characteristics of particle flow with a delay of three to four days, and due to the IMF-FEM linkage, peaks in winter, the equinoxes, and less so, the solstices. These weather phenomena indicate the Field dynamics that are also responsible for the triggering of some earthquakes and volcanic eruptions.

This electrostatic trigger has been revealed in some research. One study of major earthquakes between Bakersfield and Long Beach, California demonstrated that unexpectedly there were three out of six quakes that occurred within three days of an eclipse, and a forth within a fortnight. Furthermore, tidal or gravitational predictions were an average of 3.7 days off.[321] This three- to four-day delay is what could be expected of an electrostatic trigger, but not tidal forces. Figure 5.4 demonstrates the significance of the three- to four-day delay.

16.4 Lunar Quake Triggers

Many plate tectonics theorists dismiss lunar effects, because tides have little effect on the Earth's crust. They criticize any correlation between maximum global tidal forces and quake regions where local tides are not at a maximum, or can even be at a minimum.[318]

Meanwhile, studies of lunar phase triggers in 21 earthquakes show that fourteen occurred at the Quarter Phase, five at Full Moon, and two followed an eclipse.[332,333] Interestingly, the majority of Quarter-Phase quakes took place in the Southeast, in the region of Baton Rouge, Louisiana to Columbia, South Carolina, which surrounds the North Atlantic Field's stem.[322] In contrast, California earthquakes, which are triggered by the dynamics of the descending limb of the Field, show a peak with a three- to four-day delay.

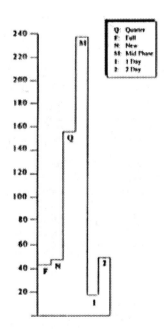

Figure 5.4. Lunar Periods in Earthquakes. Lunar periods of the same quakes (when data permitted) reveal peaks during mid-phase, the 24-hour period between three and four days after a lunar phase, and for quarter phase. [from reference 456]

Some scientists claim that the lunar effect is gravitational (i.e., tidal).[196,248,256,312,313,451] In fact, one study of 2,000 earthquakes demonstrated that they took place at times when tidal forces where over the epicenter of deep-focus earthquakes.[572] A lunar trigger was instrumental in an earthquake that took place in Parama, Italy in November 1985.[196] Lunar and solar "tidal" components were instrumental in triggering seismic activity in the volcanic areas of Hawaii, the Phlegrean Fields, and Vesuvius.[451] Earthquakes occur more often when the Sun and Moon are in opposition (opposite sides of the Earth) or in conjunction (aligned on one side of the Earth).[467] Both of these configurations have greater gravitational effects, but they are insufficient in themselves to trigger an earthquake gravitationally.

Other correlations exist between lunar phase and earthquakes, which only appear to be gravitational. A study of Nevada earthquakes revealed a close connection with variations of the gravitational forces. The Moon's active periods are 0-2 days of closest approach (*perigee*), 0-3 days of conjunction and opposition (*syzygy*), and 0-3 days of a 90° angle (*quadrature*) in relation to the Earth. These active times are not completely in accord with the gravitational effects, but indicate a delay of up to three days. A correlation of earthquakes with lunar phase and the Moon's passage through the area (*local meridian*) was demonstrated. Also a change in the polarity of the IMF triggered quakes.[593] These observations disclose the solar and lunar linkages of FEM, including a three- to four-day delay.

Another study of earthquakes in the Caucasus disclosed an increase of earthquakes four days prior to the New Moon (i.e., 3-4 days after Last Quarter).[592,593] Lunar-solar periods in quakes along the Pacific Coast were correlated with the Full or New Moon near sunrise or sunset, and also with the fortnightly ocean tides, which are regulated by lunar gravitational forces.[305] Undoubtedly, the gravitational effects are too weak, but the correlations show that there is a lunar trigger. The mechanism is revealed when it is understood that there is a lunar trigger on the occurrence of aurora.[259] In fact, sunspot cycles were investigated by using the aurora and earthquake records of ancient China.[363]

Similarly, statistical analysis indicates that there is a lunar effect on geomagnetic activity. When the Moon is within 4° of the ecliptic plane, between Full Moon and Last Quarter (on the morning side of the Earth) there are geomagnetic storms. The Moon triggers particle flow that, through electromagnetic induction of charged particles, disturbs the magnetic field.[261] The influence is electrostatic with the Moon triggering cascades of particle flow, and changing the contour of the electromagnetic characteristics of the Fields (*i.e., bow shock, plasma torus, potential gradients, electrostatic repulsion, etc.*).

Eclipses are known to have an effect on the conductivity of the atmosphere (*particularly the E-region*). As could be predicted of FEM, this affects the Earth's magnetic field.[71] In turn, the entire Field-system of FEM is affected, which can and does trigger earthquakes, as well as other geophysical phenomena.

The largest amplitudes of lunar tide on the geomagnetic field (*GMF declination*) occur in the morning around the time of the equinoxes. The effect follows the electrical conductivity of the atmosphere. The inadequacy of present models to explain the daily or diurnal effects are discussed: "We are not yet able to determine the exact height in the atmosphere or the mechanism by which this diurnal modulation is imposed. The action of different local conditions on the global lunar atmospheric tide is a complex process requiring the attention of modelers."[571]; p.309

The Earth-Moon system is a single unit that is not solely gravitational in its interactions. Shallow focus earthquakes and moonquakes vary in accord for the years 1971 to 1976.[549] Unusually large quakes in the period between 1950 and 1965 were remarkably numerous for the 20th Century. Though a lunar trigger is apparent, gravitational effects alone are not adequate for explaining the results, as a team of geophysicists state:

"Some ambiguity arises when we attempt to interpret this result within the framework of conventional gravitational models. If the pattern found is due to some physical cause, then this would seem to raise the question of the adequacy of the traditional model. There is reason to believe (on relativistic grounds) that the tidal stresses may not be the only significant stresses of external gravitational origin applied to the Earth. The underlying physical processes remain obscure."[560]; p.514

These geophysicists insightfully question the ability of the present Earth model to explain the results.

One physicist disclosed that all major earthquakes can be forecasted simply by observing the lunar surface. Seismic events on both the Earth and Moon are so precisely timed that it is "as if the Moon were in direct contact with the Earth; as if it were its seventh continent." Data from 639 major earthquakes were compared with 370 events listed in NASA's catalog of transient lunar phenomena (TLP) for the period between 1904 and 1967. TLP are luminous (high-energy charged particles) or dark (neutrons) phenomena, which are likely to be the Moon's influence on particle flow. These data made it evident that events on the Earth and the Moon are mutually registered with a time lag of up to three days.[327]

For example, consider two events. Two deep-focus earthquakes occurred on 31 March 1969, and the following day a transient red spot was noted in the Moon's Aristarchus Crater. On the 20-21 September earthquakes were followed by another spot in the same crater. Both events took place around the equinoxes, also displaying the solar linkage.[327]

The records indicated that TLP came, either or both, before and after earthquakes with a peak of two to three days prior to and after the event. Earthquakes are most likely to take place when these peaks coincide with high tides, and the Moon is at its closest approach to the Earth (*perigee*).[327] Evidently, TLP are due to the Moon's interaction with particles between the Earth and the Moon. These data reveal the existence of the electrostatic trigger and the Moon's role in the dynamics.

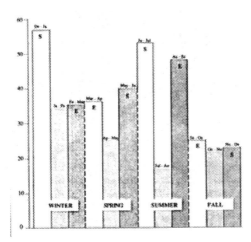

Figure 5.5. Seasonal Effects in Earthquakes. A seasonal occurrence of earthquakes is revealed in this histogram of 562 quakes of magnitude 5.0 and greater in the Northern Hemisphere, 1505-1976, and Northern California, 1901-1976. Dates were brought to the nearest mid-month to show trends related to the solstices and equinoxes. Peaks are evident in winter and around the vernal equinox, with secondary peaks around the autumnal equinox and the solstices (equinoxes: E, and solstices: S). A lesser set of data (62 earthquakes) discloses a six-month shift for the Southern Hemisphere, revealing the same peaks, as could be predicted. [from reference 456]

An eight-fold increase in earthquakes has been observed during the First and Last Quarter at closest approach (*perigee*), with the ascending node pointing toward the Sun around vernal (March) equinox, or the descending node near the September equinox. As occurs with weather, the 18.6-year lunar nodal cycle is also involved. One such example was the 2 March 1933 earthquake in Japan, which was one of the most powerful on record (8.9 on the Richter Scale).[318]

The seasonal effects can also be noted in TLP for the years of 1178 to 1982. The greatest occurrence of TLP took place around the December solstice and the March equinox, with the September equinox and June solstice represented less often. The seasonal peaks are fall and winter, as were noted for both earthquakes and weather.[118]

The evidence indicates that there is an electrostatic trigger on earthquakes. The solar wind provides particles in the form of solar plasma, the Earth's magnetosphere becomes enhanced, and the Moon triggers particle cascades along Field lines, which then interact with plate boundaries. The equinoxes are times of greater interaction between the GMF and IMF, and are also times of increased earthquake activity. Lunar phases and mid-phases (a mechanism with a 3-4 day delay) are times of greater probability for triggering particle cascades. Figure 5.5 shows the influence of these factors in histograms compiled from the analysis of 562 earthquakes. Table 5.1 displays examples of the ten worst earthquakes in history, and these factors in relation to their occurrence.

LOCATION	LIVES LOST	LUNAR PHASE AND DATE
Shensi, China	830,000	1st Full Moon after Solstice — 26 Jan 1556
Calcutta, India	300,000	Mid-phase after Equinox — 11 Oct 1737
Antioch, Syria	250,000	1st Quarter before Solstice — 20 May 526
Tangshan, China	242,000	2nd New Moon after Solstice — 28 July 1976
Nan-shan, China	200,000	1st Quarter before Solstice — 22 May 1927
Tokyo, Japan	140,000	1st Quarter before Equinox — 1 Sept 1923
Hakkaido, Japan	137,000	1st Quarter after Solstice — 30 Dec 1730
Kansu, China	100,000	1st Quarter before Solstice — 16 Dec 1920
Chihli, China	100,000	1st Quarter after Equinox — 27 Sept 1290
Aegean Sea	100,000	around Equinox — March 1201 (date unknown)

Table 5.1 The Worst Earthquakes in History. Here is a listing of the ten worst earthquakes in history, according to lives lost. All occurred around the times of the solstices and equinoxes at lunar phase or mid-phase. For example, Shensi, China had the worst earthquake in history on the first Full Moon after the solstice on 26 January 1556 (in winter), which took 830,000 lives.[456]

16.5 Other Quake Triggers?

Other, rather surprising, factors have been shown to possibly play a role in earthquake occurrence. For one, the position (115° *of meridian*) of Uranus was strongly correlated with earthquakes from 1904 to 1950. The scientist who conducted the study says that this cannot possibly due to gravitational effects.[601,602] This is quite obvious.

Earthquake occurrence in 1972 was enhanced in step with the frequency of a pulsar (CP 1133).[529] Later these data were criticized. However, one cannot help but wonder if the criticism was due to the unorthodox findings and the theoretical lack of a mechanism. Could these findings be due to action at a distance, or a system of interstellar or intergalactic fields (see Chapter 7, and sections 9.6 and 10.4)?

Another unorthodox theory was called the Jupiter Effect, when the tide raising planets are all on one side of the Sun, which was claimed to trigger earthquakes. On this basis a prediction was made that a major California earthquake would occur, but it did not come to pass. The major problem was that the assumed trigger is the gravitational forces present, but gravitational forces are too weak to be of sole importance.[231,232,274,284,396-398,435] The Jupiter Effect, as a result, has been relegated, so to speak, to the trash heap, but the suggestion that this could be an earthquake trigger shows that planetary configurations, which affect solar activity, are also correlated to earthquakes.

16.6 Volcanic Phenomena

Similar triggers that occur with earthquakes have been observed for volcanic eruptions. The global Field system is apparent in the fact that volcanoes are completely related to plate tectonics. About 80% of the world's active volcanoes are near subduction zones, another 16% are on mid-ocean ridges, and the remaining 4% are at plate-plate boundaries or hotspots in mid-plate. As already reviewed, plate motions follow solar activity, and common sense dictates volcanics do also, because volcanics are strongly related to plate dynamics.[39,340]

A solar linkage is also apparent in volcanic eruptions, such that solar motion and changes in the length of day are correlated with volcanic activity.[184,288,341,551] Volcanic eruptions are also noted at times when solar motion misses the barycenter. Two such times were 1632-1633 and 1810-1812. Many volcanic eruptions took place following the 1632-1633 period, including Hekla (1636), Roung (1638), Komagatake (1640), Awu (1641), and Gunung Adiksa (1641). Likewise, eruptions related to the 1810-1812 period were Sabrina (1811), Soufriere (1812), Awu (1812) and Roung (1817). Like other phenomena, a solar-FEM linkage is clearly evident.[551]

Lunar influences on volcanic eruptions are unmistakable. The triggering of some events took place at Full Moon and New Moon. In addition, the cycle between the closest and furthest approach to the Earth (*apogee-perigee cycle*) is observed in the timing of eruptions.[292,313,386-388,533] Likewise, micro-earthquake frequencies near an Alaskan volcano were correlated with the oceanic tides, which result from lunar effects.[387] Gravitational changes take place prior to volcanic eruptions, such as those observed at Poas Volcano, Costa Rica.[527] Like earthquakes, this effect cannot be due to gravitational forces alone, because they are far too weak.

Peaks around the equinoxes and solstices reveal the solar linkage. One study disclosed a maximum of volcanic activity for Mount Vesuvius and other volcanoes during the solstices and equinoxes.[466] Other research indicated that lava surfaces in the Halemaumau Crater at Kilauea Volcano stood higher around the solstices than around the equinoxes.[286,670] Eruptions at Kilauea and Mauna Loa clustered around the December solstice.[287,570] Fayal, in the Azores, was noted to erupt around the equinoxes.[153] Likewise, a study of 65 flank eruptions at Mount Etna, Italy from 1323 to 1980 disclosed that there were higher than average eruption rates in November just prior to the December solstice, and in March and May around the vernal equinox.[241] Another study of 4,200 eruptions showed a maximum in June, around the time of the solstice.[57] As noted for earthquakes and weather phenomena, seasonal, equinoctial and solstitial peaks in volcanic eruptions are beyond present models of the Earth, and reveal a solar linkage.

These factors are even more supportive of FEM in another study. Volcanic eruptions are more frequent for latitudes north of 30° North Latitude and south of 30° South Latitude, during April or May following the vernal equinox. For the equatorial region it is a June maximum around the solstice. When all latitudes are considered together the peak occurs in March and April, around the vernal equinox. For the period 1500 to 1700 peaks were in winter (January through March) and fall (October to December), but this correlation was not significant.[575]

Volcanic eruption frequency occurs most often in the northern mid-latitudes (30° to 60° North), and less so in the low or equatorial southern latitudes (20° North to 10° South), and each hemisphere has its own history.[485] FEM and its solar linkage is evident in the absence of synchronicity between the hemispheres. The solar linkage activates one hemisphere in one period and the other in another period, and this activation depends on which hemisphere was pointing away from the Sun and the polarity of the IMF at the time of peak solar activity; as occurs with glacial advance.

16.7 Origin of the Non-Gravitational Force

Evidence indicates that there is an interaction between gravitational and electromagnetic fields in accord with general relativity.[195,528,589] Non-gravitational forces are evident in the Earth-Moon system.[26,289,413,423-426] Furthermore, gravity has been observed to shift during solar eclipses, such as in June 1954.[14-16,41,221,319] The conclusion of the physicist who performed the original experiment has been relatively ignored. Such observations, he concludes, can be accounted for "only by the existence of a new field."[14] Once again FEM is called for.

Gravitational anomalies are known to exist along the mid-ocean ridges and faults. Anomalies also precede earthquakes and volcanic eruptions with such frequency that they have been used to predict them. Anomalous gravity measurements have been observed with depth in the oceans and in the Earth's crust, and have been attributed to a property of the Fields, known as vacuum polarization.[544]

The solar-FEM linkage is the result of the conditions that existed before the Earth became a full-fledged planet. Any modification of the distance dependence of gravity is inadequate for explaining the gravitational effects. A long-range force coupled with angular momentum, originating with the planetary nebula, is required. This external-field-effect (EFE) consists of a micro-dynamical system (particle flow) placed in an external field of dominant acceleration (FEM), which is affected above the gravitational effects. Furthermore, EFE can explain the inconsistences in conventional gravitational theory.[404]

The angular momentum versus mass relationship for the full range of astronomical objects, from asteroids to galactic superclusters, shows a remarkably tight clustering around a specific power law (*of index 2*). This shows that there is a saturation of the electromagnetic force that has not been accounted for theoretically. Most classes of systems are grouped around the mean relation than the stability argument would require.[606]

A new constant that equates angular momentum and mass must exist on theoretical grounds. There is a symmetrical property (self-similarity) that most astronomical systems share. Most systems have density and rotation laws that are scale-free (i.e., not restricted to the size and composition of the object).[76,647-649] As was shown in this and the previous tome, the Field-dynamical Model is found everywhere and is responsible for this symmetry.

Gravity is not only attractive, but posses a torsion analogous to the spin of nuclear particles. Matter is continually created near existing matter and is spinning outward due to the creation of angular momentum.

Rotational mass produces a magnetic field, which can be observed in most celestial objects.[6,7]

The usual models of dynamo processes of celestial bodies encounters difficulties, and alternatives should be considered. The protoplanetary nebula contained active plasma regions in which most of the energy exchange took place, converting gravitational and electromagnetic energy so that they were and are now self-supporting.[12] Electric fields have very long-range force, and are involved in FEM and its solar linkage. This is why we see gravity anomalies prior to earthquakes and volcanic eruptions, non-gravitational effects in the Earth-Moon system, and a solar-FEM linkage (for other discussions see Chapter 10 and "Churning Up Creation" in the **Conclusions**).

CHAPTER 17

Solstitial and Equinoctial Phenomena (The Solar-FEM Linkage in Geophysical Phenomena)

This chapter is mostly a review of observations of phenomena that occur during the equinoxes and solstices that were mentioned earlier. They provide further confirmation of the factual basis of the solar linkage. A few observations have not been presented in other chapters.

17.1 Solstices

A number of phenomena take place during the solstices aside form earthquakes, volcanics, weather and solar activity. For one, atmospheric electricity is at a minimum.[88] Pillar aurora observed at low altitude are correlated with the solstices.[114] The majority of booming "water guns" occur around the solstices, especially around the December solstice.[117] Superbolts, seen near the Japanese Field, take place from mid-December to mid-April, mostly in winter, or from solstice to equinox, with a peak around the vernal equinox.[475,590,591,615] The rotation of the Earth peaks and then drops, as El Nino develops around the solstice. El Nino of 1982-83, for example, had an accompanying strong peak in the length of day following the December solstice, especially in January-February

1983.[97] Over the Antarctica there is increased storminess from the December solstice to around the vernal equinox.[470] Following the solstice, autumnal changes indicate a considerable fall in ozone levels over Antarctica.[194] This is aside from the references to phenomena in mythology and history (see *In Defense of Nature—The History Nobody Told You About*), as well as solar activity, and biological and geophysical phenomena, that peak at these times.

A pronounced maximum of stratospheric circumpolar storms (*cyclonic vortices*) takes place in winter, particularly around the December solstice. The middle atmosphere displays an upward propagation of planetary waves that reach a maximum after the solstice, in January. An intense coupling between the basic states of the upper atmospheric layers (stratosphere, mesosphere and thermosphere) takes place during winter, following the solstice. Internal atmospheric processes in the upper region are most pronounced after the solstice, in winter.[564]

17.2 Equinoxes

A far more active time for most geophysical phenomena is around the equinoxes, particularly the March or vernal equinox. The depth of the equatorial ionization anomaly is reduced during the March (vernal), and enhanced during the September, equinoxes on days when the IMF is moving towards the Sun (*negative polarity*). There is also a modification of equatorial zonal electric fields at that time. Other characteristics of the equatorial regions also change (*in the F-region, as well as an end to the D-region winter anomaly*). These changes are primarily related to east-west, electric-field responses to IMF polarity shifts during equinoctial months at times of solar maximums (*as noted, for example, in the 19th and 20th cycles*).[401,532]

The winter stratospheric circumpolar cyclonic vortex begins in September, and lasts into April, or from equinox to equinox, though the maximum is around the December solstice. In the summer, the stratospheric-mesospheric east wind blows from April to the end of August. A transition from winter to summer conditions takes place abruptly in early April (at all heights from 30 to 150 kilometers, or 18.5 to 93 miles). This sudden change involves a shift from west to east winds in the stratosphere, an end to upward propagating planetary waves, a pronounced minimum in one (D-) region's absorption, and the onset of a mesospheric pressure maximum. Also, a meteor wind reversal takes place, and a geoelectromagnetic (Sq-) current becomes half as strong as it was.[514,564] These currents are enhanced around the September equinox, and are accompanied by atmospheric "tidal" waves (east-west component at the equator, and ionospheric plasma coupling that is ground, magnetospheric and solar influenced).[637]

A peak in a number of anomalous phenomena takes place around the equinoxes, as well. Pillar and low-level aurora peaked around the vernal (March) equinox. The majority of wheel-type luminous seas were noted around the vernal equinox, and less so, the September equinox. Milky seas were more frequent around the September equinox, while parallel bar seas occurred more often around the vernal equinox.[116] The South African event, that some claim was an atomic blast, occurred during the September equinox.[380]

Other phenomena occur around the equinoxes. A peak in atmospheric electricity also occurs around the equinox.[88] Circulation and composition of the thermosphere alters dramatically, as well.[512] Even major power blackouts take place near the vernal equinox, such as 24 March 1940, when New England, New York, Pennsylvania, Minnesota, Quebec and Ontario where hit simultaneously.[347] The largest know outbreak of tornadoes within a 24-hour period occurred shortly after the equinox on 3-4 April 1974, and involved 148 tornadoes that hit 14 states. Other extreme weather also takes place around the equinoxes, such as Hurricane Fran in September 1996, Hurricane Marilyn in September 1995, a major snow blizzard hit the East Coast in March 1993 (weather bombs most often hit around the March equinox), and so on. The electrostatic effects are evident in a study that disclosed 50% of 985 whistlers, which are lightning generated, that occurred in one year, took place in March-April.[594]

Geomagnetic storm activity displays semi-annual peaks around the equinoxes. The Sun's axis is more inclined toward the Earth just prior to the equinoxes. As a result, recurring geomagnetic storms are most prominent near the equinoxes due to interaction with the IMF.[525,608,609,651] The largest amplitudes of lunar tide in geomagnetic declination occur in the morning during the equinoctial months, with a significant peak in early April.[571]

The rotation of the Earth fluctuates around the equinox, displaying first a peak and then a drop.[97] The result is that there is consistently a vernal equinox increase in the length of day.[335] El Nino, ozone levels, earthquakes, and other geophysical phenomena are accompanied by changes in the length of day, disclosing a peak due to a global system (see Chapter 18).

As will be discussed in section 19.1, the equinoctial effects are also noted in polar jerks. The Sun's helical motion around the barycenter, or center of mass, reached a minimum in 1913 when the Earth had undergone a polar jerk. The greatest solar interactions were observed around the equinoxes.[399,525,571,608,609] Universal time variations are associated with the Earth's tilt on its axis, displaying the largest variations around the equinoxes.[525] In addition, shifts in the geomagnetic field peak *first*.[18,19]

Solar activity continually peaks around the equinoxes, particularly the vernal equinox. Examples include, April 1980 when numerous flares developed, and several large x-ray flares, particle events, and shocks took place during April and early May in 1981, and numerous peaks throughout 1990 to 1998.[213,367,534] Solar maximums occur more often around the vernal equinox, as a study of 21 maximums revealed.[399] Furthermore, chemical reactions (*Piccardi Tests*) on Earth peak *first*, then solar activity, and these chemical reactions have an annual cycle with a peak around the March (vernal) equinox.[94,473,474]

CHAPTER 18

When Time Changes
(Alterations in the Length of Day or in the Earth's Rotation)

Another phenomenon that reveals the workings of FEM, and its solar and lunar linkages, is changes in the Earth's rotation, affecting the length of day. Again, the forces involved have a non-gravitational character. Also the changes occur along with other geophysical phenomena.

18.1 Solar Effects

The length of day (LOD), the period required for the Earth to complete it rotation, has varied throughout time. Like so many other geophysical phenomena, solar activity has been correlated with changes in the length of day.[135,168,225,484] Different LOD occurred along with fluctuations in solar activity in the periods 1780 to 1960, and 1900 to 1979.[135,484]

Solar flares are known to abruptly alter the Earth's rotation.[100] For example, the great flares of 1959 and 1972 brought abrupt changes in the LOD.[231] Both long-term and short-term changes in solar activity alter the Earth's rotation.

The 11-year, 22-year and 56-year solar cycles are conspicuous in the LOD data.[124,131,183,225,361,365] Long-term (*secular*) changes were noted at times of fluctuations in solar activity in the past, and the evidence indicates that these changes were very likely abrupt.[403,415] Observations of the LOD, like so many other geophysical phenomena, reveal the solar-FEM linkage.

18.2 Lunar Effects

Like other Earthly phenomena, the Moon also exerts an influence on the length of day. The 18.6-year lunar nodal cycle is apparent in LOD changes.[281,365] During the last 2,500 years the deceleration rate of the rotation and the uniform recession of the Moon appear to be necessary to explain the variations. However, this would require a change in gravity, or in the mass of the Earth and/or the Moon, if it were to be described in conventional, gravitational terms.[595,596] Other evidence demonstrates that neither gravity nor the mass of the Earth or Moon have changed substantially within that same time frame.

This conclusion is surprising within the framework of conventional thought, to say the least. Notwithstanding, such an observation is in accord with the FEM-lunar linkage, which goes beyond gravitational factors. Furthermore, the greatest lunar tidal effect is between the 40° latitudes, which are the latitudes of the Fields of FEM.[595,596] This makes the Moon ideally located with respect to interacting with Field dynamics and the triggering of particle flow.

The "surprising" aspect of such an interpretation, that gravity or mass have changed, is bound up in the perspective that everything is explainable by gravity. Basing conclusions on a gravitational perspective (*lunar tidal torque*) forces such an enigmatic deduction. However, the real effect is non-gravitational, and involves the Moon's interaction with particle flow along the Field lines of FEM (an effect that originated in the polarization of the vacuum during the protoplanetary stage).

18.3 A Non-Gravitational Component

Confirmation of the effect's non-gravitational nature is evident in numerous observations. One is that a correlation exists between the LOD, weather and solar activity. Alterations in the Earth's rotation, variations in the upper atmosphere's density, and geomagnetic storms fluctuate with solar activity.[168] Variations in the upper atmosphere's density and geomagnetic storms would be expected from the ionization produced by particle flow.

A few scientists have claimed that the changes in rotation are due to weather phenomena acting on the Earth's surface (i.e., wind shear).[183,484,638] Meanwhile, the rotation slowed just *prior* to one of the greatest weather disturbances in some forty years, the strongest El Nino on record. A strong peak in the increase of LOD *preceded* an unusually intense El Nino in 1982-83, not the other way around.[97] A similar rotational change also preceded the El Nino of 1997-98.

Likewise, a correlation between weather and solar activity still stands up against all tests, though it remains unexplained by conventional physics. Furthermore, weather, solar activity, earthquakes, volcanic eruptions, and other geophysical phenomena are correlated with changes in the LOD. A solar-FEM linkage exists that is coupled with FEM's global Field system and dynamics, and therefore, these phenomena should be correlated.

Further confirmation of the non-gravitational component can be found in evidence that correlates the LOD with physical phenomena on Earth. For one, there is a well-established correlation in variations of the Earth's rotation and polar motion, such as the Chandler Wobble.[484] Our present understanding of the wobble has many unresolved problems, the most important of which is that it should decay and eventually disappear, but it is somehow continually reinstated. This leaves open the question of what is responsible for maintaining the wobble, as well as what mechanism is responsible for LOD alterations.[166,484,565]

Compounding this lack of a viable conventional theory are the correlations that demonstrate that the Chandler Wobble, solar activity, weather, earthquakes, and LOD fluctuate under a common influence. Two scientists studying the problem comment: "Because of several forces of varied nature acting on the Earth it seems quite natural that both fluctuations in the Earth's rotation and in polar motion may have at least in part a common geophysical cause."[484]

Variations in the Earth's rotation and continental motion (plate tectonics) correspond in such a way that the elastic deformations in the Earth's crust have the same motions (*torque*) as the annual and semi-annual shifts in the LOD.[122,459,489] This observation can only be explained by a field system controlling both LOD and plate tectonics. As in the discussion on earthquakes (see Chapter 29 of Volume One, and Chapter 16 of this volume), this represents an enigma for conventional theorists: "The good agreement between the Earth's accelerations and crust movements seems to confirm the hypothesis that we have a westward movement of the Earth's crust corresponding to a deceleration in the Earth's rotation. It is very difficult to give the reason for the mutual relations between such different plates; nevertheless the same remarkable regularities can be evinced from the analysis."[489]; pp.391-393

A relatively good correlation exists between changes in the LOD and the westward drift of the geomagnetic field.[524] Periodic components in LOD fluctuations are described as annual, semi-annual, lunar monthly, and fortnightly.[97] Evidence also reveals a spring increase in the LOD, which is activated at the vernal equinox.[335] Similar periods are noted in all geophysical phenomena.

Considering all of these facts, as evidence of a coherent whole, leaves us with the inescapable conclusion that an integrated mechanism is involved that encompasses the whole Earth. Furthermore, it involves interrelationships with the Sun and Moon as a single unit, and it is non-gravitational. This non-gravitational or non-tidal component is clearly shown in the data. "Hence, there is a non-tidal component in the past millennium which acts to decrease the length of day."[415]

18.4 <u>FEM</u>, <u>Life</u> and <u>LOD</u>

Once again, with changes in the LOD, it becomes clear that life has a stabilizing effect. The human destruction of life has increased, solar activity has increased, and the Earth's rotation has slowed down, along with a weakened geomagnetic field. This increase in the LOD brings about an increase in solar energy for photosynthesis and an extended growing season. Furthermore, a number of biological, meteorological and geophysical phenomena are closely associated with the period of the Earth's rotation.[484] In fact, the study of changes in the LOD in the past involves the utilization of fossils to uncover the data.

The destruction of life, through the solar-FEM linkage, increases solar activity, which increases the growing season. The season is longest approximately one year after solar maximum, and if unusually high there is a correspondingly higher peak. For example, the growing season was 200 days until the mid-1930s, then there were three solar maximums producing 220 days, 225 days, and in 1959's major flares, 235 days.[225] This extended growing season discloses the solar linkage, and life's response. Likewise, at these times there were also changes in the LOD.[335] Again, the evidence discloses the importance of life in FEM's dynamics.

An examination of solar activity and history shows that solar maximums take place at times of civilizations' heights when there is an increased destruction of life. The increased solar activity triggers events that lead to the downfall of these civilizations. Minimums take place after the downfall of civilizations on a global basis, and there is the renewal of life (see *In Defense of Nature—The History Nobody Told You About*). A correlation exists between these data and the LOD.

There are long-term trends in changes in the LOD. The first was between 700 BC and AD 1000, when there was an average increase of

2.4 milliseconds per century. The second, from AD 1000 onwards, displayed a similar average increase. Since AD 1600, when data is more available on a shorter timescale, the LOD fluctuated by about four milliseconds on a time scale of decades.[415] From 1650 to 1900 there was a slowing of the rotation compensated for by an acceleration comparable in magnitude.[403]

The correlation with the waxing and waning of civilization is conspicuous in the data. When civilization increases destroying life, the LOD increases. As civilization collapses with life renewed, the LOD decreases. Civilization was at its height in the 8th Century BC, followed by a worldwide collapse by the end of that century (approximately 700 BC). From that time there was the Homeric Minimum in solar activity, and an increase in the Earth's rotation (a decrease in LOD). Likewise, there were the Dark Ages following the Byzantine Maximum (6th Century AD), and with the exception of the Medieval Maximum (AD 1140-1240), it was a period of low solar activity. Minimums occurred (Wolf, AD 1280-1350, and Sporer, AD 1420-1570) when the rotation again speeded up (decreased LOD). Then came the bubonic plague and the conquest of the Americas, which was followed by the Maunder Minimum (AD 1645-1715), the most inactive period in the known history of solar activity. It is after this that we find the greatest fluctuations in the LOD. A general increase in the LOD has been taking place during the Modern Maximum (AD 1800-present) along with modern-day civilizations' destruction of life. These dates all coincide with the historical cycles that are presented in *In Defense of Nature—The History Nobody Told You About*, and where the evidence for the Earth's influence, along with life, on solar activity are addressed in more detail in terms of this causal relationship (also see sections 13.4 and 13.7).

Furthermore, the Sun itself shows similar trends, such as an increase in its rotation as the Maunder Minimum was beginning, at about AD 1642 to 1644. It was then that the equatorial velocity was 3% to 5% faster, and differential rotation was enhanced by a factor of three.[177] In both cases the rotation increased when the Sun and FEM (with flour-ishing wildlife) were functioning more efficiently. As will be shown, there are also ancient writings that confirm these relationships, and demonstrate the reality of the solar-FEM linkage and life's important in the dynamics in the historical process itself, as discussed in *In Defense of Nature—The History Nobody Told You About*

CHAPTER 19

Polar Wander, Flip-Flops and Twitches (*Mechanisms Behind Polar Wander, Reversals and Jerks*)

Other phenomena that uncovers the workings of FEM are polar wander, reversals and jerks. The solar and lunar linkages are again apparent in the data gathered on these phenomena. Both the polar jerks and reversals display evidence of a mid-latitude effect that can only be explained by FEM.

All excursions and reversals of the geomagnetic are anomalous in the sense that they cannot be explained by the dynamo theory. It is not known how or why they are initiated. For example, if the reversal occurs rapidly, then according to the dynamo theory there would have to be unrealistically rapid motions in the core. Generally speaking, polar excursions and reversals seriously challenge the dynamo theory, and they alone call for a new model of the Earth.

19.1 Polar Jerks

Reports of sharp, brief variations in the Earth's magnetic field, often referred to as polar jerks, appeared with increasing frequency beginning in the 1950s. The 1950s were also noted for changes in the Sun with the first appearance of polar faculae, and a solar magnetic field reversal, as tectonic stress on the Earth increased, and industrialization entered a new era (post World War II). The 1950s were also noted for major solar flares, changes in the LOD, and increased growing seasons.

In 1969-70 there was jerk in the geomagnetic field that is global and appears "simultaneously" at stations far apart. Expected of FEM, a geophysicist comments: "This suggests that this phenomenon is produced either by global or regional sources."[518]; p.869. Furthermore, stations along mid-latitudes, particularly the Field latitudes (and associated longitudes), display similar periods, and the 22-year solar cycle in polar jerks.[19] The source of the jerk has important implications for the origin of the Earth's magnetic field and internal structure, and the observations lead to interpretations that depart from the presently accepted model of the Earth, but conform with what would be due to the Field-dynamical Earth Model.

As with other observations there are indications of a solar linkage in jerks. The jerks, such as in 1913 and 1969, are correlated with the 11-year solar cycle.[680] The Sun's helical motion around the barycenter, or center of mass, reached a minimum in 1913, and the greatest solar interaction took place around the equinoxes.[399,525,571,608,609] Universal time (LOD) variations are associated with the Earth's tilt on its axis, displaying the largest variations around the equinoxes.[525] In addition, shifts in the geomagnetic field peak *first*.[18,19] Again, all of this indicates that the Earth is responsible for the solar transformations. Further support for this linkage is evident in theories that attribute the jerks to both internal and solar changes.

One of the most active periods of solar activity have been taking place as the new millennium approaches. As might be predicted, a geomagnetic jerk, as noted in the records of the geomagnetic observatories in Europe, took place about 1991. Again, a correlation between polar jerks and solar activity is noted.

Between 1947 and 1972 there is no evidence for short-period signals that originate in the Earth's core. Jerks were first attributed to internal processes, but were later correlated with solar activity.[18,19,120,127] The entire system can be noted in that geomagnetic activity, solar activity, changes in the LOD, and other geophysical phenomena, including climate are correlated.

Typical of other relationships, these impulses display correlations with solar activity and human activity on Earth. A geophysicist discusses a time when there were mutual changes of many solar-terrestrial relationships: "A preliminary analysis of the 1700-1870 period seems to indicate that this was a time with no dramatic geomagnetic changes, no alterations of westward drift, and also no large changes in the acceleration of the Earth's rotation."[120]; p.715. This period began near the end of a remarkable low in solar activity, known as the Maunder Minimum, and was a time prior to the Industrial Revolution with its intensified destruction of life.[175,540]

This minimum in solar activity took place after the bubonic plague or Black Death, the conquest of the Americas, and other events stopped the wholesale destruction of wilderness in the AD 1400-1600 period (see *In Defense of Nature—The History Nobody Told You About*). Following that period the Earth's magnetic field strengthened and stabilized, the Sun calmed down, and there were less fluctuations in the Earth's magnetic field and rotation (LOD). In fact, about 1800 the Sun was on the verge of lapsing into a passive state.[507] Afterward, the Earth's magnetic field began a relatively steady decay in strength in the early 19th Century, as the Industrial Revolution commenced.[101,394,540] Then, aurora displayed ever increasing peaks in the 1850s, and particularly following 1900, as the Modern Maximum in solar activity increases even more.[358] At the same time there is the 1913 polar jerk and changes in the Earth's rotation (LOD). Other correlations of this nature can be witnessed throughout history, as will be shown.

19.2 <u>Polar</u> <u>Reversals</u>

Polar reversals are being comprehended with more accuracy due to high-resolution studies of sediments and lavas. For one, reversals are very rapid, on the order of the same time scale as that of the polar jerks, such as that of the 1969 polar jerk.[487] Some scientists studying reversals admit that this timescale is far too incompatible for conventional solid Earth theories proposed for the lifetime of interior convention patterns (*mantle mixing and flux diffusion*).[333]

As with earthquake waves, which indicate very strong magnetic fields near the core, reversals demonstrate that, "Decay times of the order of a few years require implausibly large fields at the core-mantle boundary."[487]; p.233. There is no generally accepted theory of how the Earth generates its magnetic field, and what the reversal mechanism involves.[269,455] Unquestionably, this is because the Field-dynamical Earth Model (FEM) is not know. Demonstrating life's contribution to FEM is the fact that the core, as well as the magnetic field, became strong at the same time that life became well established.[618]

Observations of the sedimentary and volcanic records demonstrate that there are "transitional fields" before reversals and "aborted" reversals. Sediments and lavas laid down during times of reversals or transitional fields leave behind directional changes of the geomagnetic field in the formation of crystals in these rocks. Due to the magnetic properties of minerals, these crystals form in the direction of the magnetic field at that time. By studying these rocks and their crystals, geologists are able to determine the reversal record.

The layered lava flows of Steens Mountain preserved a record of the geomagnetic field reversal. It takes about two weeks for a two-meter thick flow to solidify, and thereby, a record of the behavior of the geomagnetic field was established. The lava flow indicates that the field's axis was rotating 3° to 6° per day. That is incredibly fast according to present thinking, and is, in fact, several thousand times faster than theorized. According to these figures (confirmed by a number of studies), the reversal would take no more than months, but theories claim that they should take thousands of years. If the dynamo theory were at work, then the molten rock of the core would have to move at speeds of several kilometers per hour, an impossible scenario.

At Steens Mountain, Oregon there were two phases when the field shifted away from the typical north-south (*axial*) symmetry, when a "transitional field" between the 30° and 40° latitudes was recorded. The first phase was a complete flip from normal to reversed polarity, while the second phase was normal-transitional-normal, and did not undergo a reversal.[487,620,621] Similar records of transitional fields are found at numerous sites in North America, Puerto Rico, Iceland, Japan, Africa, Australia and New Zealand.[269,487,622] In addition, the sedimentary record also supports the volcanic evidence for a transitional field phase.[620-622]

The transitional field is not dipolar, not north-south for example, but is instead a single pole.[264] The Coriolis Force (most active around the 35° latitudes) is said to be involved in both the two phase and other long-term (*secular*) variations, such as the polar jerk.[352] For example, a study of rocks from southwestern Puerto Rico indicated an ancient pole at about 39.4° North Latitude. This lead to the suggestion that the region had undergone a 40° to 45° anticlockwise rotation.[622]

However, this is actually the activation of a mid-latitude Field that produced the magnetization observed in the rocks. Such transitions occur at a time when the main dipole or north-south magnetic field is about one-fifth or less of its typical strength.[487] What these facts uncover is that the main dipole is weakened at the same time that a mid-latitude Field is activated, releasing hydrogen fusion by-products, and producing a non-dipole transitional field that is recorded in the sedimentary and volcanic rocks laid down at the time.

Other observations clearly support this interpretation. Reversals occur in two steps separated by an interval of relative inactivity in the dipole field.[268] Recent theories have suggested that reversals have been caused by comet or meteor impacts, but the record indicates that this is not so. A study of the Ivory Coast tektites, for example, disclosed that the reversal occurred prior to the deposition of the tektites, which are thought to be of impact origin. Moreover, other times of proposed impacts are not clearly linked with a reversal.[538]

Departures from the original (*initial axial*) dipole field to the transitional field shows the same geometry, and hence, a "memory." As could be predicted of FEM, "such a geometrical aspect may not always be associated with the core process."[268] This stability has brought the idea of a "frozen-in-flux" into the theoretical picture. Such a concept as frozen-in-flux is just another way of describing the stability of Field position when scientists don't know of its existence. In terms of FEM, this demonstrates that the Fields prevent any great departure from the original geometry, as could be predicted from a geometric arrangement of a system of fields.

This is also why the transitional field is not considered an "aborted" reversal.[620,621]

Moreover, the paths taken by the reversing poles follow similar routes with each reversal. One preferred path is a band about 60° wide running north-south through the Americas. The other preferred path is 180° away with a band running through east Asia and just west of Australia. To think that the convection and mixing of a molten core could do this is shear hallucination.

Likewise, long-term changes in sea level and reversals are closely associated, and even synchronized. That is, ridge volume and seafloor creation are correlated with polar reversals.[204] The Mesozoic-Cenozoic histories of reversals are not only accompanied by seafloor spreading, but also climate changes, black shales, tektites and geologic periods. This could be predicted of FEM with the Fields controlling climate, tektite ejection, and the ridges and seafloor spreading, as well as reversals.

The facts indicate a particular scenario. First the geomagnetic field (GMF) weakens, and around that time an increase in solar activity takes place due to the solar linkage with the Interplanetary Magnetic Field (IMF). The burst of solar activity is accelerated into the Earth's core, shortly followed by the transitional field's activation due to the triggering of one (or more) of the mid-latitude Fields. At times this causes the spewing of iridium or tektites, and shocked minerals. Finally, the polarity of the IMF changes, and through its interconnectedness with the GMF causes the reversal. Or when the IMF does not reverse its polarity there is a return to normal GMF polarity, the so-called "aborted" reversals. This is supported by studies that indicate an external field would have to play more than a transient role in reversals.[115]

Furthermore, variations in the GMF are correlated with solar activity, and relationships with the IMF are clear in both the IMF polarity and the equinoxes, which are also times of the greatest geomagnetic activity.[350,399,525,571,608,609] In fact, IMF sector boundary crossings cause changes in the Earth's spin rate (LOD), GMF, magnetosphere, ionosphere and atmosphere, displaying the solar-FEM linkage fully.[193,350,571]

The importance of life in generating electromagnetic energy to the system can be seen in a number of facts. Geomagnetic intensity increased sharply in the Proterozoic, and this has been thought to be due to the inner core consolidating (*nucleation*).[236] However, this is when life systems began to be globally established, and contribute electromagnetic energy, and conducting and semiconducting organic materials to the system, even superconductors within life itself.

Reversals take place at times when the life system on Earth is already in decline, and the geomagnetic field is weakened. In fact, there are often geological transformations and mass extinctions, during which there are reversals. Furthermore, many reversals and "aborted" reversals are followed by the blossoming of new life forms (mass speciations), and the reestablishment of a stabilized and strong geomagnetic field. Increased radiation at times of reversals is also known.[490] This could have lethal effects—extinction—and cause mutations—new species.[50,216] Moreover, chemical reactions (*colloid precipitation rates*), important in molecular evolution, are altered by both the effects of magnetic fields (or their absence) and radiation.[52,181,198] See Chapters 22, 23 and 28, and section 24.2 in Volume One for more in depth discussions. Polar reversals seem to be occurring at an ever increasing rate from the late Mesozoic (Age of Reptiles) to the present (however, this may be due to an inaccurate dating system).[269]

19.3 The Non-Gravitational Force

Jerks, changes in the length of day, and reversals, and their interrelationships can also be observed in history and prehistory. Scientists studying the long-term (*secular*) changes in the Earth's rotation note a drastic difference at about AD 700. At around the close of the 13th Century, time accelerations changed by a factor of *five!* The scientists that undertook the study comment: "We are seriously lacking in mechanisms to explain the non-gravitational force."[424] Changes in the Earth's rotation, as well as some polar wandering, jerks and reversals, appear to have taken place in each of the cycles in history discussed in *In Defense of Nature—The History Nobody Told You About.*

Observations of changes in the length of day (LOD) in modern times suggest the same. A great solar storm occurred in August 1972, and a decrease in the LOD took place.[231] The slowdown rate of the Earth's rotation has been correlated with solar activity.[223] Evidence indicates that there is an interaction between gravitational and electromagnetic fields in accord with general relativity, which are evident in the Earth-Moon system as non-gravitational forces, such as gravitational shifts during solar eclipses.[14-16,195,221,423-426,528,589] These observations can only be explained by the existence of a new field.[14-16]

For example, a torsion pendulum changed its total angular path markedly for hours before and during a solar eclipse (up to midpoint). A physicist comments on a 7 March 1970 eclipse: "These variations are too great to be explained, on the basis of classical gravitational theory, by relative change in the position of the Moon with respect to the Earth and Sun. Classical gravitational theory needs to be modified to interpret [these] experimental results."[535]

Another eclipse on 30 May 1965 showed a reduction in "galactic" radio noise.[310] Such an observation indicates that a field that absorbs radio waves is activated during the eclipse. Likewise, gravity waves were noted to change around the equinox (March) and solstice (June), and were correlated with geomagnetic disturbances and solar activity. The strongest geophysical effect was in the magnetospheric ring current (2.7 *standard deviations*).[645]

Other observations indicate that gravity appears to be decreasing.[26] Borehole and seafloor measurements have revealed deviations in gravity at greater depths, as well. These observations have led to speculations of the existence of a non-Newtonian intermediate range force, and as could be predicted, inadequate models of the Earth.[276,482] All of these data call for a new model of the Earth, for which the Field-dynamical Earth Model (FEM) is ideal.

CHAPTER 20

The Possible Outcome of Sun-Moon-FEM Linkages in History

When all of the linkages of FEM are considered there are probable events in Earth and human history. For Earth history, the linkages could bring about extinctions and speciations, polar wandering or reversal, episodes of plate tectonic activity, changes in sea level, and extreme climatic transitions, among other things. As discussed in Tomes Two and Three, this is what we find in the fossil and geological records.

For human history and prehistory, during times of solar maximums there could be mutually triggered global events. These could involve changes in the length of day, disastrous weather shifts and events, increased earthquakes and volcanic eruptions, polar wander or reversal, and increases in war and disease (due to biometeorological and geomagnetic effects). *In Defense of Nature—The History Nobody Told You About* will demonstrate that these events have taken place in human history with confirmation in geology, archeology, and ancient, mythological and historical writings.

This has extremely important ramifications for our era. As noted in this tome, the geomagnetic field began decaying about 1800, followed by a lasting ten-fold increase in aurora, leading to the Modern Maximum in solar activity, which progressively becomes more active as the present is approached. Following these changes are highly significant solar-induced weather phenomena, such as the QBO (Quasi-Biennial Oscillation), which is noted from about 1925 on, or following the increased industrialization of post-World-War I. Just prior to this there was the 1913 polar jerk. Following the even more industrialized post-World-War II era (about 1950 on) the Sun's polar faculae are first noted, and shortly thereafter there was another polar jerk (1969). In addition, there were more drastic weather phenomena, such as more frequent El Ninos, and more earthquakes and other geophysical phenomena. It is extremely important that we understand our history, and how we influence what takes place. *In Defense of Nature—The History Nobody Told You About* will illustrate this human participation in the physical world

Conclusions

It's the End of the World as We Know It!

"Man masters Nature, not by force, but by understanding."
Jacob Bronowski, p.20 in *Science and Human Values*.
London, Hutchinson, 1956.

We must consider ourselves fortunate that, in spite of our ignorance of human participation in the physical world, our expectations to the contrary, and flawed theories, science's uncovering of the facts has still managed to reveal truth. Here is one of the most interesting facets of science, it is mostly objective when obtaining facts. Interpreting those facts is quite another thing, sometimes its artistry, and at other times its dogma. Scientists admit that problems of this nature exist: "The combination of growing specialization and the peer-review system have fractured science into isolated domains, each with a built-in tendency toward theoretical orthodoxy and hostility to other disciplines."[92]; p.375. "Science is not always as objective as we would like to believe. We view scientists with a bit of awe, and when they tell us something we are convinced it must be true. We forget they are only human and subject to the same religious, philosophical, and cultural prejudices as the rest of us. This is unfortunate, for there is a great deal of evidence that the Universe encompasses considerably more than our current worldview allows."[160]; p.6

On the objective side of the factual information, however, we humans have succeeded in life itself by acquiring an understanding of the meaning of life. As a result, this knowledge leads us to a scientific basis for exemplary decision making, ethical understanding, and positive influence on the physical world. Essential ethical and physical principles that can perfect the Earth, as well as the Universe, emerge (these will be presented).

Through unconditional love and defenseless bonding, each of us can lend a helping hand and thought for developing a land of peace and plenty, in the fullest sense of these words: a veritable paradise. It will be a perfected world that touches upon the workings of the entire Universe itself, perfecting the entire cosmos. Basically, "it's the end of the world as we know it," only unlike that statement is normally contended with, we'll all feel *fantastic* about it!

The Myth of Math
(*Mathematics and Uncertainty*)

As you may have noticed, a direct application of mathematics has been left out of this book. The reason is quite simply that it is an uncertain language. In quantum mechanics one cannot simultaneously measure the momentum and position of a particle at the same time without creating some inaccuracy. This was discussed in Tome Two and is known as the Uncertainty Principle.

Mathematics involves vector analysis of geometric objects and concepts. That is, mathematics takes, for example, a circle and marks points along that circle and positions lines (vectors) through those points. The mathematical formula then does not actually allow a "particle" to "move" along the actual circle, but measures the particle's momentum and position at the same time. Hence, mathematics is an uncertain language at best, and in some cases myth has obscured reality when math is relied upon too strongly.

More importance is given to equations than to the physical plausibility of a theory, leading science, not on an errand of truth, but on the road to illusion. As a noted plasma physicist[*****] comments: "To try to write a grand cosmical drama leads necessarily to myth. To try to let knowledge substitute ignorance in increasingly larger regions of space and time is science."[2] An interdependence exists between fundamental physics and our understanding of the Universe that extends to the methods employed. Theories rely strongly on the deductive method,

---

[*****]   Dr. Hannes Alfven

deriving concepts from the symmetries of mathematics. They are not based on observation in some cases, such as theories in particle physics, and so cannot effectively be challenged by observation or experimentation. What this means is that such theories are not truly scientific, but involve belief systems.

Chaos: Order in Disguise?

The order in the Universe can be overlooked by employing mathematics the wrong way. For example, equations long used in physics (e.g., Newton's equations) can produce seemingly random results when nonlinear terms are used. As a result, we begin to see chaos when order actually exists. This does not even take into account that it could be our ignorance of what causes what looks like chaos, but instead is only an unknown factor(s).

Scientists discuss the flaw behind relying on mathematics too strongly: "Numbers, after all, are like words: they are symbols we use to describe Nature. To take mathematics as reality, to believe the Universe is formed by 'breathing fire' into mathematical equations, is to believe in magic. A search for the ultimate mathematical reality inevitably creates fairy tales, not science."[92]; p.334-335. "Usually it is necessary to resort to methods of mathematical approximation, in which small terms are omitted, leaving only the most important quantities in the equation. However, it turns out that in ignoring small terms we may be throwing out the baby with the bath water and getting results that bear no resemblance to reality."[133] Mathematical equations are a pointing finger, if you look at the finger too much, you miss what it is pointing at.

Speaking of the pointing finger, there are a number of equations that are almost identical. The equations of electromagnetism and hydrodynamics (fluid flow) are essentially the same.[33] Furthermore, the basic equation of quantum mechanics (Schrodinger's equation) is also closely related to these equations. The mathematical laws governing the motion of line vortices (i.e., plasma filaments and the Field) turns out to be a modified form of this basic equation of quantum mechanics (called the nonlinear Schrodinger equation). The pointing finger tells us that in all probability there is some interrelationship between quantum mechanics, electromagnetism, hydrodynamics and vortices; if these equations were words instead we would not doubt that conclusion. Yet, this is in clear contrast to the dominate theory: "The basic assumptions of the medieval cosmos—a universe created from nothing, doomed to final destruction, governed by perfect mathematical laws that can be found by reason alone—are now the assumptions of modern cosmology."[92; p.162]

The Cellular Cosmos
(*The Curvature of Space-Time as a Prerequisite to the Universe*)

"Empty" space holds the key to a full understanding of the forces of Nature, because the vacuum is not just empty space devoid of substance and activity. In space the vacuum is filled with what are called virtual particles. Among them are virtual electrons and positrons (the electron's antiparticle with the same mass and spin). They are referred to as virtual particles because we do not see them directly, but we do know they are there, because they leave physical traces.

Polarization of the Vacuum

If a field is present in the vacuum it will cause a net shift of charge, a phenomenon known as polarization. The fact that empty space can become electrically polarized in the presence of a field is embodied in quantum theory, and has been measured in experiments. The vacuum is not featureless and inert, but alive with vitality and energy.

The presence of a field will distort this vacuum activity, and the distortion will react back on the field. The total effects of a given particle in the presence of a field(s) will include exchange processes with other particles (e.g., messenger photons), where these other particles interact with the vacuum particles, and where vacuum particles cling to the transmitting and receiving particles. An infinite number of possible interactions will be going on all at once. The interactions between all particles produce webs of increasing complexity that have an infinite powerhouse of energy. They eventually lead to plasma and plasma filaments that produce electric and magnetic fields, which have long-range force, creating more and more distant effects.[2,61,93,121-123]

When endeavoring to understand the workings of the cosmos, it is necessary to take into account vacuum polarization of quantum particles and fields, particularly a field with a graduated series of steps (*i.e., a scalar field*).[8,41,111,138] Especially under such conditions, dynamical symmetry breaking, important for unified theory, can occur in a vacuum (*de Sitter-invariant*).[49,78,79] In addition, vacuum polarization effects produce new physics on electroweak processes that could lead to grand unification.[6,138] An important problem in quantum field theory is the contours, or topological excitations, of a given system, which has lead to the idea that the effects are due to nonlocal vortex fields.[102] What this all means is that a system of fields need only exist to begin the process of bringing about the Universe.

Some observations of the system of fields is obscured by shifts in the Doppler Effect due to the helical contouring of space-time by the Field. As you may recall, the Doppler Effect involves the assumption that the frequency of radiation observed will be higher than that of the source—blueshift—as the source and observer approach each other, and will be lower—redshift—as they move farther apart. However, these shifts in frequency do not always occur as expected. In some dispersion media certain unusual effects can occur that resemble the inverse of the Doppler Effect. In such media, a receding source may produce a blueshift and approaching source may produce a redshift. The Doppler phenomenon has an inverse relationship that is dependent on the range and polarization of the Field.

Chirality is a term used to describe the turning or twisting—the left or right handedness—of an object or field, such as a spiral, helix or vortex. Electromagnetic wave dispersal in optically active or chiral media separates a single (*monochromatic plane*) wave into two waves with two shifted frequencies (i.e., two Doppler shifts). Therefore, when motion is involved—either a moving source or a moving observer—a very narrow band of wavelengths (*monochromatic single plane wave*) in one frame can be observed as two waves with unequal frequencies and different directions. The observed frequency can become zero when the observer's velocity equals one of the velocities in the direction of the observer's motion.[44,45] Basically, Doppler shifts cannot be assumed to display velocity and/or distance relationships, because they could be indicative of a field or system of fields that produce different Doppler shifts instead.

Doppler-like frequency shifts are generated by dynamic scattering. Furthermore, the shifts may be arbitrarily large. Scatterers can imitate the Doppler Effect even though the source, the scattering medium, and the observer are all at rest relative to each other.[69] Here is the reason the Field and field system have gone unobserved; nobody ever considered this effect in astronomy, cosmology or many other disciplines in physics (except optical physics).

This explains why there are discordant redshifts in some astrophysical objects, because chiral media can involve a helical field, as occurs within active galactic nuclei, including quasars. The phenomena may also be producing the observed redshift at the solar limb, which is larger than predictions based on relativity. Significant redshifts take place before and after an eclipse. Superluminals could be due to a field with chirality that produces the illusion of faster than light velocities. Discordant redshifts in astrophysical objects were discussed in Chapter 6, Volume Two. Such an effect has not even been considered with respect to the theory of relativity nor Newtonian physics, but it is manifested in chiral forms, the most common form in Nature, particularly life forms and the Field.

Magnetic and Electric Fields Emerge

As the polarization of the vacuum produces particles, which leads to the building of plasma, other effects come into existence. The plasma will flow along field lines, and produce electric and magnetic fields reenforcing and strengthening the field system. Plasma will also begin to collapse onto itself in the pinching effect. Masses of plasma will then begin to crash into each other under their self-gravity, and the effects of the electric and magnetic fields, especially time-varying fields (see Chapter 10, Volume Two for discussion).

Tiny electromagnetic vortices penetrate plasmas, and carry electric currents. These manifestations are produced by the pinch effect. An electric current flowing through a plasma produces a cylindrical magnetic field that attracts other currents flowing in the same direction. These tiny current threads would then pinch together drawing the plasma with them leading to vortices, and eventually more massive objects. Unlike other theories offered today, that lack observational support, space is alive with a network of electric and magnetic fields filled with plasma, and laboratory experiments confirm their organizing potential.[94,95]

Any plasma with sufficient energy will create vortex filaments, which will grow without limit, as space and time allow. Filaments will then grow until they become self-gravitating. The key (*invariant*) is the velocity of matter in a magnetic field.[96] Furthermore, a structural contouring of space-time, resulting from a dynamic mechanism, such as the Field, can unify the quantum level with gravitation (*in analogy to the Casmir effect*).[86]

Using the largest pulsed generator in the world (Blackjack V), plasma filaments were produced in the laboratory. Initially the plasma filaments moved toward each other, attracted by one another's magnetic fields. Then they merged into a tight helix or vortex, and from this form the most intense X-rays radiated.[16,121-123] This pulsed generator is a time-varying accelerator as is the field system and the Fields of the Field-dynamical Model. The primordial field system can be noted in the cellular structure of the Universe, as well as gamma-ray bursters, and the effects of long-range forces discussed in Tome Four (Chapters 6, 7 and 11, and sections 9.6 and 10.4).

Large-scale, primordial magnetic fields are consistent with observations, primordial nucleosynthesis, and the formation of galaxies. In contrast, studies of galaxy clustering are placing a great deal of pressure on the standard gravitational instability picture of galaxy and cluster formation, particularly on large scales. Moreover, substantial evidence proves such fields do actually exist.[29]

Structure in the Cosmos

The Theory of General Relativity says that energy, like mass, curves space. That is, the gigantic energy density of the quantum vacuum should curve space. However, it must be guided, because if it were not, it would curve space into a sphere a few kilometers across, and the Universe as we know it would not exist.

Quantum mechanics and special relativity, likewise, have been experimentally shown to contradict each other. Hence, either or both are limited in some presently unseen way. That unseen aspect is a system of fields that curve space-time, and polarizes the vacuum.

Electrons trapped in magnetic fields produce microwaves, and the Universe shows a very uniform (homogenous) microwave background. Meanwhile, any object that emits radiation can also absorb it.[92-96,121-123] The direction of radiation absorbed, and the direction of its re-emission in a field system would be different, and thereby, scatter the microwaves to a fairly smooth, uniform background. This is especially evident in the fact that the objects continually formed would have their own field system, again scattering the microwaves. Radio brightness fades rapidly with distance. This means that the cosmic background must be generated locally by a medium that both absorbs and emits radiation.[93-95] These observations of microwaves result from the primordial field system producing objects with Field-dynamical Model structure, all of which scatter and absorb microwaves.

Because radio emissions of distant objects would be absorbed by the closer objects, distant objects should appear to have less radio emission. Galaxies, for example, are fainter and fainter radio emitters the farther they are from the Earth.[93] Something has to be absorbing the radio emission, including the microwaves, as they travel towards Earth. Furthermore, this is powerful observational evidence that the conventional explanation for the microwave background is wrong (i.e., the Big Bang scenario is flawed; see section 11.2, Volume Two for discussion).[101]

Another conformation of this effect is the fact that about 24% of the Universe is helium. Existing stars cannot have produced that much helium. At the rate that they currently produce helium it could only account for about 2%. This is direct evidence that other celestial objects and fields are generating helium, a by-product of hydrogen fusion resulting from the Field-dynamical Model.

A primordial system of fields and the resultant objects with Field-dynamical Model structure are responsible for these observations. Other cosmological models rarely, if ever, address such questions as the microwave background and the helium abundance.[172] Yet, here is a natural explanation for both of these observations.

Churning Up Creation
(*Coriolis and Centrifugal Forces in Astrophysics*)

Periodic motion or oscillation is perhaps the most widespread example of order in physics. It lies at the heart of quantum motion, electromagnetic waves carrying heat and light across the Universe, and planets, stars, galaxies and superclusters, even the Universe itself, all involve objects in periodic or time-varying motion. Furthermore, each and everyone of these objects rotate, and therefore, involve the Coriolis and centrifugal forces.

Theories on the origin of the Solar System began more than 350 years ago. Yet, no theory has been quantitatively successful in explaining its origin. As a result, "this is one of the oldest unsolved problems in modern science."[20] Plasma theory is the only concept that comes close, but it is incomplete by itself (i.e., without the existence of fields).

Dipole Fields

As the field system and the generated forces brought about more and more plasma mass, other events began to take place. After plasma collapsed under the pinching effect, and the masses of plasma gravitationally collided, they began to spin or rotate as would a rigid body.[16] As they rotated, a dipole field was generated by what is called the Coriolis force.

This induced more mass because the Coriolis coupling and centrifugal coupling contribute to (*vibrational-state*) mixing and balance, developing a dynamic core.[91] The Coriolis coupling plays an important role in the dynamical behavior in electronically excited states of (*polyatomic*) molecules.[111,112,137] Accordingly, it is active at a very basic level, and so it is referred to as a "universal coupling."[172] All of these effects are intensified by the time-varying forces produced in the interaction (*i.e., Faraday's Law applies*).[91,112]

Much of these forces' effects can be seen on Earth in a very obvious way. If one is situated at the equator, one travels at about 24,000 miles in 24 hours (1,000 mph). As one moves toward the pole less and less distance is traveled in 24 hours, while at the pole the velocity is zero. Also, the velocity is less and less as one approaches the center of the Earth with zero velocity at the center. Furthermore, the more atoms that are arranged in a certain direction the stronger their resultant magnetic field, thereby making these arrangements magnetic, and guiding mineral arrangements.[50] Here we see, in a very plain way that the Earth, as well as other celestial objects, must have a dynamic core and a dipole field with a series of graduated steps (*i.e., a scalar field*).

The role of centrifugal and Coriolis forces in producing the dipole magnetic field, and the confinement chamber of the core is evident in these facts alone, though there is more. The centrifugal force (due to angular momentum) would pull mass away from the spin axis bringing into existence the dipole field (*initially a toroidal field*). Likewise, mass would be pulled away from the center creating a lower mass core. This is also true of gravity's effect. Furthermore, because the velocity is greatest along the equator, it would bring about the equatorial bulge. The bulge has always existed perpendicular to the axis, even when polar wander had taken place, and it cannot be considered a hold over from a faster spin in the past. This set of structuring forces all began with the planetary nebula, and brought about the Field-dynamical Earth Model, as well as other celestial Field-dynamical Models (e.g., the central bulge and spiral arms of galaxies).

On other scales this same effect brought about other objects, such as galaxies and their magnetic fields.[16,49,57,78, 79,175] However, the early Universe is an unlikely place for the generation of strong magnetic fields without some initial mechanism (*e.g., equilibrium parity-violating currents, Coriolis forces, etc. are needed*).[165,175] There had to have been a system of fields.

Consider as an example the formation of a ring around stars. Its formation is the result of the ring being pushed away from the star. This causes the gas inside the ring to rotate slower than the surrounding gas. The difference produces a Coriolis force on the ring in the direction toward the rotational axis. The ring then attains an equilibrium position in which the magnetic buoyancy and tension, and Coriolis and drag forces are brought into balance.[164] This balance brings into existence the dipole field, as well as the mid-latitude fields. Star formation is an example of how other celestial objects form (see Chapter 10, Volume Two).

Mid-Latitude Fields

As the nebula spins and becomes more massive, other effects contribute to the dynamics. Vortex structures develop due to the Coriolis force, particularly if the nebula is elliptical.[33] Particles above the nebula's surface not only experience a downward vertical motion due to gravity, but also experience a horizontal deflection due to the Coriolis force. Two forces act on the particle equally, gravity and the Coriolis force.[38]

When an object is moving relative to the surface of a spinning object, such as a nebula, it experiences an acceleration to the right in the Northern Hemisphere, and to the left in the Southern Hemisphere, generating vortices. Two components of velocity form, one in the plane perpendicular to the axis of rotation (ring or disk), and a second parallel to the axis (dipole), bringing about vortices along mid-latitudes (mid-latitude Fields) and ring formation. These motions give rise to forces of acceleration that have radial components giving an object a tendency to lift off or press against the surface (i.e., acceleration towards and away from the core).[30,41,71,83,114,119,128,141,158,161]

As a massive particle races in, it interacts with the nebula via the Coriolis force in such a way that it produces a gravitational lift. In general relativity, the gravitational, centrifugal and Coriolis forces are put on an equal level, making the latter two fields in every way equivalent to the gravitational field. This is known as the Einstein Equivalence Principle (EEP).

EEP is the foundation of all gravitational theories that describe gravity as a manifestation of curved space-time, including general relativity.[172] An electromagnetic field is produced, and its source is the motion of the massive particles.[28,42,115,144,148] This force can be noted in the experiments done with the pendulum, and the other observations of anomalous gravity discussed in earlier (see Chapter 29 of Volume One, and Chapters 15, 16 and 17 of Volume Two).

The Coriolis force acting on a particle, and the magnetic force on a charge are both proportional to, and act perpendicular to, the velocity. As a result, the action of the Coriolis force on rotating mass and the action of the magnetic force on rotating charge are formally identical.[115] The whole interconnected series of events can be noted in the fact that vacuum polarization is known to have an influence on gravity.[145] Matter does not couple with these fields, but rather they mediate the manner in which matter and the Fields generate gravity (*and produce the metric*). Furthermore, these influences act back on matter in the way prescribed by EEP. In terms of the Field-dynamical Model, this brings into existence a dipole field, a disk- or ring-like phenomenon, and the mid-latitude fields (other discussions can be found in Chapter 10 of Volume Two).

Newton's Folly
(*Anomalous Gravity*)

This is why gravity measurements have revealed fluctuations in the so-called gravitational constant. Theories on the formation of celestial objects have not included the centrifugal and Coriolis forces until recently, and then separately and incompletely. As a result, the generation of the Field-dynamical Model that forms from the interactions has not been recognized.

Not Very Attractive

The homogenous Universe is *assumed* to be dominated by gravity, according to the present interpretation of general relativity. The Universe must then contract to, or expand from, a single point, known as a singularity. However, recent studies of the cosmological constant (*lambda*) have shown that there is an anti-gravity force pervading the Universe, and suggest that the expansion of the Universe is speeding up, rather than slowing down. Moreover, observations show that the Universe is arranged in an ordered cellular structure, and therefore, would not warp all of space. In fact, the evidence would seem to say that gravity conforms to, and is generated within, a curved space, not that gravitational mass is solely responsible for warped space (which seems to be a pervasive mind-set at this time).

The formation of the Universe by gravity alone would, in fact, take longer than the age of the Universe. Also, in this framework, the Universe we do see would only be one percent of the total mass, the other 99% would have to be invisible, exotic and unobserved dark matter (see section 5.8, Volume Two for discussion).[37] Present interpretations of general relativity have the events backwards in most cases; in the early Universe it was not gravity that warped space, but warped space that led to mass and gravity.

Consider the reason for this theoretical inversion in a metaphor. Imagine that there is a curved plane that is shaped in the form of a flattened funnel (something like a satellite dish) with a hole at the center. This represents the curvature of space-time by the Field. Magnetized ball bearings (plasma) are then released (accelerated) toward the center, and ends up in the central hole as a single mass of balls stuck together (mass and gravity emerge). When this final state is observed, it is easy to mistakenly *assume* that the aggregate mass curved space-time, but not consider the opposite (i.e., that curved space-time generated mass), which seems to be the present theoretical perspective.

Another indication of curved space-time leading to mass is that gravity is the weakest force. The gravitational attraction between two electrons is 10^{42} weaker than their electrical repulsion. However, gravity does have an infinite range and is always attractive. Electromagnetism is much stronger and also has infinite range, but is both attractive and repulsive. Both the weak and strong forces have extremely short ranges, which are about the diameter of a atomic nucleus. The weak force is about 100 million times weaker than electromagnetism, while the strong force is 1,000 times stronger than electromagnetism. The strong force is both attractive and repulsive, while the weak force is neither, and is mainly involved in the decay of atomic nuclei. Even at this very fundamental level one could argue that today's observations of gravity are secondary, after the Universe's formation as the result of a primordial-field system.

The direct application of Newtonian dynamics to astrophysical objects on the scale of galaxies and clusters produces a paradox.[29] The conventional reply to this problem is that there are large quantities of dark matter. However, some have cast blame for the problem on the Newtonian theory itself.[11]

Many observations of anomalous and non-Newtonian gravity exist (for example, see Chapter 29 of Volume One, and Chapter 15, 16 and 17 of Volume Two). A non-Newtonian component of gravity can give rise to a unification theory. Strong evidence, which is much more than circumstantial, indicates the existence of a possible non-Newtonian component.

There appears to be a short-range force that is directed at a substantial angle to normal gravity. This anomalous vertical gradient is observed with depth in the sea, and in boreholes and mines.[62,63,65,84,156,157,174] Interestingly, and in accord with the discussion here, the apparent anomalies have been blamed on inadequate Earth models, and can be resolved by considering a vacuum polarization influence on gravity.[84,145] A group of scientists conclude: "Geophysical measurements of the gravitational constant and a recent reexamination of experiments performed early in the century on the equivalence principle and gravitational mass may have provided either observational clues to the unification of gravity with electroweak and strong nuclear forces or an indication of a new, previously unrecognized force."[156]

This non-Newtonian gravitational potential can be explained by a structured "twisting" (*scalar-tensor model with torsion*) in the weak field and an intermediate-range force that is repulsive.[48,65,174] The results support the existence of an intermediate-range force coupling to particles, and the strong and electromagnetic interactions (*baryon number or hypercharge*).[48] It is for this reason that we find non-gravitational forces in the Earth-Moon system noted in accelerations of the Earth's spin and the Moon's angular velocity.[110] Gravity is known to change during eclipses, indicating the existence of a new field and/or unification due to vacuum polarization (i.e., the Field).[3-5,8,53,145,156] It is probably for this reason that an interaction between gravitation and electromagnetic fields exists.[46] In addition, the departure from the gravitational law (*Inverse Square Law*) is particularly noticeable parallel to the Earth's spin axis, and it is time-varying.[7]

Two tests of general relativity are the motion of bodies within the Solar System, and direct terrestrial searches for composition-dependent forces.[23] Hence, this evidence indicates that a fundamental aspect of theory has been overlooked. The Field-dynamical Model is clearly apparent in the observations of gravity alone, and lend to a reinterpretation of general relativity where curvature precedes gravitation (at least on some scales).

Another test of general relativity is the rate of the period decrease in binary pulsars. The orbital precession in binary-star systems is consistent with a very weak, long-range force. Observations are compatible with a repulsive force (e.g., the electromagnetic effects of the Field) that couples to the electrically neutral bulk matter through a (*linear*) combination of neutron and proton number with a fraction of the strength (10^{-5} to 10^{-4}) of gravity (and a range of 3-6 x 10^6 km).[23]

The deep relationship between the weak interactions and the electromagnetic interactions involves the difference in apparent strengths due to the large mass of the particle exchanged in weak interactions. Weak and electromagnetic interactions can be described by a unified (*guage-symmetric*) field theory so that the strong force could be brought into the picture. Effective interaction depends on the energies and/or distances involved. Scientists comment on this: "Perhaps the intrinsic interaction strength of the strong interactions is really of the same order of magnitude as that of the weak and electromagnetic interactions and only appears stronger because our present experiments happen to be carried out at relatively low energies and large distances."[166]

In general relativity, orbital angular momentum gives rise to a magnetic-like gravitational field in which particles are dragged along the rotational motion of the source. Because mass has gravitational effects, there should also be another spin-induced gravitational interaction. Theoretically (*i.e., the Einstein-Cartan theory*) there is a coupling of spin to the "twisting," known as torsion, and energy to curvature. This allows torsion to have indirect distant gravitational effects, although torsion itself cannot propagate.[149] Again, the facts suggest that a field system curved space-time leading to the Universe (at least on a large scale, and then on a finer scale due to long-range forces and EEP), and the merging of the centrifugal and Coriolis forces.

Forces Coupling

Gravity is warped space-time, a distorted emptiness. The other three forces can be represented by fields of force extending throughout space-time. Meanwhile, empty space can become electrically polarized in the presence of an electric field, leading to warped space-time and gravity.[34]

The sources of all the various fields are screened by vacuum polarization. The electromagnetic force gets stronger at short range, while the strong force gets weaker, and so there is a tendency for the two to merge. The weak force also gets weaker closer in. At ultra-high energy or minute distances the electromagnetic, weak and strong forces merge into a single force. These very same conditions existed when the system of fields polarized the vacuum, after which the time-varying aspects accelerated the particles to ultra-high energy causing collisions (minute distances). Hence, unification is a product of curvature and polarization.

Electrons can tunnel through an insurmountable barrier, and can disappear and re-materialize. The electron behaves as both a wave and a particle. These are properties shared by all microscopic particles and occur at relatively low energy. At high energy there can be the abrupt appearance of a particle.

Electrons and neutrinos emanate from unstable nuclei due to the weak force, but irrefutable evidence indicates neither existed inside the nucleus. It has been proposed that they did not exist prior to ejection, but that they were instantaneously created in some way out of the energy present in the radioactive nucleus. As noted, planets and galaxies eject objects, and here we see that this is also true at the atomic level.

The weak force does not produce push or pull, except in supernova. It is restricted to the driving transmutations involved in the identity of particles, often propelling them to high speeds. The weak force's activities are limited to extremely restricted region of space confined to individual subatomic particles (10^{-16} *cm of the source*).

Neutrons and protons are subject to the strong force, but electrons are not, nor are neutrinos or photons. Only the heavier particles are affected, and then the strong force acts as a pulling force that holds the nucleus together. The strong force behaves as though it were an amalgam of many forces with differing properties.

Two prominent scientists[******] used non-symmetric connections in their attempts at a grand unified theory of gravitation and electromagnetism. More recently, non-symmetric connections have reappeared in the natural extension of general relativity, known as the Einstein-Cartan theory. With regard to the properties of elementary particles, it is natural to think that their helicities or spins are related to the interconnections. Perhaps, the phenomenon of parity violation can be explained by a geometrical model in a spherical space-time with torsion. Considerations imply a kind of spontaneous symmetry breaking

---

[******]   Einstein and Schrodinger

that could unify the forces. Torsion is a pre-assigned field in space-time, existing in a spin vacuum, which (*together with the metricity condition*) completely pre-determines the connections.[12,87,118,150] Torsion is the twisting motion (helical, chiral, spiral, etc.) produced by the field system's contours. As a result, the Solar System and galaxies should "feel" the "contact" forces, and torsion can be responsible for variations in gravity.[150]

Supersymmetry is needed for gravity to be renormalizable, and it centers on the concept of spin, like a tiny ball rotating about an axis (i.e., EEP, and the Coriolis and centrifugal forces can apply). The lagrangian function is the expression for kinetic minus the potential energy in the conservation of energy. Nature tends to equalize the mean potential and kinetic energies during motion. For a dynamical system of fields its integral over time is a maximum or a minimum with respect to infinitesimal variations of the fields, providing the initial and final fields are held fixed (i.e., a field system).

Lagrangian points are five locations in space where a small body can maintain a stable orbit despite the gravitational influence of two much more massive bodies, orbiting about a common center of mass. As a result, supersymmetry guarantees that the chiral form of the lagrangian is preserved in higher orders, particularly with super fields. This implies that it could be possible for locally supersymmetric theories to unify the gravitational force with the strong and electroweak forces.[10,40,90,109,163] Hence, the mystery of grand unification is also solved with a system of fields.

Are They Really Coincidences?
(*The Fine Tuning of the Physical Constants and the Anthropic Principle*)

The basic features of galaxies, stars, planets and the everyday world are essentially determined by a few physical constants. Several aspects of our Universe, which seem to be requirements for the existence of life, depend rather delicately on what appear to be "coincidences" among the physical constants. But, are they really coincidences?

Interesting relationships exist between the different scales. For example, the dimensions of a planet are the geometric mean of the size of the Universe and the size of an atom. The mass of a human is the geometric mean of the mass of a planet and the mass of a proton. Relationships such as these might be considered coincidences if one did not acknowledge that they can be deduced from known physical laws.[24] The possibility of life as we know it emerging in the Universe depends on the values of a few basic physical constants that are remarkably sensitive to mathematical values.

Life is in the Cards

The same set of laws give rise to simple crystals, as well as permit systems as complex and intricate as living organisms. It is easy to envision a Universe with objects such as stars and planets. However, to extend the laws to include complex structures, such as polymers and DNA, requires an exceptionally fine tuned set of laws.

Moreover, the evidence indicates that the variety and complexity of the real world is even more extraordinary with a hierarchy of living things. Life permeates all of space, it is built into the very substance of the Universe, and has even brought about its own self-consciousness— we humans. Yet, we have done little, in the scientific realm, to ask one question: Why? And the reason is that most scientists are afraid to admit that the Universe is purposeful and fundamentally biological. A scientist comments on the mind-set involved: "Every advance in fundamental physics seems to uncover yet another facet of *order*. A professional scientist is so immersed in unraveling the laws of Nature that he forgets how remarkable it is that there are laws in the first place. Because science presupposes rational laws, the scientist rarely stops to think about why these laws exist."[34]; p.223

Consider some of these laws and what they mean to life. If electromagnetism did not exist then there would be no atoms, no chemistry, no life, and no heat and light from the Sun. If there were no strong force then nuclei would not have formed, and therefore, nothing would be. Likewise, if the weak force and gravity did not exist, then you would not be reading this, nor would any form of life be here. Yet, these four very different forces (and no others), each vital to all of the complex structures that make up the Universe, are so fine tuned that they all combine to make a single superforce.

The complex structures throughout the Universe are the direct result of the balance or competition of these four forces. Where a tightly interlocking competition occurs, the structure of the system depends delicately on the strengths of these forces (i.e., the fundamental constants). If these constants differed by only a few parts in say a million trillion little, if anything, would survive, particularly life, and especially, human life.

The charges of the proton and electron have been measured in experiments, and have been found to be *precisely* equal and opposite. If this were not so, then the charge imbalance would cause *every* object in the Universe to explode. The Universe would consist of an ether, something like thin gas, and nothing else.

In order for matter to remain electrically neutral the electron and proton must exactly balance in opposite charges. If the charges between the electron and the proton where to differ by as little as one part in 100 billion, everything on Earth would fly apart. Larger objects like the Earth and the Sun would require an even more precise balance of one part in a billion billion. Yet, Nature has adjusted things even more accurate than that. If it had not, the cosmos would have been uninhabitable throughout all space and time. The fine-tuned laws of Nature are directly responsible for our, as well as everything's, existence!

The heat of the Sun keeps the Earth from rapidly cooling down to the near zero temperature of space. It is the source of all food (photosynthesis), and nearly all energy on Earth. Yet, the Sun's existence hangs by a thread.

The Sun is possible only because certain physical properties are present. The neutron outweighs the proton by a fraction of a percent. If it did not, the Sun or any other star like it would not shine for more than a mere few hundred years. The strong force is just strong enough to hold the deuteron together. However, if it were only slightly weaker, the deuteron would not be, and stars like the Sun could not generate nuclear energy. Conversely, had the strong force been slightly stronger the generation of energy would be violently unstable.

These precise properties bring about the Sun, and the temperature of the Sun matches the absorptive properties of chlorophyll. If this were not so, then no chemical reaction capable of utilizing the Sun's energy would exist, making life impossible. All of these conditions balance perfectly, and make life possible.

If gravity's force were to differ from what it is by only one part in 10^{40}, stars like our Sun would not form. If it differed by only one part, there would only be giant stars. Such giant stars would very likely not support life.

If a certain energy level in the oxygen nucleus (how fast the nucleus vibrates) were 0.5% higher than it is, life could not exist. That is, all the carbon produced in a star would burn immediately to oxygen. Thereby, no carbon would be left to make living things.

Likewise, water would not be. Moreover, water has properties that are strikingly unusual and not found in any other liquid. These characteristics make it indispensable to every living creature science calls living.

The properties of water are miraculous. Water is extremely adept at dissolving and transporting substances. It is essential to the ultimate source of all food on Earth, photosynthesis, making it the primary source of oxygen in the atmosphere. Water's ability to retain great amounts of heat while only undergoing slight changes in temperature (i.e., high specific heat) keeps the climate at a level comfortable for life. Its peculiar ability to expand upon freezing is what prevents most of it from permanently freezing solid. Water is the best solvent and has the largest dipole moment of any commonly occurring molecule as well, making it ideal for biological reactions.[54] Yet, had the constants not been so fine tuned, water could not exist.

One constant, omega, defies present physical theory. Omega is the ratio of the average density of the Universe to the force that is needed to gravitationally contain the Universe's (Hubble) "expansion" or growth. However, observations show that as objects get larger their density drops.[39] This particular relation and the observed cosmic density discloses an omega of 0.0002, which contradicts the Big Bang theory that says omega should be 1. If the average density drops as size increases, an average density for the whole Universe cannot be defined. In a Universe with that little matter, gravitation will be so weak that the difference between Newtonian gravity and general relativity would be of little account; yet, the difference is real.

Particle physicists suggest that dark matter must be present in order to reach a satisfying explanation that would create an omega of 1. Heavy neutrinos, axions, and WIMPs (Weakly Interacting Massive Particles) could provide the mass needed. Again, the problem is that there is no real evidence that these particles actually exist. They are simply derived from imagination and mathematical formulas. For the Universe to exist under the tenets of the Big Bang as it is normally envisioned, then omega had to be 1 and dark matter is required.

Such fine tuning is hard to imagine within the context of the early universe. Science generally refers to this enigma as "the problem of pre-arrangement." How could these limits have "known" to adjust themselves so precisely that when everything settled the cosmological constant would turn out to be virtually zero? This is a question science is now asking, but with hesitation and hypothetical theorems. The answer may be what are called wormholes, which are connections between two large and otherwise smooth regions of space-time.[19,117,143,167] The Unified Field is a wormhole, but unlike present assumptions they are not only microscopic (the mathematics even indicate that large visible wormholes could form). The small value of the cosmological constant is telling us that a remarkably precise and totally unexpected relationship exists, and also unknown physics.

Life requires chemical elements heavier than hydrogen and helium—biochemistry requires such heavier elements. Likewise, hydrogen and helium are not capable of forming solids and liquids at physiological temperatures. Stars synthesize these heavier, crucial elements.

For a long time it was known that there is a barrier in the way of these reactions, and no one knew how Nature found its way around that barrier. Finally, it was discovered that these reactions occur because there are two separate resonances between nuclei in red giant stars. Were it not for this double correspondence, the cosmos would consist solely of hydrogen and helium, making life impossible. Likewise, resonances between helium, beryllium and carbon were discovered because we knew we were here and so they had to match, though the original data did not reveal the resonance.

The Earth can be thought of as a niche in the Solar System, a niche in the Milky Way—like a flower finding the right place to sprout its seeds. But, how did life find out that this planet was a proper niche? If this time is right, how did life's organization arise "now"? How did the seeds of life get here, and what are the "seeds" of life?

Energy and matter, the seeds of life, are everywhere. That is how the seeds of life got here. They have been here all along. Contemplating this fantastic Earth one might consider it an immense seed that has bore fruit. These fine tuned constants made it all possible.

The Earth abides in a region of space where the stars are not packed too densely so that there is the danger of stars colliding. Many safe regions such as this can be found throughout space. Why do such niches exist at all? If stars were more densely packed everywhere throughout the Universe, life would not exist.

The characteristic size of galaxies allows for such roominess, had the stars been greater in number the Universe would be uninhabitable. A guiding mechanism, such as a system of fields, and fine tuned constants are essential. Furthermore, the dimensions of the cosmos are very large, had it been smaller the galaxies would overlap and again the Universe would be uninhabitable. Indeed, it is a symbiotic Universe—a Universe of, by and for life.[54] One scientist expresses this:

> "A major revolution in thought is in the offing. Whatever the explanation turns out to be for that massive series of coincidences whereby life arose in the Universe, it is not going to be simple. Each and every one of them flows from the laws of Nature, and it is to these principles themselves that our thinking must turn. Life obeys the laws of physics—this much is a truism. What is new and incomprehensible here is that in some extraordinary way the reverse seems also to be true— that the laws of physics conform themselves to life."[54]; p.27

The Cosmos' Self-Awareness

In order to obtain an age for the Universe we need to work with truly fundamental units of time. Such a unit is called the "jiffy." A jiffy is the time required to traverse a fundamental building block, the proton, at the most fundamental speed, the speed of light. Because light is fast and the proton small, that time works out to be 0.000000000000000000000009 of a second. As a result, the age of the Universe expressed in jiffies turns out to be 66,000,000,000,000,000,000,000,000,000,000,000,000,000 (6.6×10^{40}).

Another cosmic number compares the strengths of two of the fundamental forces of Nature: gravitation and electromagnetism. The number involves two of the most fundamental subatomic particles: the electron and the proton. That cosmic number is electromagnetism divided by the gravitational force between an electron and a proton. The answer is remarkable in the sense that it is almost exactly equal to the first cosmic number, the age of the Universe in jiffies!

Another cosmic number involves the aging of a star. It is the formula for a star's lifetime compared to a jiffy. The answer turns out to be equal to the second cosmic number!

The questions abound when those numbers are considered. What does the ratio of two forces have to do with the Universe's age? What relationship could there be between electromagnetism and gravitation, and the size of the proton, the speed of light and the age of matter? Why does the age of a star equal the lifetime of the Universe? This mystery remains unsolved in contemporary science.

The anthropic principle (AP) is a concept that deals with the consideration that we observe around us not some arbitrary Universe, but one compatible with our existence therein. The improbable fine-tuning of the physical constants turns out to be a necessary precondition for the existence of our orderly world. The strong anthropic principle rules out our existence in an unfavorable Universe. The weak anthropic principle says they are merely the result of our existence as observers. Still, some perplexing questions remain as to why we find ourselves in such a favorable Universe.[1,9]

We now understand some of the deeper, more intricate levels of order in the Universe, but we still do not know what is the need of humans, in the form of four billion participators and observers. And how many other similar creatures are there in the Universe? Still, the key word is "participators"; we are here to participate in making it work its best and observe its magnificence.

Building Stability With Life
(*Entropy and Life*)

One of the most prominent astronomers in 1928 revised all theories about the fate of the Universe. The Second Law of Thermodynamics, he reasoned, forces us to conclude that the Universe began at some finite point in time, and must become more and more disordered and break-down. The Universe must, therefore, move more from matter to energy, leading to the death of the Universe.

The Big Bang initially arose from the ideas embodied in thermody-namics, particularly the Second Law. The Second Law says that the Universe is continually running down, approaching the heat death of equilibrium. Therefore, it was reasoned, the Universe could not have endured forever, and must have begun some time in the past. The Second Law justifies the Big Bang and the Big Bang justifies the Second Law, a rather circular reasoning that really has no place in truly scien-tific thinking.

According to all the laws in physics there is no distinction between the future and the past, no direction to time. Meanwhile, the Second Law says that entropy increases with time, and therefore, the past and future differ. Collectively this means that the Second Law is contra-dicted. In relativity, time is simply the fourth dimension with no differ-ence between the past and present. Also, Newton's laws and the laws of quantum mechanics are time reversible. If the laws of physics have no direction in time, why does the Second Law of Thermodynamics?

The Optimistic Arrow

The problem is that the Second Law of Thermodynamics has been overgeneralized and applied to the Universe as a whole. The Second Law states that disorder can never spontaneously give rise to order. Yet, the Universe is not running down, but is becoming more and more ordered. It is moving toward increasingly complex structures, and faster rates of organization and evolution. Quite simply, the Universe we observe is not decaying or becoming disordered, and therefore, the gen-eralization of the Second Law is not supported by the observations.

Natural processes reuse and recycle energy, creating new energy flows. As a result, there is no limit to organization as long as the process increases the efficiency with which it recycles the energy. Life is particularly adept at doing just that. Life changes the environment in such a way that it increases the captured energy flows. Furthermore, as uncovered in this book, more is alive than we have previously admitted.

The processes we see in Nature must form the basis of our understanding of the Universe. There cannot be a perfect microscopic realm and an imperfect realm of the everyday.[55,126,127] It is a remarkable result—in the conventional sense—of the past few decades of research in non-equilibrium thermodynamics that non-equilibrium or irreversible processes may be a source of order in a system.[32,173] This orderliness is especially true of the interaction of a large system with time-varying fields.[32] In other words, there is an optimistic arrow that points to increasing order, stability and complexity that is particularly focused on life itself.

Demons or Angels?

A physicist over a century ago proposed a concept in thermodynamics that has come to be known as Maxwell's demon. He imagined a microscopic demon that sat at a tiny door between two compartments. One compartment contained gas, while the other was empty. The demon watches each gas molecule that approaches and opens the door only if the molecule is moving fast—that is, hot. As a result, the demon separates the hot molecules from those that are cold, thereby heating one compartment and cooling the other. If this scheme were utilized it would provide an limitless supply of energy.

Scientists have long struggled with Maxwell's demon. They cannot prove the demon does not exist. Yet, if it does, then they must give up on the laws of thermodynamics.

In information theory, the demon can extract energy from a gas by waiting for rare temperature fluctuations in the gas that put the molecules into unusually simple arrangements. A crude, theoretical demon is called a Szilard engine. This engine consists of a box with a piston at each end, and contains only one molecule. The demon creates energy by first noting which side of the box the molecule is on, then drops a partition and moves the piston in the empty half of the box up to the partition. When the screen is removed, the molecule pushes back the piston transferring energy to it.

The energy gain is not canceled out in making the decision if there are a number of engines. The amount of energy that can be gained by many engines turns out to be more than the amount of energy required to store the information. This energy gain is the result of the arrangement of the molecules, which can be described by less than the amount of energy required to store the information. As, for example, when the molecules are all on one side of the box it requires only one bit of information.[25,97] In other words, the demon can overcome the laws of thermodynamics by creating work without a compensating loss of energy.

If we think of the demons as life forms, their ecosystems and the overall biosphere, we begin to see the potential that a life-filled planet has. Life can lower the entropy of its surroundings by an amount equal to the difference between the maximum entropy of responding to the information (i.e., recording it) and its initial entropy, without generating a compensating entropy increase. Thus, mutual information can be used to reduce entropy, particularly if the system is like life (*i.e., ferromagnetism, and the Josephson Junction Effect, which involves biological superconductivity*).[97] Life systems in states of low entropy can get a certain amount of information about other systems without increasing entropy overall, even acquiring the information nonlocally.[15,74,75]

The perception of reality by biological systems is based on different, and in some ways more effective, principles than those used by the more formal procedures of science. As a result, what appears as a random pattern to the scientific method can be a meaningful pattern of reality to living organisms. The existence of this mutually perceived reality makes it possible for organisms to have direct connections between separated objects that are not anywhere nearby.[75]

Evidence demonstrates the existence of direct connections between separated biological objects, while no real physical manifestations of these interconnections actually exist or can be observed. Scientists assert: "From the view of quantum mechanics, these connections appear to be unphysical [and] associated specifically with the activities of living things, in terms of which the interconnections may be very concretely real, and capable of being put to practical use. From the point of view of a biosystem itself, this possibility translates into one that biosystems can have more discriminative knowledge of Nature than is obtainable by quantum measurement."[75]

The motion of particles in a quantum system is governed by an interaction determined by nonlocal phenomena in life systems in a way that is unlike other physical systems. Quantum formalism does not apply in any obvious way to natural situations, such as those involving life, because their actions are not fully governed by scientific design. The common belief that it should be possible to apply quantum mechanics to natural systems just as readily to controlled experiment is one that appears to evolve out of an extrapolation that cannot be justified under closer examination.

The meaning of an information pattern requires concepts that cannot be reduced to physical properties of elementary particles. Only life can derive meaning. Meanwhile, there are two sides or aspects of the act of measurement: an instructive and a destructive side. The instructive (or constructive) side increases one's information. The destructive side produces an uncontrollable change in the values of other, complementary, observables, and it decreases information and produces entropy. Scientists expound on this:

> "These arguments lead us to the conclusion that, because of the different kind of perceptual and interpretative processes characteristic of life compared with those of science, living organisms can possess knowledge that is more detailed in certain aspects than is the knowledge specified by the quantum

theory. One may talk in terms of higher discrimination and selectivity, which improvements can be compared to a process that makes contact with Nature. By way of analogy, it can be compared to a process that makes contact with individual atoms, relative to one that makes contact with the macroscopic aspects of a system only."[75]

Just as the eye developed to perceive objects at a distance, so too can the Earth's life system as a whole react at a distance to perturbations of other habitats, the living Earth as a whole, and those objects reacting with the Earth in the Universe. Within the nervous systems of living things forms of organization of microphysical organization exist capable of selecting, amplifying and analyzing perceptions through nonlocal connections of individual objects (microphysical or otherwise). This is very much like the nervous system itself, which is an interaction of many individual subunits. As a result, life can have perceptions of distant objects and events, just as we can perceive the external environment with our senses. Yet, life can have perceptions of nonlocal events or objects that do not involve sensory input. Such connections are fragile and easily broken by disturbances, such as by those created by humans, and so the actual undisturbed situation has not yet been observed, and shows every potential of bettering and stabilizing the physical world itself (see Chapters 5 and 15 of Volume One other discussions).

The Earth's life system can be considered a many-electron system in a superconducting state, which could bring far more order than we presently give it credit for. Such order could reduce, possibly to near null levels, the Second Law of Thermodynamics' disorder and decay. A state of highly organized and higher order physical reality could exist. This is described: "In the superconducting state of a many-electron system, there is a stable overall organized behavior, in which the movements are coordinated by the quantum potential so that the individual electrons are not scattered by obstacles. One can say indeed that in such a state, the quantum potential brings about a coordinated movement which can be thought of as resembling a 'ballet dance.'"[15]

The superconducting state provides an example of a setting where different organisms can be highly correlated. Such a context is relevant to the origin of life and the organism Earth. This provides a mechanism by which the amount of linking of an individual organism to other systems through nonlocal interconnections is adjustable. Life existed from the beginning as a cooperative whole directly connected at a distance by (*Bell-type*) nonlocal connections. Modifications throughout organisms' life histories (including evolution) allow them to be interconnected directly with each other and with physical objects to the extent of being able to adapt to circumstances. This is in contrast to the "conventional wisdom" that all knowledge can be reduced to quantum mechanical knowledge. Rather, these capabilities allow life to have its own possibilities, beyond the constraints of what the "good scientific method" will allow, for knowing and acting on such knowledge.[15,74,75] In short, a whole, complete and complex life system can perform "miracles" with the Earth and its interconnections with the cosmos: a Cosmic Ballet of "Angels"!

Chirality

Chirality is a term used to describe the direction in which something turns or twists. It is derived from the word Chiro meaning hand. Chirality is the handedness of spiral and helical forms, which are the most common forms in Nature.

Chirality is one of the clearest distinctions between living and non-living things. Processes in Nature are asymmetrical. For example, right-handed or dextral shells dominate both hemispheres on Earth, while left-handed ones are mutations. This is not a universal observation, but a very dominate one. Also conditions may change the situation, such as *Bacillus sutilis* bacteria are usually right-handed spiral colonies, but if temperatures rise the spiral becomes left-handed.[59] Bats come out of caves in a helical pattern. Narwhal, the "horned" whale, always has a horn with a counterclockwise spiral. The human umbilical cord is a triple helix of two veins and one artery. A variety of sugar arrays, amino acids, DNA and organic polymers are among the natural chiral objects. The examples of chirality in life forms is literally endless.

Individual collagen molecules have chirality in the form of a triple helix. Collagen (*three alpha-chains*) is right-handed, hence, fibrils and most tissues, layers and bundles must be handed. Handedness may even alternate from left to right in consecutive layers.

Only living tissues have asymmetric substances composed of just one type of asymmetric molecule, making it optically active. For example, microorganisms change optically inactive material to optically active material. Time-harmonic electromagnetic fields are known to exist in chiral objects. This all means that life organizes and directs electromagnetic energy in much the same way that optic fibers and computers enhance communication, only with life it is done much more efficiently, and even nonlocally, without any observable physical connection.

Helices are perhaps the most common chiral objects found in Nature. They are the basis for the helical structure of many organic molecules: "Indeed, it appears that the helix as well as its two-dimensional counterpart, the spiral, may be considered as the canonical chiral structures. All other manifestations of chirality stem from the helix. Thus, the scattering response of a simple helix assumes a fundamental significance. Additionally, it is important therefore to validate the observation that electromagnetic waves do distinguish between chiral scatters and their mirror images."[59] ; p.6

Achiral objects are those that are an identical mirror image of the chiral forms. Mirror-image molecules are referred to as enantiomers. L-enantiomers are left-handed (L: left or levo) and D-enantiomers are right-handed (D: right or dextro). Enantiomers are present in all molecules essential to life, such as proteins and DNA. Synthetic molecules cannot form the regular helix present in natural enzymes (i.e., proteins).

All amino acids with the same handedness enhance the catalytic activity of proteins, and without uniform chirality the activity would be extremely poor or nonexistent.

This uniformity of handedness is referred to as homochirality. Homochirality in biomolecules must have arisen before life existed. However, this is so difficult to achieve it had to have arisen in the first cell, because no set of conditions would have allowed it otherwise.[59]

Chemical reactions are essentially the result of the electromagnetic interaction of atoms. In terms of the electromagnetic force, the given process should produce a mirror image just as frequently, this is referred to a parity-conservation. Thus, the L- and D-enantiomers should be in equal numbers, but they are not.[59] The question is why?

Chemical reactions or interactions between atoms and light appeared to conserve parity. However, the electroweak force (the electromagnetic and weak forces together) between an atom's electrons, protons and neutrons in the nucleus does not conserve parity. Chiral asymmetry at the subatomic level is fundamentally connected to not conserving parity.

The electroweak force distinguishes between left and right through weak charged currents and weak neutral currents. The strength of these currents—referred to as the W and Z forces—depends on the distance and charge of any elementary particle. The weak W charge is nonzero for left-handed electrons, but zero for right-handed ones. Consequently, the right-handed electrons do not "feel" the W force.

The W force is a fundamental property of the weak force. One result of this asymmetry is nuclear beta decay, which is governed by the W force and produces mostly left-handed electrons. The Z force produces both left- and right-handed electrons, and have Z charges of opposite signs with approximately equal magnitudes. The different signs cause the right-handed electrons to be attracted to the nucleus by the Z force, while the left-handed ones are repelled. This is strictly valid only when electrons are at high energy, such as near the speed of light. The weak force is chirally asymmetric and parity is not conserved.

Another important consequence of the weak Z force between electrons and the nucleus is that all atoms are chiral. When the electron is near the nucleus its direction of motion is partially aligned with its spin axis because of the weak Z force. There is spin orbit coupling. This makes it a right-handed helix—left-handed helices do not exist—and this helical motion has been measured and observed. Through symmetry breaking the weak nuclear force could bring about asymmetry (i.e., it favors L-enantiomers).[17,59] Chiral asymmetry at the level of elementary particles causes chiral asymmetry at the higher level of atoms.

Beta particles produce left-handed electrons far more than right-handed electrons. During beta decay, neutrinos and antineutrinos are produced, and travel at the speed of light. The neutrino and antineutrino are always right-handed. In fact, neutrinos are almost pure nothing, except for the property of spin. No one knows why chiral asymmetry exists at such a fundamental level.[59]

The electron's current density has a helical structure due to the parity-violating weak neutral current. The interaction between the nucleus and the electron in atoms is chiral, and therefore, optically active (see Figure 1). The mirror image (i.e., enantiomers) does not exist in Nature.[60] The nucleus of an atom is like a spinning cone, actually a tetrahedron.[50] This means that the Coriolis and centrifugal forces may apply (EEP) at a very basic atomic and subatomic level, and can also be linked to the Field-dynamical Model. While, the Coriolis and centrifugal forces are very weak at the atomic level, there is a structural arrangement between the nucleus and the electrons so that the electrons miss the "poles" and create a ring-like arrangement (*actually toroidal*).

Until recently the atom was regarded as a system governed only by the electromagnetic force (non-chiral), which conserves parity. The forces at work in the atom are gravity, and the strong (nuclear) and weak forces. Of the three, gravity is negligible on the atomic scale, and the strong force does not affect electrons (whose properties determine the properties of all atomic interactions). It is the weak force between each electron and the atomic nucleus that causes the electron's orbit to perturb. The weak force and the electromagnetic force can be understood as different manifestations of a single underlying force, known as the electroweak force.

The greater the mass of a particle the shorter the range of the force it mediates. Photons, which mediate the electromagnetic force, have zero mass, which makes their range infinite.[17] A photon is a quantum of radiant light energy.

A trend in life systems is towards increasing diversity, which indicates that the rate of energy flow increases through an ecosystem. This may be interpreted as the inherent tendency to return to a near-instantaneous release of energy with the emission of photons. This may be why living things can perceive and communicate with other objects or beings over large distances without there being any known interaction between them.

The electromagnetic field between particles involves a force of electrons and other charged particles in terms of force carriers called photons. Life, in terms of diversity, actually enhances the flow of light through a system (*Fermat's Principle of Least Time*). In contrast, individual species tend to increase the time-delay and tie-up nutrients, making them unavailable for other species (e.g., human society). Light travels at speeds that are relative to diversity (space-time relationships). The greater diversity of life at near equilibrium can reach a state of energy flow without resistance (superconductivity; see Chapters 5, 12 and 15 of Volume One), leading to a relativistic relationship that could transcend some assumptions in, and laws of, physics, such as entropy.

Biological systems can show phenomena not accounted for by purely physical theories (see Tome Two, especially Chapter 25 of Volume One). Yet, we have not fully explored biological systems, because part of the problem when studying life systems physically involves the Uncertainty

Principle. The question arises, what probabilities would exist if the boundaries included the entire biosphere or beyond, producing continuous chiral material. This is especially relevant in undisturbed ecosystems, as plants and animals tend to align with the Earth's magnetic field when undisturbed. This uncertainty is in addition to the confounding effect of our participation in the biosphere and not taking that into account either; i.e., our practice of disturbing ecosystems.

Time-harmonic fields apply not only to chiral objects composed of helices, but also to any lossless, reciprocal, chiral media composed of chiral objects of arbitrary state, such as life systems, including the biosphere. Chirality and its effects on optical activity began to attract attention in electromagnetics with microwave experiments. In the microwave regime, they obtained results that are somewhat similar to those for optical frequencies. In chirowaveguides—material with uniform chirality—there are integrated optical and electric effects.[120] Optical and electrical activity occurs in organic molecules due to its chirality; that is, life acts as a chirowaveguide.[29,159]

Life generates electricity and magnetism, while having properties of superconductivity and chirality making it a unique physical force. Life can respond to energy shifts from nonlocal sources allowing energy to follow in ways that are far, far superior to those at work in supercomputers and optic fibers. Science has noted this is the case, and plans to use life-like processes and materials in electronics, known as nanotechnology.

Balance

Throughout this book we have seen the stability that life brings to the physical world. This stability and balance was witnessed at every level from the subatomic and quantum levels, to the Earth and its interconnections with the Sun, stars, galaxies, clusters, superclusters, and the large-scale structure of the Universe. The physical world is shaped by and for life.

In Tome One, the Earth as an organism was addressed, and later discussions served to support that interpretation (also see *In Defense of Nature—The History Nobody Told You About*). Many life-like homoeostatic mechanisms were discussed that adjust the environment for and by life. These mechanisms allowed for the regulation of ocean salinity, planetary temperature, atmospheric composition, element cycles, electric currents, geomagnetic field strength, the Earth's electrical environment, soil dynamics, nutritious foods, evolution, planetary dynamics (i.e., plate tectonics, etc.), solar-terrestrial linkages, and weather dynamics, among other things. *In Defense of Nature—The History Nobody Told You About* will show our interaction with the living Earth and cosmos in human history and prehistory. At every level, life has brought about stability and reduced entropy, while life itself has been becoming more and more complex, and stable.

This is, of course, in reference to life other than humans who have often destroyed other life, and caused instability and increased entropy. However, we are a conscious species, and therefore, can reverse this trend. If we are to be truly in touch with the realities about living, then we must complete the experiment and work with life. Then and only then will we know how different living can be in terms of a new physical reality. It will be the first time ever that a globally united effort in support of Nature will have been consciously established; not going back to Nature, but toward a unified, holistic and intelligent way of life.

Science and human history have taught us about the potentials of working with Nature. There is every possibility that much of human suffering can be ended. World peace appears to be achievable only with this life-centered holism in mind. The effects reach far beyond the Earth and touch upon the workings of the cosmos. True to the quantum tenet that we are participators, we are at the controls of the entire cosmos, and must see what different actions—intelligently working with Nature—can produce.

A Purposeful Universe
(*The Teleological Cosmos*)

The Living Universe

In many of our modern activities we are on a "collision course" with the living Earth.[21,22,142] Scientists assert: "Taken as a whole, the planet behaves not as an inanimate sphere, but more as a biological superorganism—a planetary body—that adjusts and regulates itself. Geophysicists warn that if the planet does function as a body, the Earth may have the equivalent of vital organs and vulnerable points. Once destroyed, these planetary organs can debilitate the entire system, much as an injury to the spine can cripple one's body up through the neck and down to the toes."[73; pp.1-2]

At the same time we are ignorant of the living Earth. "Our knowledge of the Earth is insufficient; it is impossible to predict how the system will react to the enormous changes. We need a grand interdisciplinary research effort on the interaction of the physical, chemical, biological and cultural factors affecting the dynamism of our planet. Ultimately, a drastic societal reorganization will be required. If somehow the change can be kept under control, new and unprecedented opportunities may emerge for humanity and Nature alike!"[169]

Life, so vital to the stability of the entire planet, is in trouble, because of our misguided decisions.

"Entire ecosystems are now disintegrating as the result of the changes that have been taking place in the post World War II era. The process of evolution has worked towards optimizing conditions and bringing about the regulation of physical phenomena so that such processes in turn will better support life. We, as one species, are now threatening that state and inevitably are threatening our own survival. The dying forests, the spread of deserts, the death of the Baltic seals, the cycles of droughts and floods are all indications of perturbations. Such perturbations may have underlying natural causes but their accentuation to the point of catastrophe is surely our own doing."[21]

One of the major problems hindering our transition to a new and better way of life is the unwillingness of humans to give up on so many material possessions. Can you imagine the effects of another billion cars or air conditioners on the Earth? This is not an issue of choice or a democratic right, but rather an ethical matter.

"We can keep our comforts for some time, and let the world's poor continue to rot; but that would be evil, and would sooner or later become destructive in physical as well as moral terms to us all. Thus, [the living Earth], as a sharpening of an ecological perspective, provides us with two philosophical issues arising out of the destructiveness of the ordinary operations of our modern industrial technology. In the long run (which may not be very long by planetary standards or even by human ones) the disruptive effects of our material culture will be producing vast destructive changes; so that our own status as beings endowed with some superior qualitites, is called into question. Then, even in the short run, the impossibility of extending the current material benefits of our industrial system to all of humanity means that we are the Rich and they are the Poor; and the evil of injustice on a planetary scale is enforced not merely by consciously selfish politics, but by the exigencies of our productive machine."[131]

The real question is "Shall we survive?"

"Natural selection has not equipped us with a long-term sense of self-preservation. Our population cannot continue to expand at its present rate for much longer, and the examples of many other species suggests that expansion can end in catastrophic collapse. Survival beyond the next century in a tolerable state seems most unlikely unless all religions and economies begin to take account of the facts of biology. Scientists must learn to argue their case in aesthetic terms— and indeed in religious terms. Religion doesn't have to be a lot of theological and mystical airy-fairiness. In essence, it is simply an attitude to the Universe; and the proper attitude is that of respect, [love and] reverence."[162]

The discipline of deep ecology is beginning to become a widespread concern. New terms are being employed, like ecoethics and geophysiology.[99-100] Some scientists are pleading for what might be called biogeology, which is a merging of biology and geology at all scales from microbes to plate tectonics.[168] Others are calling to heal the living Earth by changing our perspective of dealing with the symptoms of environmental distress to one that focuses on the underlying causes to eliminate them; a kind of preventive medicine for the living Earth.[43] Contemporary methodologies of science, mathematics and computation are deeply flawed, and therefore, are inadequate for resolving complex problems of a global nature in the necessary holistic manner.[27] As a result, we cannot rely on mathematics and computers to predict the outcome. Rather, it is essential that we establish co-operation and community in a social symbiosis with the living Earth.[72] This would necessarily include what has been referred to as "bio-logical architecture," which involves human dwellings in a symbiotic relationship *with* Nature (i.e., life incorporated into the architectural design).[140]

Meanwhile, the full scope of the matter does not only involve the Earth. "Far from being an inert lump of matter, the Earth behaves as a giant organism [and] the presence of life anywhere in the Universe is a signal that the whole of reality is, in a sense, alive."[136]; pp.3-4 . True to the tenets of quantum mechanics, what we do here on Earth affects the entire *living* cosmos! And let us not forget that living things always have built-in mechanisms for self preservation, so let's not go against ourselves by mismanaging the living cosmos, for in the long run it will stop us if we do.

Physical Laws: Biological, Ethical and Spiritual Laws

The "Life-Giver"

Could it be that suddenly, without anybody's looking for it, we have found evidence of something supernatural at work? The question has to be asked due to the immense significance of the orderliness of the Universe. Why do the laws of physics nurture life so carefully? Why do the laws of physics harmonize themselves with life so completely? Nothing in physics, or science in general, explains why the laws conform so well to life's requirements. A scientist comments: "The more I read, the more I became convinced that such 'coincidences' could hardly have happened by chance."[54]; p.25. When you see evidence of a masterpiece than you begin to understand that there has to be a master at work, and you begin to really understand that master.

These laws had to be laid down at the moment of the creation of the cosmos. And creation takes intelligence. In the fitness for life of the cosmos we are witnessing the effects of a giant symbiosis between the Universe and life. And yet, the Universe brought forth life in order to exist, because it requires being observed to truly exist, and here we are, the billions of us (and whatever other intelligent beings there are in the cosmos), observing it!

"If Nature is so 'clever' it can exploit mechanisms that amaze us with their ingenuity, is that not persuasive evidence for the existence of intelligent design behind the physical universe? If the world's finest minds can unravel only with difficulty the deeper workings of Nature, how could it be supposed that those workings are merely a mindless accident, a product of blind chance?"[34]; pp.235-236

"The very fact that the Universe *is* creative, and that the laws have permitted complex structures to emerge and develop to the point of consciousness—in other words, that the Universe has organized its own self-awareness—is for me powerful evidence that there is 'something going on' behind it all. The impression of design is overwhelming. Science may explain all the processes whereby the Universe evolves its own destiny, but that still leaves room for there to be a meaning behind existence."[35]; p.203

One of the more interesting considerations that come from this orderly Universe is why the Field exists and not some other form of establishing order. The answer may simply be that it is a wormhole in space-time. That is, it serves as a useful test bed for ideas about causality, the role of participation and observation (the quantum theory of measurement), as well as free will.[108] It is a matter of knowing right and wrong by interaction with the cosmos.

Certainly, if there were a supernatural "Life-Giver" then such a condition would need to exist, and the reason is diverse. It would serve in testing what our will generates, and the cause and effect of our action. That is, if we were to be brought back into the world through space-time and experience what we created. After all, the theory of relativity tells us that time is merely another dimension that is traversable in any direction. This is a great deal like many religious teachings, such as "an eye for an eye," "being born again," "justice," "kharma" and so forth. Let us not get the wrong idea, this discussion is not an exercise in religion, but rather, a spiritually sustained scientific understanding.

Consider the reports of those people who have died for a short time only to be revived. They uniformly describe their experience. The accounts relate that they were traveling within a tunnel-like structure at the end of which light is observed.[88] Certainly, this could be a description of the Field, which transcends time, space, matter and energy.

We must fully grasp that the quantum world is uncertain, it is intrinsic and irreducible. The simplistic classical relationship between the whole and its parts is totally inaccurate. Quantum factors force us to perceive particles of matter only in relation to the whole. That is, there is a network of relationships that reach out to other things, and the mind is connected to these objects in a nonlocal way.[75]

The new physics restores mind to a central position in Nature. The quantum theory as it is usually interpreted is meaningless without introducing an observer of some sort. Likewise, the paradigm supporting quantum mechanics insists that we are participators, not just observers.

The mind has been shown to influence random physical systems.[75,130] Scientists state: "It is difficult to avoid the conclusion that under certain circumstances, consciousness interacts with random physical systems."[130] Virtually all of the founders of quantum mechanics had considered this problem in depth.[36,67] At present only certain interpretations of quantum mechanics convincingly explain or predict this conscious control on physical systems.[31,36,66,75,155]

One of the most important contributions to psychology is insight into the workings of the collective unconscious.[77] Many symbols and works of art have continually reappeared in different or the same points in time, and at different locations around the Earth. Historians cannot explain the diffusion of these symbols through intercultural contacts or migrations. Other studies have clearly demonstrated that minds are connected in unseen ways and that single, or collective minds can overcome time, space, matter and energy.[13,15,36,51,68,76,77,80-82,113,125,134,135,146,147, 160,171] To put that in the perspective on an old cliche, it involves restoring the truism: "Mind over matter." Basically, this translates into the fact that what is conscious and unconscious can collectively become a causal force, making the world what it is. Above all else we must make what is ethically and morally, even physically, right known to ourselves and others, otherwise we are lost to our own unconscious self-destructiveness.

Some statements of scientists reflect this very well, such as this one made by a scientist studying the role of consciousness in the physical world: "Industrialized society is presently in severe trouble. It is not necessary to be a doomsayer to recognize this—only to read the signs of the times. As with individuals whose lives are in trouble, the critical spot to look—either for diagnosis or remedy—is at what guides decisions. The way a society makes its choices is key both to its dilemmas and to their possible resolution."[56]; pp.121-123. While this appears to be a complex undertaking, and current psychology would tend to state it so, it is the purpose of this book to simplify that view.

Many will respond to the role of their individual actions with statements about how they do not see what good one person is going to do. Statements like this obviously arise from dependence, as it indicates powerlessness and helplessness. They may even want the present socioeconomic-political structure to do it for them, but the importance of the situation dictates otherwise. "To me this means that it is more important to try to live in harmony with the Earth at a personal level than to allow any of the numerous human collectives and parties to take that responsibility away from us."[98]

Even having this state of mind—the love of living things—itself is important. A scientist examining the role of consciousness in the mind's electromagnetic field states: "Since science is concerned with ultimates, the ultimate consciousness in the electromagnetic field is Love, which is the same as enlightenment espoused by many religious teachings and has been placed in an empirical and experimental scientific framework."[132]; pp.48-49.

It is only by fully welding ourselves to the purpose of love that full human potentials can be realized. The key to developing this is in the understanding of all-encompassing love through definition, thereby creating an interchange between intellect and emotion, between the left and right brains. This is where genius comes from, as in the self-actualized who make great contributions to society through the arts and sciences. Their lives are directed towards bettering the human and natural worlds. That is, facts are gathered in the expression of love, bringing about genius, as well as an improved world.[103-106]

It is the extent of knowledge that alters our behavior by dictating certain choices, and love is the most powerful and intelligent choice. The wakeful mind listens for the truth of the outer world, and then conforms to that world. As it has been stated: "The mind bows down before reality."[116] However, as has been the course throughout history, love has been poorly understood, and "reality" too dictated by collective opinion, pervasive social misperceptions, and actions taken against living things. Ultimate reality rests upon all-encompassing love, and it is this mind-set that unleashes the unseen powers of the mind, love's expression, and literally changes the physical world.

Understanding love can unleash the mind's full potentials, and there is no doubt that the mind is an active force shaping the world around us.[18,36,51,76,77,80,81,135,146,147,160] Each of us, everyday, by our thoughts alone, are subtly influencing our environment, our reality, our universe without consciously knowing it. A scientist indicates that we are not merely passive spectators: "We have to cross out the old word 'observer' and replace it by the new word 'participator.' In some strange sense, the quantum principle tells us that we are dealing with a participatory universe."[170]

Experimental evidence indicates that thoughts directed towards a specific goal (mental activity) can produce measurable changes in the normal operation of external physical devices or apparatus.[36,64,68,74,75,129,130,134,146,147,160] Reality itself is dictated by the observer, and his or her limited view of what it is expected to be. **And this is exactly the problem with today's world: we see what we expect to see—But is it *really* reality?**

This effect of mind is twofold, with collective choice shaping "reality," and collective mind with its electromagnetic energy changing the physical world itself through belief about what exists. Other relationships of time, space, matter and energy are also factors that govern physical law and are controlled by electromagnetic fields. Mind and especially minds joined have power over the physical world, which has been proven experimentally (as well as the fact that the mind functions as an electromagnetic field). Finally, this brings us to the point of how minds joined, people together in common belief and action, can participate in the physical world and reality.

Far greater than was ordinarily assumed, expectation, image and suggestion influence perception and behavior as a result. For example, the outcome of experiments have actually been changed by the expectations of the experimenters. The ability to "know" unconsciously is far more widespread and accurate than is ordinarily accounted for, but we do not learn this fact in schooling nor exercise it. In fact, like many human capabilities, we are encouraged to deny it even exists. "The phenomenon of repression indicates that a person 'knows' how to hide information or distort perceptions that might be disturbing."[56]; p.116. Because of the various aspects of this repression, individuals, whole cultures, and even most people on Earth are not only susceptible to self-deception, but *have and continue to be* engaged in it.

Cultural and individual beliefs about potentials are powerfully limiting influences, and actual potentialities are far greater in many ways than has been ordinarily realized.[56] The mind is not limited by the physical brain, and is also not limited by distance (space), matter, energy or time.[56,76,77,80,81,151-154] Ultimately, the mind is predominant over the physical world.

This is because minds are joined and the perception of separateness of an individual from other persons and the Universe is, as stated in an American Association for the Advancement of Science (A.A.A.S) publication, and elsewhere, an "illusion."[18,36,47,56,68,76,77,80,81,125,146,147] One of the biggest problems is that science insists that ultimate knowledge is only measurable physically, with the perspective that we have nothing to do with what takes place in the physical universe.[14,56] In sum total, this means that we are basically living an illusion in which our most powerfully limiting force is our perception of ourselves in the Universe, and our ability to influence reality.

The mind is capable of transcending physical laws, and so, with proper direction can create a new world, a new universe. Our current understanding of the mind's ability to repair the brain and influence the physical world has been stated: "It is characterized as placing 'mind back over matter,' and as a scheme that idealizes ideas and ideals over physical and chemical interactions, nerve impulse traffic, and DNA"[153]; p.204. All encompassing love is the strongest of all forces that the mind can grasp, and thereby, is the greatest human influence on reality, especially when collectively undertaken by all humans. This is, of course, not a new idea, but what is new is that this has been proven on an empirical, scientific level, especially if love is defined as the nurturing of all life, including the living Earth.

Included in all the "objects" of the physical universe are the mind and minds joined. And as far as the Universe is concerned: "The violation of separability seems to imply that in some sense all these objects constitute an indivisible whole."[36]; p.181. Because the mind is definitely influenced by love—even perfected and enhanced—then so is the Universe, because it is interconnected with the mind. Meanwhile, this holds true even without the mind.

The mind is actually an electromagnetic field, which when combined with others can change time, space, matter and energy.[129,132] This is not only on a minute scale, but is universal as well. Electric and magnetic forces control matter, and the mind's most enhanced electromagnetic functioning is accomplished with love. Furthermore, we interact, both through thoughts and actions, particularly with the quantum and molecular world, which through unifying forces and physical law effects the entire Universe, as well as enhances mental functions—it is a truly symbiotic relationship.[14,56]

As the A.A.A.S. tells us, the implications of recent research makes our choice explicit. In a book titled, *The Role of Consciousness in the Physical World*, a scientist tells us what those plans should include: "The 'higher' unconscious mind, as we have noted earlier, displays astounding problem-solving and creative abilities. If minds are joined, does it not seem plausible that plans and problem solutions appropriate for a group of minds—or society—might be available? And if mind has dominance over the physical, might it not be that resources as needed to carry out these plans would also be available? There must be a willingness to perceive differently. Minds need to be joined in unconditional love and defenseless bonding."[56]; pp.129-130. This statement is the conclusion published by the most world-renowned scientific association, and based on scientific and experimental evidence obtained worldwide.

It is only with unconditional love and defenseless bonding, combined with an understanding of the cosmos, that any of us will find true freedom, self-actualization, and ultimate reality. A scientist examining the power of consciousness comments: "On this view, psychical freedom is not mere absence of constraint by other persons or circumstances, but positive power to create, to add to the definite character of the cosmos, to turn antecedently nebulous causal possibility into definite actuality. It is power to do this, not in appearance only, relative to our ignorance of causal laws and conditions, but in reality. Freedom is not simply absence of complete control by others, but of complete control by anything, even one's own past self plus the surrounding universe."[58]; p.124

Implicit in the foregoing is that our choices must be unselfish, even altruistic. Making choices in accordance with one's own wishes and desires does not make one free from what those choices create.[52,58,139,153] As a scientist indicates: "A person may be relatively free in this view from much that goes on around him, but he is not free from his own inner self."[153]; p.200. True consciousness is without ego boundaries, striving for the care of all living things, and is at one with the entire cosmos.

Ultimate consciousness is, in fact, definable in this way. "It belongs to the very definition of consciousness to be related to that which is not itself."[52]; p.311. "When we speak of consciousness we are referring to the sum total of events in awareness."[58]; p.225. The more we consider and understand, the more conscious we become. Awareness of what constitutes the nurturing of life is supreme love and true consciousness—it is the ultimate in reality itself.

In this book we have witnessed much of the unconscious at work in the world around us. We often express a love for Nature, but in the choices we make, we unconsciously destroy it. We grasp at seeking protection from the Earth by changing it, rather than loving it the way it is. We take the Earth and fashion it into something else, when working with Nature would take care of all our necessities. Or, worse yet, we do not believe in the capabilities that lie there because we do not give ourselves credit for our own capabilities and influences, or some of us even claim it is environmental hysteria to concern ourselves with protecting and nurturing life.

We even overindulge in matter by being overly involved with time and material possessions. When people focus outside themselves, as occurs when growing up under the influence of dependence—fulfilling expectations within the family and society—they begin to externalize for identity. This externalizing brings the focus of the mind onto the physical world by focusing on objects and others' acceptance for identity. Or, in other words, matter is recognized and perceived more often than had the externalizing not existed. The mind is an electromagnetic field, and such a field has power over physical laws. Therefore, the mind's focus on matter, as occurs with externalizing, makes the mind a participator in physical laws, either becoming at one with, or changing them (which we have). This effect of mind is twofold, with collective choice shaping reality, and collective mind with its electromagnetic energy changing matter itself through belief about what exists.

In the laws of physics there is the general theory of relativity, which very basically states that the laws governing the physical world involve the relationship of where an observer is situated in relation to matter itself. For example, time goes by at different rates for a person who is on Earth than for that same person, if he or she were out in space past the "edge" of the cosmos, away from matter (if this were possible). The difference in time involved is drastic, possibly something like a minute passing by in outer space, away from matter, while a million years go by on Earth. Other relationships of time, space, matter and energy are also factors that govern physical law and are controlled by electromagnetic fields, including the collective mind.

Evidence of this in relationship with regard to human problems is supported by a myriad of studies. What we have seen is relativity in relation to the mind's focus on matter. Externalizing, an aspect of dependence along with self-destructiveness, yields decay and disorder (entropy) within human populations and civilization itself. For those whose focus is on physical objects or externalizing (dependence; matter), time goes by more quickly within themselves (decay; entropy). Those civilizations that focus on urbanization and industrialization (matter) eventually lead to the downfall (decay; entropy) of those civilizations (see *In Defense of Nature—The History Nobody Told You About*), and increased rates of disease and new diseases in urban-industrial populations.

Other aspects of dependence disclose this relationship too, such as when types of externalizing are considered. Materialism, an aspect of externalizing, shows that the wealthy who lose their wealth (material objects; matter) often commit suicide, or develop cancer or another disease (entropy). That is, as a result of focusing on material objects for identity, and having those things fail (matter decaying), self-destructiveness occurs (mind causing internal decay). This is again shown in the fact that the lower socioeconomic levels, who are people who fail altogether to obtain those material objects, have shorter life spans; a professor, for example, lives longer than a laborer. Likewise, when property or status is lost, regardless of what socioeconomic levels, it often leads to cancer, heart disease, suicide, depression and other diseases. Civilization, the collective effect of externalizing, shows higher rates for many of the more serious human problems, stress lessens human abilities, historical renewing cycles remove civilization in favor of life (see *In Defense of Nature—The History Nobody Told You About*), and so forth.

When we look at another aspect of this externalizing, consumerism, the collective impact of individual choice, more becomes evident. Consumerism, especially as it stands today, leads to the wholesale exploitation of natural resources. Given enough time, species become extinct, environmental degradation sets in, the physical world becomes imbalanced and other problems are created. Money flows towards those industries (chemical, technological, etc.) that exploit natural resources, making them the focal point of the world economy along with the major lending institutions. Such a focal point leads to the necessity of continually exploiting natural resources, and in order to make profits maximal, waste is not controlled, leading to further environmental degradation. When this gets out of hand, controls need to be employed, lessening profits, thereby causing loans to be defaulted. Eventually the entire world economy hangs in the balance, with war or total exploitation of the natural world as the only ways to revitalize the economy. Again, the focus on matter—the self and self-aggrandizement—leads to decay (entropy)—eventual self-destruction.

Division can be understood as the decay of the whole. This is the basis of nationalism, racism, and so on, as opposed to a unified human race. Such division (of matter; humans) is at the root causes of war, riots, and civil unrest that cause the loss of human life (decay). Likewise, when life systems are destroyed it brings disaster, and thereby, loss of human life.

There's much more, and it all represents a focus on matter as the result of dependence leading to externalizing. Therefore, in relation to relativity and the Second Law of Thermodynamics, time goes by faster, within individuals themselves and collectively, thus decay (entropy) is also speeded up. In contrast, if we eliminate this externalizing focus (focus on love instead), we will be less focused on matter itself, and therefore, experience less aging and loss of life (decay; entropy). What it all comes down to is that mind matters, but matter does not. **No matter where we look, this increased entropy (decay) is evident whenever love and life are *absent*.**

Today we have more knowledge than ever, but still we see what we want to see and disregard the rest. It is an extremely difficult task to separate what we see from what we know, because what we call seeing is invariably colored and shaped by our knowledge or belief of what we see. Collectively, this creates a mess, because knowledge is limited and individuals are usually selfish. In this work (both Volumes One and Two, and *In Defense of Nature—The History Nobody Told You About*) we have seen that the world, bitter, strife-filled and war-torn, is but a product of unloving choice by the masses. An abundance of life, which requires love, will change that.

A hologram is basically a part of a larger picture from which one can reproduce the entire picture. Likewise, if we take a small portion of the Universe, such as the atom, and understand it fully, it should allow us the ability to "reconstruct" the entire cosmos from that knowledge. This is a law of physics, that is within the general theory of relativity, and is often referred to as the "unified principle."

Humans are part of the Universe, and therefore, are a hologram of it. If we understand the human being well enough then we should see that unified principle at work. Since we humans are born as lovers, perfected by love, and imperfect without love, then so is the cosmos. Meanwhile, if we totally disregard humans in our observations, as prior discussions have shown, this unifying principle of love—nurturing life—still holds true. Most importantly, those observations were not gathered with the intention of finding such things, and therefore, indicate facts that go far beyond human thoughts and expectations.

It is because we have not achieved an all-encompassing consciousness that the paradise within our grasp has not yet come to exist. One of the most basic problems of consciousness is understanding how things "fit into the ordered array of Nature."[70]; p.2. In order for this consciousness to take place, there must be adequate information sources about the world, and we must not be engaged in self-deception, not be subject to wishful thinking, and not be biased nor distorted in our perception. Getting in touch with feelings is no guarantee that we will draw the correct conclusion about the kind of life we should lead. We must become deliberately aware of the way we are engaged in the world, and how this awareness can become distorted, grow and diminish.

Emotions serve to motivate and are aroused by the mind's processing of information from the environment or memory. It is true that "subjective values can be treated in principle as causal agents in the objective world"[152]; p.121. And if we are involved in self-deception, our world itself becomes misleading—we *are*, and it *has*.

Like the ancient belief that the Earth was in the center of the Solar System with the stars, planets and Sun revolving around it, we are still human-centered in our thinking. The natural world is perceived of as something to exploit for our "benefit," but our mind-set does not allow us to perceive, nor the extent of our knowledge allow us to know, that much of what we do is really to our detriment. No human culture has attempted to be givers, only takers, and if species become extinct in the process, some would even argue that it is natural selection in the evolution of life, but instead, *it is <u>unn</u>atural selection in the evolution of ignorance!*

It is this absence of nurturing life that the paradise available to us remains to be experienced. At no time in all of human history has humanity sought to work with Nature, and enhance and perfect its functioning. All-encompassing love necessitates this, even self-love necessitates this. "The overall 'image' (of the way things are, should be, can be), which is to a great extent influenced by the individual's history of experience and his ability to schematize, process, or incorporate current stimulation into that image, is a powerful determinant of what makes a given environment 'dull' or 'exciting' for a given individual."[124]; p.189

Industrialized society has ideals that are material, and therefore, anti-life, which has led us to believe self-aggrandizement is exciting. For this reason, some, or even many, will say a world blossoming with life, brought about by our nurturing, would be "dull." However, all we need to contemplate is one simple fact to know that the conclusion is totally unfounded. That fact is simply: *We have never yet given it a chance!*

A noble prize winner for his research on the mind describes what is needed to acquire full capacity of the mind, including our participation in the Universe.

> "Our top social priority today is to effect a change worldwide in man's sense of value. This translates by hierarchic value theory into a change in what is held most sacred. What is needed, more specifically, is a new ethic, ideology, or theology that will make it sacrilegious to deplete natural resources, to pollute the environment, to overpopulate, to erase or degrade other species, or to otherwise destroy or defile the evolving quality of the biosphere. This is exactly what is found from our current approach to the theory and prescription of human values."[154]; p.9

In our understanding of a living Earth, humans take on the most essential position. We can understand the needs of each and every species upon this organism Earth. Like the nucleus of a cell, we are capable of reproducing the components of the cell or another cell (habitat) itself. This objective is something that has *never* been undertaken by *any* human society, so let us leave out the stigmatism of "going back." Instead, it is a time like no other in which we become *positive* participators in the physical world!

As we examined in Tome One, the plant and animal life on Earth perform similar functions to the components of a cell. By increasing life to its greatest abundance and diversity we enhance the cellular and nervous system (electromagnetic) functioning of the organism Earth. In this way, ecosystem (plant, animal and environmental interrelationships) complexity is established, which makes the life system more stable. This results in perfecting the environment, which, in turn, perfects the so-called "evolutionary" destiny of *all* species, because a stable environment makes adaptation stable. More life will generate more electrical current on the Earth's surface, strengthening the Fields, thereby perfecting weather, life's capacity for abundance and health, and our relationship with the cosmos (i.e., solar-terrestrial linkages, solar activity, stability in the relationships between celestial bodies, etc.). Ions become balanced in the air, soil becomes super fertile, plant growth is enhanced, and in fact, all of the physical world becomes perfected and stabilized.

It may sound far-fetched, but the entire physical world becomes so perfected, in fact, that aging, disease, mental illness, violence and even death will be lessened (for example, see sections 14.9, 14.10, 15.9, and Chapters 16, 17,18, 19, 20, 21, and 25 of Volume One, and Chapters12, 14, 15, and 17 of Volume Two, as well as *In Defense of Nature—The History Nobody Told You About*). As the electromagnetic forces are brought to perfection by our efforts of working *with* Nature, entropy will be reduced (see earlier discussion and Chapter 15 of Volume One), not only on this planet, but it will also have its effects throughout the cosmos. That is, strong and integrated electromagnetic forces stabilize matter and have long-range force. A self-stabilizing, maintenance-free environment will be established: *our* **leisure paradise**. Replenishing the Earth will bring stability, and so, there will be an end to the disastrous historical events we have continually brought upon ourselves (see *In Defense of Nature—The History Nobody Told You About*).

Currently, though, we are something akin to a cancer on the face of the Earth, because of our poor self-image leading to externalizing (self-aggrandizement). It has been, and still is, an exercise of our will, though we have chosen negative will in the form of an exaggerated independence (rebellion) away from the natural world. However, it is mind over matter, and love and life over mind, that bears the fruits of the tree of knowledge. This means we must give up our material self-aggrandizement. The violence that is already beginning to strike us is like that of our own body attempting to survive a cancerous disease. The Earth must survive for love and life to continue, for humanity to continue. The major transformation that is at the door is not of some vengeful "god," but our own violence coming back on us. What follows is *not* the end of the human race, but the beginning of Heaven on Earth, if we take our participation seriously and ethically.

Learn of this as it happens, or better yet, before it happens. A paradise awaits us like none ever imagined if we just learn of our profound place in the Universe. Our actions can make this Earth a veritable Heaven, or a "hell," but certainly it is Heaven that awaits. Make yourself at one with it, for Heaven is within. This is not a religious statement, as Heaven is defined as a place or condition of utmost happiness, and one of harmony with all. Hell, on the other hand, sounds much like our present world, as it is defined as a place or state of turmoil, destruction, and misery—of being out of harmony.

To Love and Be Loved in Return

It seems completely fitting that if the Universe is fundamentally biological then nurturing life—love—would create a totally different world. It would be the best way of caring for ourselves and those we love. Tomes One through Five, and particularly *In Defense of Nature— The History Nobody Told You About*, effectively have demonstrated the negative repercussions of *not* nurturing life, as well as the positive impact of caring for life.

This understanding brings us to two principles that are truly all encompassing and will give us an extremely powerful, positive participation in the physical world. They are multifaceted principles, and therefore, need to be understood completely. If we apply these principles in our decision making, we will create a far-sweeping positive influence throughout the world and the cosmos.

I. Respect, Revere and Nurture Life. Never destroy, degrade or defile any life form (other than disease) or life system, but help it prosper and flourish.

In Tome One, we examined the components of habitats and how they resembled the components or organelles of a cell. Humans are the only organism on Earth that is capable of regulating and establishing life systems. Basically, humans can best maintain the Earth-cell in much the same way that the nucleus of a cell sustains the entire cell. We are the most influential in regulating the homoeostatic mechanisms and cycles in the biosphere. In our understanding of a living Earth, humans take on the most essential position. We can understand the needs of each and every species upon this organism Earth.

We should, therefore, work with Nature by understanding what it is cultivating. On Earth the life system or ecosystem Nature works most to establish on land is the forest (see Chapter 17 of Volume One). Forests can also rectify most of the environmental problems facing us today. They generate ozone, trap greenhouse gases, stabilize and enhance the fertility of soils, prevent soil erosion, and can supply our needs by providing food, fuel, fiber and fodder. Contributing greatly to the Earth's electrical environment, forests also bring about stronger and more stable electric and magnetic fields, and in turn, stabilize the geomagnetic field. In addition, these forces, as well as the generation of negative air ions, strengthens and enhances the physiological and mental functioning of plants, animals and humans, and increases the abundance and strength of all life.

Wetlands are the second most abundant land-based life system. However, wetlands are mostly affected by what takes place on adjacent lands, and wetlands are not always completely land-based, but occur along coast lines and rivers. Beaver meadows, swamps, marshes, bogs, fens and riparian (river bank) habitats are some of the more land-based, while estuaries form where a river and the sea meet. Estuaries, swamps and riparian habitats are more species diverse than the other wetlands. Notwithstanding, wetlands are profoundly affected by what is done to surrounding lands, particularly erosion and runoff, which is often a problem when forests are cut down. So even these extremely important life systems are affected by what we do to forests and grasslands. All of the benefits associated with forests are also provided by wetlands.

Moreover, life has the capability of reversing decay and disorder. It may even curb the normal operation of the Second Law of Thermodynamics, entropy (see Chapter 25 of Volume One). In other words, we will have a more secure and balanced world in every respect, if we nurture life.

Because all things in the Universe are interconnected and fundamentally biological, we will be touching upon the entire workings of the cosmos (see Chapter 12 and Tome Five of Volume Two). This is especially true when we consider that the Universe is still evolving and interconnected, and consequently, we will be experiencing an improved situation through time.

Therefore, we should put nothing above respecting, revering and nurturing life. This means that we should end our material self-aggrandizement. The Earth and its biosphere cannot survive a human population that has billions of people exploiting its natural resources. Can we imagine the consequences of a billion more cars or air conditioners alone? This, of course, will be more and more important as populations increase, because the environmental hazards will become more and more devastating.

What this all comes down to is we must enact a new mind-set that assimilates total respect for living things. Wise consumerism is essential, and should always be based on life's preservation and enhancement. Our dwellings should have life incorporated into their structure. That is, our dwellings should be somewhat subterranean, and covered with soil and plant life, and also have plant life inside. Or, we should build tree houses and cave houses.

Some people who are religious may have a hard time putting life as a primary concern. Yet, surely loving God (or by whatever name you may refer to the Life-Giver, Supreme Being or Creator) with all your heart would entail respecting and maintaining what God has established as the primary objective of the Universe.

II. Love Life as You Would Yourself. Because of the interconnections, what you do to other life affects your life and those you love. So, loving life as you would yourself is actually loving your neighbor and yourself.

Numerous examples of this principle are throughout this book. Wild foods are more nutritious, and nutrition is intimately connected to health and longevity (see Chapter 16 of Volume One). Not establishing wilderness systems in favor of more artificial systems has lead to soil erosion, ozone depletion, pollution, acid precipitation, the greenhouse effect, and other forms of environmental instability. These effects end up causing problems for humans.

This was obvious in the role of human participation and life in the causes of diseases and in the maintenance of health (see Chapters 18 and 19 of Volume One). Also, the principle was clear in the role of human participation and life in the causes of war, civil unrest, and "natural" disasters (see Chapter 14 of Volume One, and *In Defense of Nature—The History Nobody Told You About*). The care and maintenance of life is also essential to the stability of the geomagnetic field, electrical environment and the physical environment (see Chapters 15, 20, and 21 of Volume One, and *In Defense of Nature—The History Nobody Told You About*). Due to the solar-lunar-FEM linkages, and the system of fields, stability in the cosmos is also affected (see Chapters 13 through 20 of Volume Two, and *In Defense of Nature—The History Nobody Told You About*).

Most importantly, we must face the fact that life will continue regardless of what we do (see Chapters 22, 23, 24, and 26 through 30 of Volume One, and *In Defense of Nature—The History Nobody Told You About*). We shall be risking the loss of peace, and the high level of our culture with its knowledge and progress. For life is an all-powerful and all-encompassing *LAW*, governing the cosmos.

On a purely physical level alone, humans are the only creature we know of that is on the mean ratio between the atom and the star. Psychically, physically and spiritually, humanity is the only creature to understand the microcosm (the subatomic world) and the macrocosm (the Universe), and their interrelatedness. It effectively follows that through pure and simple will, we can be the most affective and effective link between these two worlds. This is the human place in Nature: *to love and be loved in return*.

SELECTED REFERENCES

Tome Four

1. Acterberg, A., et al. (1982) Spoke Electrodynamics. pp.549-550 in (ed) A. Brahic, Anneaux des Planetes, Planetary Rings. Toulouse, France, Cepadues-Editions.
2. Acuna, M.H., et al. (1981) Topology of Saturn's Main Magnetic Field. Nature 292:721-24.
3. Adams, E.M., Schlesinger, F. (1910) Pwdre Ser. Nature 84:105-106.
4. Adams, F.C. (1990) Eccentric Spiral Modes in Disks Associated with Young Stellar Objects. pp.85-91 in (eds) L.A. Willson, R. Stalio, Angular Momentum and Mass Loss for Hot Stars. Boston, Kluwer.
5. Adelman, B. (1986) Question of Life on Mars. J Brit Interplanet Soc 39(6):256-62.
6. Aglietta, M., et al. (1988) Neutrino Burst from SN1987A Detected in the Mont Blanc LSD Experiment. pp.119-129 in (eds) M. Kafatos, A.G. Michalitsianos. Supernova 1987A in the Large Magellanic Cloud. NY, Cambridge Univ.
7. Ahluwalia, D.V., Sirag, S.P. (1979) Gravitational Magnetism. Nature 278:535-538.
8. Ahluwalia, D.V., Wu, T.Y. (1978) On the Magnetic Field of Cosmological Bodies. Lettere al Nuovo Cimento 23:406-408.
9. Albritton, C.C. (1989) Catastrophic Episodes in Earth History. NY, Chapman & Hall.
10. Alexander, J.K. Desch, M.D. (1984) Voyager Observations of Jovian Millisecond Radio Bursts. J Geophys Res 89:2689-97.
11. Alfven, H. (1981) Cosmic Plasma. Boston, D Reidel.
12. Alfven, H. (1984) Magnetospheric Research and the History of the Solar System. Eos 65:769-770.
13. Alfven, H. (1988) Comments on H. Arp "The Persistent Problem of Spiral Galaxies" pp.213-218 in (eds) F. Bertola, et al. New Ideas in Astronomy. NY, Cambridge Univ.
14. Allaby, M., Lovelock, J. (1984) The Greening of Mars. London, Andre Deutsch.
15. Allen C.C., et al. (1982) Hydrothermally Altered Impact Melt Rock and Breccia: Contributions to the Soil of Mars. J Geophys Res 87B:10083-101.
16. Allen, D.A., Crawford, J.W. (1984) Cloud Structure on the Dark Side of Venus. Nature 307:222-224.
17. Allen, D.A., Wickramasinghe, D.T. (1987) Discovery of Organic Grains in Comet Wilson. Nature 329:615-616.
18. Allen, M., et al. (1980) Titan: Aerosol Photochemistry and Variations Related to the Sunspot Cycle. Astrophys J 242: L125-L128.
19. Allen, M., et al. (1987) Evidence for Methane and Ammonia in the Coma of Comet P/Halley. Astron Astrophys 187:502-512.

20. Aller, H.D., et al. (1987) Evidence for Shocks in Relativistic Jets. pp.273-279 in (eds) J.A. Zensus, T.J. Pearson, Superluminal Radio Sources. NY, Cambridge Univ.

21. Amato, I. (1989) Meteorite May Carry Organic Martian Cargo. Sci News 136:53.

22. Anders, E. (1962) Two Meteorites of Unusually Short Cosmic-Ray Exposure Age. Science 138:431-433.

23. Anders, E. (1973) Organic Compounds in Meteorites. Science 182:781-790.

24. Anderson, I. (1986) Uranus Stays Shrouded in Mystery. New Sci 109(1493):21-22.

25. Anonymous (1846) Sci Amer 2:79.

26. Anonymous (1861) Rpts Brit Assoc Adv Sci:30.

27. Anonymous (1867) Shower of Sulphur. Symons's Mon Meteorol Mag 2:130.

28. Anonymous (1870) Shower of Shell-Fish. Sci Amer 22:386.

29. Anonymous (1872) Black Worms Fall from Sky. Nature 6:356.

30. Anonymous (1877) Snake Rain. Sci Amer 36:86.

31. Anonymous (1879) Extraordinary Phenomena in Fifshire. Symon's Mon Meteorol Mag 14:136.

32. Anonymous (1882) Curious Appearance of the Moon. Sci Amer 46:49.

33. Anonymous (1883) Symon's Mon Meteorol Mag 18:120.

34. Anonymous (1887) [Pilot Chart of the North Atlantic Ocean] Nature 37:187.

35. Anonymous (1893) [Das Wetter] Nature 47:278.

36. Anonymous (1894) Bright Projections on Mars. Observatory 17:295-296.

37. Anonymous (1897) Phenomenal Weather in Victoria. Symons's Mon Meteorol Mag 32:27.

38. Anonymous (1901) Rain of Small Fish. Mon Weather Rev 29:263.

39. Anonymous (1901) Sci Amer 84:179.

40. Anonymous (1908) Brilliant Sky Glows. Nature 78:228.

41. Anonymous (1910) Bright Projection on Saturn. Nature 84:507.

42. Anonymous (1910) Projection on Saturn's Outer Ring. Nature 85:248.

43. Anonymous (1912) Brilliant White Spots on Mars. Nature 89:17.

44. Anonymous (1912) Possible Changes in Saturn's Rings. Nature 88:388-389.

45. Anonymous (1915) [During a Thunderstorm Near Gibraltar] Nature 95:378.

46. Anonymous (1932) Sunspots and Comet Activity. Nature 130:371.

47. Anonymous (1933) White Spot on Saturn. Science 78:6.

48. Anonymous (1963) Why Did Mariner II Find Venus Non-Magnetic? New Sci 17:10.

49. Anonymous (1964) Remarkable Eclipse of the Moon. Sky Telesc 27:142-146.

50. Anonymous (1970) Gold and the Glassy Craters. Nature 226:598.

51. Anonymous (1970) Jupiter's Radiation Mysteries. Sci News 97:577.

52. Anonymous (1973) Lunar Terminator Phenomenon. Sky & Telesc 46:146.

53. Anonymous (1973) Venus Breathes in Steady Fashion. New Sci 58:72.

54. Anonymous (1974) Changing Micrometeoroid Influx. Nature 251:379-380.

55. Anonymous (1974) Mercury's Moon That Wasn't. New Sci 63:602.

56. Anonymous (1974) Mystery of the Hemispheres. Sci New 105:241.

57. Anonymous (1976) Normal Galaxy Emits Jets. New Sci 71:694.

58. Anonymous (1977) Venus Light Confirmed. Astronomy 5:65.

59. Anonymous (1985) Tunguska: Comet or Asteroid? Sky & Telesc 72:577-578.

60. Anonymous (1987) New Evidence Unearthed at Tunguska Site. Sky & Telesc 74:459.

61. Antonello, E. (1990) Rotation, Pulsation and Atmospheric Phenomena in A-Type Stars. pp.97-121 in (eds) L.A. Willson, R. Stalio, Angular Momentum and Mass Loss for Hot Stars. Boston, Kluwer.

62. Antonucci, R.R.J. (1985) Extended Radio Emission Associated with Blazars. pp.98-102 in (ed) J.E. Dyson, Active Galactic Nuclei. GB, Manchester Univ.

63. Armstrong, T.P., et al. (1981) Low-Energy Charged Particle Observations in the 5-20 RJ Region of the Jovian Magnetosphere. J Geophys Res 86A:8343-55.

64. Arp, H. (1971) Observational Paradoxes in Extragalactic Astronomy. Science 174:1189-1200.

65. Arp, H. (1976) Ejection from the Spiral Galaxy NGC 1097. Astrophys J 207:L147-L150.

66. Arp, H. (1987) Quasars, Redshifts and Controversies. Berkeley, Ca, Interstellar Media.

67. Arp, H. (1987) Spiral Arms as Ejection Phenomena. pp.65-70 in (eds) R. Beck, R. Grave, Interstellar Magnetic Fields—Observation and Theory. NY, Springer-Verlag.

68. Arp, H. (1988) Galaxy Redshifts. pp.161-172 in (eds) F. Bertola, et al. New Ideas in Astronomy. NY, Cambridge Univ.

69. Arp, H. (1989) Redshift Controversy—Further Evidence Against the Conventional Viewpoint. Comments Astrophys 13:57-65.

70. Arp, H., Burbidge, G. (1990) Pecular Hydrogen Cloud in the Virgo Cluster and 3C 273. Astrophys J 353:L1-L2.

71. Arp, H., et al. (1986) Future Needs for the Observations of the Virgo Cluster: Panel Discussion. pp.439-470 in (eds) O.-G. Richter, B. Binggeli, The Virgo Cluster of Galaxies. Proc ESO Conf & Workshop No.20. Fed Rep Germany European So Observatory.

72. Arp, H., et al. (1990) Extragalactic Universe: An Alternative View. Nature 346:807-812.

73. Arrhenius, G., Arrhenius, S. (1988) Band Structure of the Solar System: An Objective Test of the Grouping of Planets and Satellites. pp.357-372 in (eds) C.-G. Falthammer, et al. Plasma and the Universe. Boston, Kluwer Acad [and] Astrophys Space Phys 144:357-372.

74. Artyukh, V.S. (1989) Physical Conditions in Active Galactic Nuclei. pp.535-536 in (eds) D.E. Osterbrock, J.S. Miller. Active Galactic Nuclei. Boston, Kluwer.

75. Ashour-Abdalla, M., et al. (1981) Acceleration of Heavy Ions on Auroral Field Lines. Geophys Res Lett 8:795-98.

76. Asseo, E. (1987) Magnetic Fields, Black Holes and Accretion Disks. pp.194-202 (eds) R. Beck, R. Grave, Interstellar Magnetic Fields: Observations and Theory. NY, Spring-Verlag.

77. Astapovich, I.S. (1934) Air Waves Caused by the Fall of the Meteorite of 30 June 1908 in Central Siberia. Q J Roy Meteorol Soc (2).

78. Astapovich, I.S. (1935) New Investigations of the Fall of the Great Siberian Meteorite of 30 June 1908 [In Russian]. Priroda (9):70-72.

79. Astapovich, I.S. (1951) Great Tunguska Meteorite [In Russian]. Priroda (2):23-32.

80. Atreya, S.K. (1984) Modification of Planetary Atmospheres by Material from the Rings. Adv Space Res 4(9):31-40.

81. Atreya, S.K., et al. (1982) Copernicus Measurement of the Jovian Lyman-Alpha Emission and Its Aeronomical Significance. Astrophys J 262:377-87.

82. Atreya, S.K., et al. (1984) Theory, Measurements, and Models of the Upper Atmosphere and Ionosphere of Saturn. pp.239-280 in (eds) T. Gehrels, M.S. Matthews, Saturn. Tucson, Univ Arizona.

83. Avigliano, D.P. (1956) Mars, 1954—Unusual Observations. Strolling Astron 10:26-30.

84. Azar, M., Thompson, W.B. (1988) Magnetic Confinement of Cosmic Clouds. pp.587-614 in (eds) C.-G. Falthammer, et al. Plasma and the Universe. Boston, Kluwer [and] Astrophys Space Phys 144:587-614.

85. Backer, D.C. (1987) Superluminal Expansion in NGC 1275. pp.76-82 in (eds) J.A. Zensus, T.J. Pearson. Superluminal Radio Sources. NY, Cambridge Univ.

86. Backer, D.C., Kulkarni, S.R. (1990) New Class of Pulsars. Phys Today 43(3):26-35.

87. Bagenal, F., Mc Nutt, R.L. (1989) Pluto's Interaction with the Solar Wind. Geophys Res Lett 16:1229-1232.

88. Baggaley, W.J. (1990) Incidence of Equinoctial Es Related to Meteoric Influx. Planet Space Sci 38:145-147.

89. Bajkov, A.D. (1949) Do Fish Fall from the Sky? Science 109:402.

90. Balick, B. (1989) Shapes and Shaping of Planetary Nebulae. pp.83-92 in (ed) S. Torres-Peimbert. Planetary Nebulae. Boston, Kluwer.

91. Barbosa, D.D., et al. (1984) On the Acceleration of Energetic Ions in Jupiter's Magnetosphere. J Geophys Res 89A:3789-3800.

92. Barthel, P.D. (1987) Feeling Uncomfortable. pp.148-154 in (eds) J.A. Zensus, T.J. Pearson, Superluminal Radio Sources. NY, Cambridge Univ.

93. Bassani, L., et al. (1985) Gamma-Rays from Active Galactic Nuclei. pp.252-280 in (ed) J.E. Dyson, Active Galactic Nuclei. GB, Manchester Univ.

94. Basu, D. (1969) Relationship Between the Visibility of Jupiter's Red Spot and Solar Activity. Nature 222:69-70.

95. Bates, J.F. (1981) Origin of Microscopic Planetary Ring Particles. Geophys Res Lett 8:835-836.

96. Baum, R.M. (1956) On the Observed Appearance of a Remarkable Light Spot on the Night Side of Venus. Strolling Astron 10:30-32.

97. Baum, R.M. (1978) Maedler Phenomenon. Strolling Astron 27:118-119.

98. Beasley, W.H., Tinsley, B.A. (1974) Tungus Event Was Not Caused by a Black Hole. Nature 250:555-556.

99. Beck, R. (1987) Interstellar Magnetic Fields: Past, Present and Future. pp.2-7 (eds) R. Beck, R. Grave. Interstellar Magnetic Fields—Observation and Theory. NY, Springer Verlag.

100. Belcher, J.W., et al. (1980) Low Energy Plasma in the Jovian Magnetosphere. Geophys Res Lett 7(1):17-20.

101. Belcher, J.W., et al. (1989) Plasma Observations Near Neptune: Initial Results from Voyager 2. Science 246:1478-1483.

102. Benford, G. (1983) Jets, Magnetic Fields and the Central Engine. pp.271-279 in (eds) A. Ferrari, A.G. Pacholczyk, Astrophysical Jets. Boston, D Riedel.

103. Ben-Menahem, A. (1975) Source Parameters of the Siberian Explosion of June 30, 1908, from Analysis and Synthesis of Seismic Signals at Four Stations. Phys Earth Planet Interiors 11:1-35.

104. Bergrun, N.R. (1986) Ringmakers of Saturn. Edinburgh, Pentland.

105. Bertaux, J.L., et al. (1985) Venus EUV Measurements. Adv Space Res 5(9):119-124.

106. Biemann, K., Lavoie, Jr., J.M. (1979) Some Final Conclusions and Supporting Experiments Related to the Search for Organic Compounds on the Surface of Mars. J Geophys Res 84B:8385-90.

107. Bignami, G.F. (1988) Two Results on PSR 0833-45:(1) Optical Data on its Birthplace, and (2) Possible Polarization in its High Energy Gamma-Ray Emission. Adv Space Res 8(2):681- 689.

108. Binder, A.B. (1977) Fission Origin for the Moon: Accumulating Evidence. Lunar Planet Sci 8:118-120.

109. Binney, J. (1975) Oddballs and Galaxy Formation. Nature 255:275-276.

110. Binzel, R.P. (1989) Pluto-Charon Mutual Events. Geophys Res Lett 16:1205-1208.

111. Biretta, J.A., Cohen, M.H. (1987) Investigations of 3C345. pp.40-47 in (eds) J.A. Zensus, T.J. Pearson, Superluminal Radio Sources. NY, Cambridge Univ.

112. Birgham, F. (1881) Discovery of Organic Remains in Meteoritic Stones. Pop Sci 20:83-87.

113. Birmingham, J.J. (1984) Pitch Angle Diffusion in the Jovian Magnetodisc. J Geophys Res 89:2699-2707.

114. Blamont, J. (1985) Exploration of the Atmosphere of Venus by Balloons. Adv Space Res 5(9):99-106.

115. Blandford, R.D. (1977) Super-Luminal Expansion in Extragalactic Radio Sources. Nature 267:211-213.

116. Blanford, R.D. (1985) Theoretical Models of Active Galactic Nuclei. pp.281-299 in (ed) J.E. Dyson, Active Galactic Nuclei. GB, Manchester Univ.

117. Blandford, R.D. (1987) Grand Unified Models. pp.310-327 in (eds) J.A. Zensus, T.J. Pearson, Superluminal Radio Sources. NY, Cambridge Univ.

118. Blandford, R.D. (1987) Quasar Evolution and the Growth of Black Holes in the Nuclei of Active Galaxies. pp.233-240 in (eds) D.E. Osterbrock, J.S. Miller. Active Galactic Nuclei. Boston, Kluwer.

119. Bland-Hawthorn, J., et al. (1991) Ultramassive (1011 [Solar Masses]) Dark Core in the Luminous Galaxy NGC 6240? Astrophys J 371:L19-L22.

120. Blank, D.L., Soker, N. (1988) Evolutionary Sequence of Seyfert Galaxies. pp.427-429 in (eds) H.R. Miller, P.J. Wiita. Active Galactic Nuclei. NY, Springer-Verlag.

121. Blum, J. (1986) Discovery of a New Jupiter Ring. Sterne Weltraum 25(17):8.

122. Bodenheimer, P. (1989) Stellar Structure and Evolution. pp.689-721 in (ed) R.A. Meyer, Encyclopedia of Astronomy and Astrophysics. NY, Academic.

123. Bogard, D.D., Johnson, P. (1983) Martian Gases in an Antarctic Meteorite? Science 221:651-654.

124. Bolton, S.J., et al. (1989) Correlation Studies Between Solar Wind Parameters and the Decimetric Radio Emission from Jupiter. J Geophys Res 94A:121-128.

125. Bone, D.W. (1934) Lunar Ray. Marine Observer 11:49.

126. Bonsignori-Facondi, S.R. (1983) Rapid Radio-Variability at 408 MHz in SS433. pp.161-164 in (eds) A. Ferrari, A.G. Pacholczyk. astrophysical Jets. Boston, D. Riedel.

127. Borderies, N., et al. (1984) Unsolved Problems in Planetary Ring Dynamics. pp.713-734 in (eds) R. Greenberg, A.Brahic, Planetary Rings. Tucson, Univ Arizona.

128. Botley, C.M. (1976) TLPS and Solar Activity, and Other Phenomena. Brit Astron Assoc J 86:342.

129. Boyden, C.J. (1939) Shower of Frogs. Meteorl Mag 74:184-5.

130. Brace, L.H., et al. (1990) Response of Nightside Ionosphere and Ionotail of Venus to Variations in Solar EUV and Solar Wind Dynamic Pressure. J Geophys Res 95A:4075-84.

131. Brace, L.H., et al. (1983) Ionosphere of Venus: Observations and Their Interpretations. pp.779-840 in (eds) D.M. Hunten, et al., Venus. Tucson, Univ Arizona.

132. Brand, P.W.J.L. (1985) Infra-Red and Optical Photopolarimetry of Blazars. pp.215-219 in (ed) J.E. Dyson, Active Galactic Nuclei. GB, Manchester Univ.

133. Brandt, J.C., Niedner, M.B. Jr. (1987) Plasma Structures in Comets P/Halley and Giacobini-Zinner. Astron Astophys 187:281-286.

134. Branduardi-Raymont, G., Mittaz, J.P.D. (1988) Flux Variability of the Seyfert Galaxy NGC 6814 as Observed with EXOSAT. Adv Space Res 8(2/3):(2)61-(2)64.

135. Brauner, B. (1908) Recent Nocturnal Glows. Nature 78:221.

136. Bridle, A.H., Perley, R.A. (1983) Physical Properties of the Jet in NGC 6251. pp.57-66 in (eds) A. Ferrari, A.G. Pacholczyk. Boston, Riedel.

137. Briggs, M.H. (1961) Organic Constituents of Meteorites. Nature 191:1137.

138. Briggs, M.H. (1962) Meteorites and Planetary Organic Matter. Observatory 82:216-218.

139. Briggs, M.H., Kitto, G.B. (1962) Complex Micro-Structures in the Mokoia Meteorites. Nature 193:1126.

140. Broadfoot, A.L., et al. (1981) Overview of the Voyager Ultraviolet Spectrometry Results Through Jupiter Encounter. J Geophys Res 86A:8259-84.

141. Broadfoot, A.L., et al. (1989) Ultraviolet Spectrometer Observations of Neptune and Triton. Science 246:1459-1466.

142. Bronstein, V.A. (1961) Problem of the Movement of the Tunguska Meteorite in the Atmosphere [In Russian]. Meteoritika (20):72-86.

143. Brooks, E.M. (1964) Why Was Last December's Lunar Eclipse So Dark? Sky & Telesc 27:346-348.

144. Brosche, P. (1980) Mass-Angular Momentum-Diagram of Astronomical Objects. pp.375-382 in P.G. Bergmann, V. De Sabatta (eds) Cosmology and Gravitation: Spin, Torsion, Rotation and Supergravity. NY, Plenum.

145. Browne, I.W.A. (1987) Extended Structure of Superluminal Radio Sources. pp.129-147 in (eds) J.A. Zensus, T.J. Pearson, Superluminal Radio Sources. NY, Cambridge Univ.

146. Browne, P.F. (1987) Magnetic Vortex Tubes and Charge Acceleration. pp.211-221 in (eds) R. Beck, R. Grave. Interstellar Magnetic Fields—Observation and Theory. NY, Springer Verlag..

147. Buie, M.W., Tholen, D.J. (1989) Surface Albedo Distribution of Pluto. Icarus 79:23-37.

148. Burbidge, G. (1971) Was There Really a Big Bang? Nature 233:36-40.

149. Burbidge, G. (1985) Violent Events in Extragalactic Objects. pp.369-372 in (ed) J.E. Dyson. Active Galactic Nuclei. GB, Manchester Univ.

150. Burbidge, G. (1988) Observations of Unusual Objects. pp.101-110 in (eds) F. Bertola, et al. New Ideas in Astronomy. NY, Cambridge Univ.

151. Burbidge, G. (1988) Problems of Cosmogony and Cosmology. pp.223-238 in (eds) F. Bertola, et al. New Ideas in Astronomy. NY, Cambridge Univ..

152. Burbidge, G., Hewitt, A. (1989) Ejection of QSOs from Galaxies. pp.562-564 in (eds) D.E. Osterbrock, J.S. Miller. Active Galactic Nuclei. Boston, Kluwer.

153. Burns, J.A. (1985) New Things About Rings. Bull Amer Astron Soc 17(3):10.3

154. Burns, J.A., et al. (1984) Ethereal Rings of Jupiter and Saturn. pp.200-272 in (eds) R. Greenberg, A. Brahic, Planetary Rings. Tucson, Univ Arizona.

155. Burns, J.O. (1983) Bent Jets and Tailed Radio Galaxies. pp.67-79 in (eds) A. Ferrari, A.G. Pacholczyk. Astrophysical Jets. Boston, Riedel.

156. Burns, J.O., et al. (1987) Radio-Interferometric Imaging of the Subsurface Emissions from the Planet Mercury. Nature 329:224-226.

157. Bursa, M., Sima, Z. (1985) Dynamic and Figure Parameters of Venus and Mars. Adv Space Res 5(8):43-46.

158. Cabrit, S., et al. (1990) Forbidden-Line Emission and Infrared Excesses in T Tauri Stars: Evidence for Accretion Driven Mass Loss? Astrophys J 354:687-700.

159. Caldwell, J., et al. (1986) Infrared Images of Jupiter and Saturn. Bull Amer Astron Soc 18(3):789.

160. Calvani, M., Nobili, L. (1983) Jets from Supercritical Accretion Disks. pp.189-199 in (eds) A. Ferrari, A.G. Pacholczyk, Astrophysical Jets. Boston, Riedel.

161. Cameron, W.S. (1972) Comparative Analyses of Observations of Lunar Transient Phenomena. Icarus 16:339-387.

162. Campbell, P.M. (1969) Spin-Orbit Resonance of the Inner Planets. Science 165:930.

163. Campbell, W.H. (1960) Magnetic Micropulsations Accompanying Meteor Activity. J Geophys Res 65:2241.

164. Campins, H., et al. (1987) Thermal Infrared Imaging of Comet P/Halley. Astron Astrophys 187:601-604.

165. Carbary, J.F., et al. (1982) Spokes in Saturn's Rings: A New Approach. Geophys Res Lett 9:420-422.

166. Carilli, C.L., et al. (1989) Disturbed Neutral Hydrogen in the Galaxy NGC 3067 Pointing to the Quasar 3C232. Nature 338:134-136.

167. Carlqvist, P. (1988) Cosmic Electric Currents and the Generalized Bennett Relation. pp.73-84 in (eds) C.-G. Falthammar, et al. Plasma and the Universe. Boston, Kluwer [and] Astrophys Space Sci 144:73-84.

168. Carswell, R.F. (1985) Emission-Line Regions of Active Galactic Nuclei and Quasars. pp.157-170 in (ed) J.E. Dyson, Active Galactic Nuclei. GB, Manchester Univ.

169. Chakrabarti, S.K. (1985) Some Exactly Solvable Models of Thick Discs and Radio Jets Near the Black Hole. pp.346-350 in (ed) J.E. Dyson, Active Galactic Nuclei. GB, Manchester.

170. Chambers, K.C., et al. (1987) Alignment of Radio and Optical Orientations in High-Redshift Radio Galaxies. Nature 329:604-606.

171. Chapman, S., Ashour, A.A. (1965) Meteor Geomagnetic Effects. Smithsonian Contrib Astrophys 8(7):181-197.

172. Cheng, A.F. (1980) Effects of Io's Volcanoes on the Plasma Torus and Jupiter's Magnetosphere. Astrophys J 242:812-27.

173. Cheng, A.F. (1984) Magnetosphere, Rings, and Moons of Uranus. NASA Conf Pub 2330:541-56.

174. Cheng, A.F., Hill, T.W. (1984) Do the Satellites of Uranus Control its Magnetosphere? NASA Conf Pub 2330:557-558.

175. Chu, Y.-H. (1989) Multiple Shell Planetary Nebulae. pp.105-115 in (ed) S. Torres-Peimbert. Planetary Nebulae. Boston, Kluwer.

176. Ciardullo, R., et al. (1988) Morphology of the Ionized Gas in M31's Buldge. Aston J 95:438-444.

177. Clark B.C., et al. (1982) Chemical Composition of Martian Fines. J Geophys Res 87B:10059-67.

178. Clark B.C., et al. (1987) Systematics of the "CHON" and Other Light-Element Particle Populations in Comet P/Halley. Astron Astrophys 187:779-784.

179. Clarke, J.T., et al. (1981) IUE Detection of Bursts of H Lyman Alpha Emission from Saturn. Nature 290:226-227.

180. Clarke, J.T., et al. (1981) Observations of Polar Aurora on Jupiter. NASA-CP-2171 Universe at Ultraviolet Wavelengths. N81-25902:45-8.

181. Clarke, J.T., et al. (1986) Continued Observations of the Hydrogen Lyman Alpha Emission from Uranus. J Geophys Res 91:8771-81.

182. Clavel, J, et al. (1991) Steps Toward Determination of the Size and Structure of the Broad-Line Region in Active Galactic Nuclei. I. An 8-Month Campaign of Monitoring NGC 5548 with IUE. Astrophys J 366:64-81.

183. Cloutier, P.A., Russell, C.T. (1982) International Conference on Venus. Solar Wind Interaction. Nature 296:20.

184. Clube, S.V.M. (1988) Dark Matter and the 15 Myr Galacto-Terrestrial Cycle. pp.343-346 in (eds) F. Bertola, et al. New Ideas in Astronomy. NY, Cambridge Univ.

185. Cobb, J.C. (1966) Iron Meteorites with Low Cosmic Ray Exposure Ages. Science 151:1524.

186. Cochran, W.D., Barker, E.S. (1979) Variability of Lyman-Alpha Emission from Jupiter. Astrophys J 234:L151-L154 [and] NASA-CR-158756.

187. Connerney, J.E.P. (1986) Magnetic Connection for Saturn's Rings and Atmosphere. Geophys Res Lett 13:773-776.

188. Connerney, J.E.P., et al. (1981) Saturn's Ring Current and Inner Magnetosphere. Nature 292:724-7.

189. Connerney, J.E.P., et al. (1984) Magnetic Field Models. pp.354-377 in (eds) T. Gehrels, M.S. Matthews, Saturn, Tucson, Univ Arizona.

190. Connerney, J.E.P., Ness, N.F. (1988) Mercury's Magnetic Field and Interior. pp.494-513 in (eds) F. Vilas, et al. Mercury. Tucson, Univ Ariz.

191. Conrath, B.J. (1981) Planetary-Scale Wave Structure in the Martian Atmosphere. Icarus 48:246-255.

192. Conrath, B.J., et al. (1981) Thermal Structure and Dynamics of the Jovian Atmosphere. 2. Visible Cloud Features. J Geophys Res 86:8769-8775.

193. Conrath, B.J., et al. (1989) Infrared Observations of the Neptunian System. Science 246:1454-1459.

194. Conway, R.R. (1983) Multiple Flourescent Scattering of Nitrogen Ultraviolet Emissions in the Atmospheres of the Earth and Titan. J Geophys Res 88A:4784-92.

195. Cook II, A.F., et al. (1981) Visible Aurora in Jupiter's Atmosphere? J Geophys Res 86A:8793-6.

196. Corliss, W.R. (comp) (1983) Handbook of Unusual Natural Phenomena. Garden City, NY, Anchor Press/Doubleday.

197. Corliss, W.R. (comp) (1985) Moon and the Planets: A Catalog of Astronomical Anomalies. [and] (1986) The Sun and Solar System Debris. Glen Arm, Md, Sourcebook Project.

198. Cowan, C. (1965) Possible Anti-Matter Content of the Tunguska Meteor of 1908. Nature 206:861-865.

199. Cowan, C., et al. (1965) Possible Antimatter Content of the Tunguska Meteor of 1908. Nature 206:861-865.

200. Cowen, R. (1990) Mapping New Features of the Milky Way's Bulge. Sci News 137:340.

201. Craven, J.D., Frank, L.A. (1987) Atomic Hydrogen Production Rates for Comet P/Halley from Observations with Dynamics Explorer 1. Astron Astrophys 187:351-6.

202. Cravens, T.E., et al. (1984) Evolution of Large-Scale Magnetic Fields in the Ionosphere of Venus. Geophys Res Lett 11:267-270.

203. Cross, F.C. (1947) Hypothetical Meteorites of Sedimentary Origin. Pop Astron 55:96-102.

204. Croswell, K (1986) Pluto: Enigma on the Edge of the Solar System. Astronomy 14:6-10, 12-24.

205. Crowther, J.G. (1931) More About the Great Siberian Meteorite. Sci Amer 144(5):314-17.

206. Cruikshank, D.P. (1963) A Review of Some Apollo Venus Studies. Strolling Astron 17:202-208.

207. Cruikshank, D.P. (1966) Possible Luminescence Effects on Mercury. Nature 209:701.

208. Cugnon, P. (1987) Interstellar Polarization and Magnetic Fields. pp.100-109 in (eds) Interstellar Magnetic Fields—Observation and Theory. NY, Springer Verlag.

209. Cuzzi, J.N., et al. (1984) Saturn's Rings: Properties and Processes. pp.73-199 in (eds) R. Greenberg, A. Brahic, Planetary Rings. Tucson, Univ Arizona.

210. D'Alessio, S.J.D., Harms, A.A. (1989) Nuclear and Aerial Dynamics of the Tunguska Event. Planet Space Sci 37:329-340.

211. Danielson, G.E., Porco, C.C. (1982) Kinematics of Spokes. pp.219-222 in (ed) A. Brahic, Anneaux des Planetes. Planetary Rings. Toulouse, France, Cepadues-Editions.

212. Davies, P.C.W. (1974) How Special is the Universe? Nature 249:208-209.

213. Davies, P.C.W. (1984) Superforce: The Search for a Grand Unified Theory of Nature. London, Heinemann.

214. Davies, P.C.W. (1995) Cosmic Blueprint. London, Penguin.

215. Davies, P.C.W. (1988) Cosmological Horizons and Entropy. Class Quantum Gravity 5:1349-1355.

216. Davies, R.D. (1985) Radio Emission from the Nuclei of Sbc Galaxies. pp.81-86 in (ed) J.E. Dyson, Active Galactic Nuclei. GB, Manchester Univ.

217. De Bernardis, P., et al. (1990) On the Dipole and Quadropole Kinematic Anistropy in the Brightness of the Cosmic Background Radiation. Astrophys J 353:145-148.

218. de Lapparent, V., et al. (1986) A Slice of the Universe. Astrophys J 302:L1-L5.

219. Delsemme. A.H. (1987) Galactic Tides Affect the Oort Cloud: An Observational Confirmation. Astron Astrophys 187:913-8.

220. Denning, W.F. (1903) White Spot on Saturn. Sci Amer 89:79.

221. DePater, I., et al. (1989) Uranus Deep Atmosphere Revealed. Icarus 82:288-313.

222. Dermott, S.F., Gold, T. (1978) On the Origin of the Oort Cloud. Astron J 83:449.

223. Dermott, S.F., Murray, C.D. (1980) Origin of the Eccentricity Gradient and Apse Alignment of the E Ring of Uranus. Icarus 43:338-349.

224. Desch, M.D., et al. (1989) Impulsive Solar Wind-Driven Emission from Uranus. J Geophys Res 94A:5255-5263.

225. Dessler, A.J., Chamberlain, J.W. (1979) Jovian Longitudinal Asymmetry in Io-Related Europa-Related Auroral Hot Spots. Astrophys J 230:974-81.

226. Djorgovski, S., Sosin, C. (1989) Warp of the Galactic Stellar Disk Detected in IRAS Source Counts. Astrophys J 341:L13-L16.

227. Dolgov, Y.A., et al. (1971) Chemical Composition of Silicate Spherules in Peats of the Tunguska Meteorite Fall Region. Doklady Akademii Nauk USSR 200(1):201-4.

228. Dollfus, A. (1984) Saturn Ring Particles from Optical Reflectance Polarimetry—A Review. In (eds) R. Greenberg, A. Brahic, Planetary Rings. Tucson, Ariz, Univ Arizona.

229. Dollfus, A. (1985) Photopolarimetric Sensing of Planetary Surfaces. Adv Space Res 5(8):47-58.

230. Donahue, T.M. (1979) Pioneer Venus Results: An Overview. Science 205:41-44.

231. Drapatz, S., et al. (1987) Search for Methane in Comet P/Halley. Astron Astrophys 187:497-501.

232. Dressler, A. (1987) Large-Scale Streaming of Galaxies. Sci Amer 257(9):46-54.

233. Dressler, A. (1989) Observational Evidence for Supermassive Black Holes. pp.217-232 in (eds) D.E. Osterbrock, J.S. Miller, Active Galactic Nuclei. Boston, Kluwer.

234. Dulk, G.A. (1965) Io-Related Radio Emission from Jupiter. Science 148:1585-1589.

235. Durrance, S.T., Clarke, J.T. (1984) Lyman-Alpha Aurora. NASA Conf Pub 2330:559-72.

236. Dutsch, H.U. (1983) Ozone Variability. Planet Space Sci 31(9):1053-64.

237. Dyal, P., et al. (1970) Apollo 12 Magnetometer: Measurement of a Steady Magnetic Field on the Surface of the Moon. Science 169:762-764.

238. Eades, H.L. (1867) Yellow Rain. Sci Amer 16:233.

239. Eberhart, J. (1985) Towering Dust Devils Discovered on Mars. Sci News 127:197.

240. Eberhart, J. (1985) Two More Sun-Grazing Comets Discovered. Sci News 128:326.

241. Eberhart, J. (1986) Mars-to-Earth Rock-Throwing Method. Sci News 130:246.

242. Eberhart, J. (1986) Voyager 2's Uranus: 'Totally Different'. Sci News 129:72-73.

243. Eberhart, J. (1988) Sun-Grazers: A Hot Road to the End. Sci News 134:39.

244. Eberhart, J. (1989) R.I.P. Solar Max: The Satellite's Last Days. Sci News 136:357.

245. Eberhart, J. (1990) Within a Galaxy and Outside a Supernova. Sci News 138:151.

246. Edmunds, M.G. (1976) Mysterious Meteorites. Nature 263:95-6.

247. Edmunds, M.G. (1978) More Puzzles About the Early Solar System. Nature 273:337-338.

248. Edwards, D.A., Pringle, J.E. (1987) Orbital Eccentricity of Classical Novae. Nature 328:505.

249. Elliot, J., Kerr, R. (1987) Rings: Discoveries from Galileo to Voyager. Cambridge, Mass, MIT Press.

250. Elliot, J.L., Nicholson, P.D. (1984) Rings of Uranus. pp.25-72 in (eds) R. Greenberg, A. Brahic, Planetary Rings. Tucson, Univ Arizona.

251. Ellis, R.W. (1933) Observation of a New Mexico Meteor from the Air. Science 78:58.

252. Elson, L.S. (1982) International Conference on Venus. Dynamics of the Atmosphere. Nature 296:17.

253. Epiktetova, L.E. (1976) New Evidence of the Eyewitnesses of the Tunguska Meteorite Fall. [in Russian] pp.20-34 in Voprosy Meteoritiki. Tomsk Univ.

254. Eplee, Jr., R.E., Smith, B.A. (1982) Dynamics of Spokes in Saturn's B Rings. pp.223-224 in (ed) A. Brahic, Anneaux des Planetes, Planetary Rings. Toulouse, France, Cepadues-Editions.

255. Epstein, E.E. (1966) Mercury: Anomalous Absence from the 3.4-Millimeter Radio Emission of Variation with Phase. Science 151:445-447.

256. Esposito, L.W. (1985) Long Term Changes in Venus Sulfur Dioxide. Adv Space Res 5(9):85-90.

257. Esposito, L.W., et al. (1983) Clouds and Hazes of Venus. pp.484-564 in (eds) D.M. Hunten, et al. Venus. Tucson, Univ Arizona.

258. Esteban, E.P., Ramos, E. (1988) Rotating Black Hole in an External Electromagnetic Field. Phys Rev 38D:2963-2971.

259. Eugster, O. (1989) History of Meteorites from the Moon Collected in Antarctica. Science 245:1197.

260. Evans, D.R., et al. (1981) Impulsive Radio Discharges Near Saturn. Nature 292:716-718.

261. Evans, I.M., Dopita, M.A. (1985) Are Radio-Quiet Active Nuclei the Result of Thermal Winds Driven by Bubbles? pp.323-329 in (ed) J.E. Dyson, Active Galactic Nuclei. GB, Manchester Univ.

262. Eviatar, A. (1984) Plasma in Saturn's Magnetosphere. J Geophys Res 89A:3821-8.

263. Eviatar, A., Barbosa, D.D. (1984) Jovian Magnetosphere Neutral Wind and Auroral Precipitation Flux. J Geophys Res 89A:7393-8.

264. Fabian, A.C. (1985) X-Rays from Active Galactic Nuclei. pp.221-232 in (ed) J.E. Dyson, Active Galactic Nuclei. GB, Manchester Univ.

265. Falle, S.A.E.G., Wilson, M.J. (1985) Internal Shocks in Jets. pp.342-345 in (ed) J.E. Dyson, Active Galactic Nuclei. GB, Manchester Univ.

266. Farrell, W.M., Calvert, W (1989) Source Location and Beaming of Broadband Bursty Radio Emissions from Uranus. J Geophys Res 94A:217-225.

267. Feast, M.W. (1989) Planetary Nebulae and the Galactic Bulge. p.167 in (ed) S. Torres-Peimbert. Planetary Nebulae. Boston, Kluwer.

268. Fechtig, H., et al. (1979) Micrometeoroids Within Ten Earth Radii. Planet Space Sci 27:511-531.

269. Feigelson, E.D. (1983) X-Rays from Jets and Lobes. pp.165-172 in (eds) A. Ferrari, A.G. Pacholczyk, Astrophysical Jets. Boston, Riedel.

270. Feitzinger, J.V., Spicker, J. (1987) Stochastic Star Formation, Magnetic Fields and the Fine Structure of Spiral Arms. pp.171-178 in (eds) R. Beck, R. Grave. Interstellar Magnetic Fields—Observation and Theory. NY, Springer-Verlag.

271. Ferris, J.P., Ishikawa, Y. (1987) HCN and Chromophere Formation on Jupiter. Nature 326:777-778.

272. Fesenkov, V.G. (1949) Atmospheric Turbidity Caused by the Fall of the Tunguska Meteorite. [In Russian]. Meteoritika (23):3-29.

273. Festou, M.C., et al. (1987) Periodicities in the Light Curve of P/Halley and the Rotation of its Nucleus. Astron Astrophys 187:187:575-580.

274. Fielder, G. (1965) Distribution of Craters on the Lunar Surface. Roy Astronom Soc Mon Not 129:351-361.

275. Filliys W., et al. (1980) Trapped Radiation Belts of Saturn: First Look. Science 207:425-431.

276. Fiore, F., et al. (1990) X-ray Spectral Variations in NGC 4151. Mon Not Roy Astron Soc 243:522-528.

277. Fisher, D.E. (1963) Ages of the Sikhote Alin Iron Meteorite. Science 139:752-753.

278. Florensky, K.P. (1965) Preliminary Results from the 1961 Combined Tunguska Meteorite Expedition. Meteoritika 23:3-37.

279. Fomalont, E.B. (1983) Summary of Properties of Radio Jets. pp.37-46 in (eds) A. Ferrari, A.G. Pacholczyk, Astrophysical Jets. Boston, Reidel.

280. Fox, J.L. (1986) Models for Aurora and Airglow Emissions from Other Planetary Atmospheres. Can J Phys 64:1631-1656.

281. Fox, J.L. (1986) Studies of the Aurorally Induced Ultraviolet Emissions on the Nightside of Venus. NASA-CR-177004 [and] NTIS Sci Tech Aerosp Rept 1986, 24(22) Abstr N86-31478.

282. French, R.G., et al. (1986) Structure of the Uranian Rings. II. Ring Orbits and Widths. Icarus 67:134-63.

283. Fricke, K.J., Kollatschny, W. (1989) Relationships of the Active Nucleus, Galaxy and Environment. pp.425-444 in (eds) D.E. Osterbrock, J.S. Miller. Active Galactic Nuclei. Boston, Kluwer.

284. Friend, D.B. (1990) Rotational Evolution of Hot Stars Due to Mass Loss and Magnetic Fields. pp.199-203 in (eds) L.A. Willson, R. Stalio, Angular Momentum and Mass Loss for Hot Stars. Boston, Kluwer.

285. Fujimoto, M. (1987) Bisymmetric Spiral Magnetic Fields in Spiral Galaxies. pp.23-30 in (eds) R. Beck, R. Grave. Interstellar Magnetic Fields—Observations and Theory. NY, Springer-Verlag.

286. Fukui, Y., et al. (1989) Molecular Outflows in Protostellar Evolution. Nature 342:161-163.

287. Furst, E. (1987) Magnetic Fields in Supernova Remnants—Results from Radio Continuum Observations. pp.179-184 in (eds) R. Beck, R. Grave. Interstellar Magnetic Fields — Observations and Theory. NY, Springer-Verlag.

288. Furst, E., et al. (1989) Nature of the Central Source of the Supernova Remanent G179.0 + 2.7. Astron Astrophys 223:66-70.

289. Galeev, A.A. (1987) Encounters with Comets: Discoveries and Puzzles in Cometary Plasma Physics. Astron Astrophys 187:12-20.

290. Gamaleldin, A.I. (1990) Do Elliptical Galaxies Suffer from Warp? Astophys Space Sci 168:89-101.

291. Ganaoathy, R. (1983) Tunguska Explosion of 1908 Discovery of Meteoritic Debris Near the Explosion Site and at the Pole. Science 220:1158-1160.

292. Gaskell, C.M., Keel, W.C. (1988) Another Supernova with a Blue Progenitor. pp.13-15 in (eds) M. Kafatos, A.G. Michalitsianos, Supernova 1987A in the Large Magellanic Cloud. NY, Cambridge Univ.

293. Gathier, R. (1987) Radio Observations of Young Planetary Nebulae. pp.371-383 in (ed) S. Kwok, S.R. Pottasch, Late Stages of Stellar Evolution. Boston, Reidel.

294. Gault, D.E., et al. (1970) Lunar Theory and Processes: Post-Sunset Horizon "Afterglow". Icarus 12:230-232.

295. Gear, W.K., et al. (1985) Infra-Red to Millimetre Continuum Emission of Blazars. pp.152-156 in (ed) J.E. Dyson, Active Galactic Nuclei. GB, Manchester Univ.

296. Geballe, T.R. (1987) Organic Chemicals in Comets. Nature 329:583.

297. Gehrels, N., Stone, E.C. (1983) Energetic Oxygen and Sulfur Ions in the Jovian Magnetosphere and Their Contribution to the Auroral Excitation. J Geophys Res 88A:5537-5550.

298. Geller, M.J., Huchra, J.P. (1989) Mapping the Universe. Science 246:987-903.

299. Genet, R.M., et al. (1987) Supernova 1987A: Astronomy's Explosive Enigma. Mesa, Az, Fairborn.

300. Genova, F. (1987) Auroral Radio Emissions of the Planets. Ann Phys 12(2):57-107.

301. Gibson, Jr., E.K., Moore, G.W. (1973) Volatile-Rich Lunar Soil: Evidence of Possible Cometary Impact. Science 179:69-71.

302. Giddings, N.J. (1946) Lightning-Like Phenomena On the Moon. Science 104:146.

303. Gilman, D.A., et al. (1986) Upper Limit to X-Ray Emission from Saturn. Astrophys J 300:453-55.

304. Giommi, P., et al. (1988) X-ray Time Variability and Luminosity Correlations in BL Lacertae Objects. Adv Space Res 8(2/3):(2)79-(2)83.

305. Giovanelli, R. Haynes, M.P. (1989) Protogalaxy in the Local Supercluster. Astrophys J 346:L5-L7.

306. Giovannini, G. (1985) Properties of the Radio Cores in Elliptical Galaxies. pp.93-97 in (ed) J.E. Dyson, Active Galactic Nuclei. GB, Manchester Univ.

307. Glaisher, J., et al. (1874) Mass of Burning Sulphur Falls. Rpts British Assoc, p.272.

308. Gnedin, Y., et al. (1988) Astrometry, Photometry and Spectropolarimetry of SN1987A. Adv Space Res 8(2/3):(2)691-(2)694.

309. Goddard, A.V. (1932) Unusual Lunar Phenomenon. Pop Astron 40:316-317.

310. Godfrey, D.A., Hunt, G.E. (1985) Periodic Features in the North Pole of Saturn. Bull Amer Astron Soc 17(3):3.1.

311. Goertz, C.K., Ip, W.H. (1982) On the Structure of the Io Torus. Planet Space Sci 30(9):855-64.

312. Gold, T. (1969) Apollo 11 Observations of a Remarkable Glazing Phenomenon on the Lunar Surface. Science 165:1345-1349.

313. Gold, T. (1976) Accretion of the Moon. Lunar Planet Sci 7:304-306.

314. Goldstein, B.E., et al. (1981) Mercury: Magnetosphere Processes and the Atmospheric Supply and Loss Rates. J Geophys Res 86A:5485-5499.

315. Golenetskii, S.V. (1988) On the Spatial Distribution of Gamma-Ray Burst Sources. Adv Space Res 8(2):653-7.

316. Gonzalez-Serrano, J.I., et al. (1988) CCD Photometry of the Jet in M87: New Features Revealed. Adv Space Res 8(2/3): (2)635-(2)637.

317. Goodrich, R.W., Bianchi, L. (1989) The Shocking Truth About Some "Proto-PN". p.447 in (ed) S. Torres-Peimbert. Planetary Nebulae. Boston, Kluwer.

318. Gopal-Krishna, Witta, P.J. (1989) Interaction of the Beams of Active Galactic Nuclei with Their Environment at High Redshifts. pp.469-471 (eds) Active Galactic Nuclei. Boston, Kluwer.

319. Gorenstein, P., Bjorkholm, P. (1973) Detection of Radon Emanation from the Crater Aristarchus. Science 184:792.

320. Gott (III), R.J., et al. (1990) Topology of Microwave Background Fluctuations: Theory. Astrophys J 352:1-14.

321. Graham, J.R., et al. (1990) Double Nucleus of Arp 220 Unveiled. Astrophys J 354:L5-L8.

322. Graf, E.R., et al. (1968) Correlation Between Solar Activity and the Brightness of Jupiter's Great Red Spot. Nature 218:857.

323. Greeley, R. (1987) Release of Juvenile Water on Mars: Estimated Amounts and Timing Associated with Volcanism. Science 236:1653-1654.

324. Green, D.W.E., Morris, C.S. (1987) Visual Brightness Behavior of P/Halley During 1981-1987. Astron Astrophys 187:560-568.

325. Greiner, J. (1991) Distribution of Turned-off Pulsars and Consequences for Gamma-ray Burst Sources. Astron Astrophys 242:417-424.

326. Grewing, M. (1989) Wind Features and Wind Velocities. pp.241-250 in (ed) S. Torres-Peimbert. Planetary Nebulae. Boston, Kluwer.

327. Greyber, H.D. (1989) Importance of Strong Magnetic Fields in the Universe. Comments Astrophys 13:201-213.

328. Gribbin, J. (1974) High Energy Radiation from White Holes. Nature 251:590.

329. Gribbin, J. (1975) White Holes—A Coming Fashion? New Sci 68:199.

330. Grieve, R.A.F., Robertson, P.B. (1979) Terrestrial Cratering Record, I. Current Status of Observations. Icarus 38:212-29.

331. Grigoryan, S.S. (1979) Nature of the Tunguska Meteorite. Soviet Phys Doklady 21:603-5.

332. Gringauz, K.I. (1983) Bow Shock and the Magnetosphere of Venus According to Measurements from Venera 9 and 10 Orbiters. pp.980-993 in (eds) D.M. Hunten, et al., Venus. Tucson, Univ Arizona.

333. Grun, E., et al. (1984) Dust-Magnetosphere Interactions. pp.275-332 in (eds) R. Greenberg, A. Brahic, Planetary Rings. Tucson, Univ Arizona.

334. Gurnett, D.A., et al. (1989) First Plasma Wave Observations at Neptune. Science 246:1494-1498.

335. Halthore, R., et al. (1986) Thermal Equilibrium Model for Jupiter's North Polar Hot Spot. Bull Amer Astron Soc 18(3):773.

336. Hamilton, D.C., et al. (1983) Energetic Atomic and Molecular Ions in Saturn's Magnetosphere. J Geophys Res 88A:8905-22.

337. Hammel, H.B. (1986) Methane-Band Imaging of Neptune and Uranus. Bull Amer Astron Soc 18(3):764.

338. Hammel, H.B., et al. (1989) Neptune's Wind Speeds Obtained by Tracking Clouds in Voyager Images. Science 245:1367-69.

339. Hammond, A.L. (1972) Lunar Research: No Agreement On Evolutionary Models. Science 175:868-870.

340. Hammond, A.L. (1976) Presolar Grains: Isotopic Clues to Solar System Origin. Science 192:772-773.

341. Hartle, R.E., Grebowsky, J.M. (1990) Upward Ion Flow in Ionospheric Holes on Venus. J Geophys Res 95A: 31-38.

342. Hartmann, W.K. (1974) Geological Observations of Martian Arroyos. J Geophys Res 79:3951-3957.

343. Harwit, M. (1968) "Spontaneously" Split Comets. Astrophys J 151:789-790.

344. Hasegawa, I. (1980) Catalogue of Ancient and Naked-Eye Comets. Vistas Astron 24:59-102.

345. Haywood, J. (1884) Auroral Glow on the Moon. Sidereal Messenger 3:121.

346. Heap, S.R., Stecher, T.P. (1981) Discovery of the Molecular Hydrogen Ion (H2+) in the Planetary Nebulae. pp.657-661 in The Universe at Ultraviolet Wavelengths, NASA Conf Pub NASA-CP-2171.

347. Hedgepath, J. (1949) Rainfall of Fish. Science 110:482.

348. Heidmann, J. (1988) MEGA SETI, A Major Step in Bioastronomy. pp.9-22 in (eds) F. Bertola, et al., New Ideas in Astronomy. Cambridge Univ.

349. Hill, T.W., Michel, F.C. (1976) Heavy Ions from the Galilean Satellites and the Centrifugal Distortion of the Jovian Magnetosphere. J Geophys Res 81:4561-5.

350. Hillebrandt, W. (1987) A Few Summarizing Remarks. pp.264-288 in (eds) R. Beck, R. Grave, Interstellar Magnetic Fields—Observation and Theory. NY, Springer-Verlag.

351. Hillebrandt, W. (1987) Stellar and Extragalactic Jets: An Introduction. pp.208-210 in (eds) R. Beck, R. Grave, Interstellar Magnetic Fields: Observation and Theory. NY, Springer-Verlag.

352. Hodges, M.W., Mutel, R.L. (1987) Are Compact Doubles Misaligned Superluminals? pp.168-173 in (eds) J.A. Zensus, T.J. Pearson, Superluminal Radio Sources. NY, Cambridge.

353. Hofstadter, M.D., Muhleman, D.O. (1988) Latitudinal Variations of Ammonia in the Atmosphere of Uranus: An Analysis of Microwave Observations. Icarus 81:396-412.

354. Holden, E.S. (1878) Moon's Zodiacal Light. Amer J Sci 3(15):231.

355. Hoppe, M.M., Russell, C.T. (1982) Particle Acceleration at Planetary Bow Shock Waves. Nature 295:41-42.

356. Horne, K., et al. (1991) Echo Mapping of Broad H-Beta in NGC 5548. Astrophys J 367:L5-L8.

357. Horgan, J. (1989) Galactic Center. Sci Amer 260(3):22.

358. Houpis, H.L.F., Flammer, K.R. (1985) Asteroids, Comets and the Electrostatic Mechanism. Bull Amer Astron Soc 17(3):1.22.

359. Hoyle, F. (1988) Is the Universe Fundamentally Biological? pp.5-8 in (eds) F. Bertola, et al., New Ideas in Astronomy. NY, Cambridge Univ.

360. Hoyle, F., Narlikar, J.V. (1974) Action at a Distance in Physics and Cosmology. SF, WH Freeman.

361. Hoyle, F., Tayler, R.J. (1964) Mystery of the Cosmic Helium Abundance. Nature 203:1108-1110.

362. Hoyle, F., Wickramasinghe, N.C. (1985) Living Comets. So. Glamorgan, Univ College Cardiff.

363. Huchra, J.P. (1987) Galactic Structure and Evolution. pp.203-220 in (ed) R.A. Meyer, Encyclopedia of Astronomy and Astrophysics. NY, Academic.

364. Hudec, R., et al. (1988) Optical Transient Searches. Adv Space Res 8(2):665-688.

365. Hughes, D.W. (1974) Meteor Rates, Volcanoes and the Solar Cycle. Nature 252:191-192.

366. Hughes, D.W. (1974) Super-Rotation of the Upper Atmosphere. Nature 249:405-406.

367. Hughes, D.W. (1975) Hydrous Minerals in Meteorites. Nature 256:697.

368. Hughes, D.W. (1976) Brightness Variations of Saturn's Rings. Nature 261:191.

369. Hughes, D.W. (1976) Meteor Swarms Colliding with the Moon. Nature 262:175-176.

370. Hughes, D.W. (1976) Tunguska Revisited. Nature 259:626-627.

371. Hughes, T.M. (1910) Pwdre Ser. Nature 83:492-494.

372. Huguenin, R.L., et al. (1986) Injection of Dust into the Martian Atmosphere Evidenced from the Viking Gas Exchange Experiment. Icarus 68:99-119.

373. Hummel, E., et al. (1987) Magnetic Field in the Anomalous Arms in NGC 4258. pp.61-64 in (eds) R. Beck, R. Grave, Interstellar Magnetic Fields—Observation and Theory. NY, Springer-Verlag.

374. Hunt, G.E., James, P.B. (1985) Martian Cloud Systems: Current Knowledge and Future Observations. Adv Space Res 5(8):93-99.

375. Hunt, G.R., et al. (1968) Lunar Eclipse: Infrared Images and an Anomaly of Possible Internal Origin. Science 162:252-4.

376. Hunt, H. (1977) Super-Rotating Atmosphere of Venus. Nature 266:15-16.

377. Hunt, J.N., et al. (1960) Atmospheric Waves Caused by Large Explosions. Philo Trans Roy Soc Lond 252A:275-315.

378. Hunten, D.M., et al. (1988) Mercury Atmosphere. pp.562-612 in (eds) F. Vilas, et al. Mercury. Tucson, Univ Ariz.

379. Impey, C. (1987) Infrared, Optical, UV, and X-Ray Properties of Superluminal Radio Sources. pp.233-250 in (eds) J.A. Zensus, T.J. Pearson, Superluminal Radio Sources. NY, Cambridge Univ.

380. Ingersoll, A.P. (1987) Uranus. Sci Amer 256(1):38-45.

381. Ingersoll, A.P., et al. (1984) Structure and Dynamics of Saturn's Atmosphere. pp.195-238 in (eds) T. Gehrels, M.S. Matthews, Saturn. Tucson, Univ Arizona.

382. Ingersoll, A.P., Tryka, K.A. (1990) Triton's Plumes: the Dust Devil Hypothesis. Science 250:435-437.

383. Intrillagator, D.S. (1985) New Results on the Pioneer Venus Orbiter February 10-11, 1982 Events: A Solar Wind Disturbance Not a Comet. Geophys Res Lett 12(4):187-190.

384. Ip, W.-H. (1982) On Planetary Rings as Sources and Sinks of Magnetospheric Plasmas. pp.575-596 in (ed) A. Brahic, Anneaux des Planetes. Planetary Rings, Toulouse, France, Cepadues-Editions.

385. Ip, W.-H. (1986) [Plasma Processes in Planetary Rings] Plasma-Vorgange in den Ringen Planeten. Sterne Weltraum 25(4):188-194.

386. Ip, W.-H. (1986) Sodium Exosphere and Magnetosphere of Mercury. Geophys Res Lett 13:423-6.

387. Ip, W.-H. (1987) Magnetospheric Charge-Exchange Effect on the Electroglow of Uranus. Nature 326:775.

388. Jaakkola, T. (1988) Tests of the Cosmological Expansion Hypothesis. pp.333-336. in (eds) F. Bertola, et al. New Ideas in Astronomy. NY, Cambridge Univ.

389. Jackson, A.A., Ryan, M.P. (1973) Was the Tunguska Event Due to a Black Hole? Nature 245:88-89.

390. Jedrzeiewski, R., Schechter, P.L. (1988) Evidence for Dynamical Subsystems in Elliptical Galaxies. Astrophys J 330:L87-L91.

391. Jefferys, W.H. (1967) Nongravitational Forces and Resonances in the Solar System. Astron J 72:872-875.

392. Jenkins, B.G. (1878) Luminous Spot on Mercury in Transit. Roy Astron Soc Mon Notices 38:337-340.

393. Jenkins, A.W., et al. (1960) Observed Magnetic Effects from Meteors. J Geophys Res 65:1617.

394. Jewitt, D.C., Danielson, G.E. (1981) Jovian Ring. J Geophys Res 86A:8691-7.

395. Jewitt, D., Meech, K. (1985) Rotation of the Nucleus of Plareno-Rigaux. Bull Amer Astron Soc 17(3):1.24.

396. Johnson, H.M. (1941) White Spot on Saturn's Rings. British Astron Assoc J 51:309-312.

397. Johnson, R.E., et al. (1987) Radiation Formation of a Non-Volatile Comet Crust. Astron Astrophys 187:889-892.

398. Johnstone, A.D., et al. (1987) Alfvenic Turbulence in the Solar Wind Flow During the Approach to Comet P/Halley. Astron Astrophys 187:25-32.

399. Jones, D.L. (1987) Intrinsic Asymmetry in NGC 6251. pp.162-167 in (eds) J.A. Zensus, T.J. Pearson, Superluminal Radio Sources. NY, Cambridge Univ.

400. Jones, E.M. Kodis, J.W. (1982) Atmospheric Effects of Large Impacts: The First Few Minutes. Spec Pap Geol Soc Amer 190:175-186.

401. Judge, D.L., et al. (1980) Ultraviolet Photometer Observations of the Saturnian System. Science 207:431-4.

402. Kaiser, M.L., et al. (1984) Saturn as a Radio Source. pp.378-415 in (eds) T. Gehrels, M.S. Matthews. Saturn. Tucson, Univ Ariz.

403. Kaiser, M.L., et al. (1987) Sources of Uranus' Dominant Nightside Radio Emissions. J Geophys Res 93A:15169-15176.

404. Kaiser, M.L., et al. (1989) Radio Emission from the Magnetic Equator of Uranus. J Geophys Res 94A:2399-2404.

405. Kaper, L., et al. (1990) Long-Term Study of Stellar-Wind Variability of O Stars. pp.213-218. in (eds) L.A. Willson, R. Stalio, Angular Momentum and Mass Loss for Hot Stars. Boston, Kluwer.

406. Kar, J. (1990) On the Implications of an Intrinsic Magnetic Field in Early Mars. Geophys Res Lett 17:113.

407. Kassim, N.E., Weiler, K.M. (1990) A Possible New Association of a Pulsar with a Supernova Remnant. Nature 343:146-148.

408. Kaufmann, W. (1984) Jupiter: Lord of the Planets. Mercury 13:169-184.

409. Kawaler, S.D. (1990) Angular Momentum Loss in Pre-Main Sequence Objects and the Initial Angular Momentum of Stars. pp.55-63 in (eds) L.A. Willson, R. Stalio, Angular Momentum and Mass Loss for Hot Stars. Boston, Kluwer.

410. Kazes, I, et al. (1988) Magnetic Field in the Bipolar Nebula S106. pp.291-295 in (eds) R.E. Purditz, M. Fich. Galactic and Extragalactic Star Formation. Boston, Kluwer.

411. Kazanas, D., Ellison, D.C. (1986) Proton Acceleration in [Gamma]-Ray Bursts. Adv Space Res 6(4):81-84.

412. Keller, H.U., et al. (1987) Comet P/Halley's Nucleus and its Activity. Astron Astrophys 187:807-823.

413. Kennicutt, Jr., R.C. (1989) Star Formation Law in Galactic Disks. Atrophys J 344:685-703.

414. Kerr, R.A. (1983) Lunar Meteorite and Maybe Some from Mars. Science 220:288-289.

415. Kerr, R.A. (1986) The Most Complex Magnetic Field. Science 232:1603.

416. Kerr, R.A. (1986) Volcanism on Mercury and the Moon, Again. Science 233:1258-1259.

417. Kerr, R.A. (1988) Another Asteroid has Turned Comet. Science 241:1161.

418. Kerr, R.A. (1989) Neptune System in Voyager's Afterglow. Science 245:1450-1451.

419. Kerr, R.A. (1989) Triton Steals Voyager's Last Show. Science 245:928-930.

420. Kerr, R.A. (1989) Why Neptunian Ring Sausages? Science 245:930.

421. Khachikian, E. (1988) Galaxies with Double Nuclei. pp.115-118 in (eds) F. Bertola, et al., New Ideas in Astronomy. NY, Cambridge Univ.

422. Khalfin, L.A. (1989) Inflationary Fallacies. Intl J Theoret Phys 28:1109-1123.

423. Kilore, A.J. (1985) Recent Results on the Venus Atmosphere from Pioneer Venus Radio Occultations. Adv Space Res 5(9):41-49.

424. Kirsch, E., et al. (1981) X-Ray and Energetic Neutral Particle Emission from Saturn's Magnetosphere. Nature 292:718-21.

425. Kissel, J., Krueger, F.R. (1987) Organic Component in Dust from Comet Halley as Measured by the: Puma Mass Spectrometer on Board Vega 1. Nature 326:755-760.

426. Kjaergaard, P., et al. (1989) Search for Companions to High Redshift (z >= 3.0) Quasars. pp.53-54 in (eds) D.E. Osterbrock, J.S. Miller. Active Galactic Nuclei. Boston, Kluwer.

427. Kleep, H.B. (1964) Terrestrial, Interplanetary and Universal Expansion. Nature 201:693.

428. Knollenberg, R.G. (1982) International Conference on Venus. Clouds and Hazes. Nature 296:18.

429. Kollatschny, W., Fricke, K.J. (1989) Activity of Interacting Galaxies Mkn 673: A Close-By "E + A" Galaxy. pp.449-451 in (eds) D.E. Osterbrook, J.S. Miller. Active Galactic Nuclei. Boston, Kluwer.

430. Komitov, B. (1985) Ozone Vertical Distribution in Mars Polar Atmosphere. Adv Space Res 5(8):101-4.

431. Kondratyev, K.Ya., Moskalenko, N.I. (1985) The Atmospheric Greenhouse Effect and Climates on Various Planets. Adv Space Res 5(8):37-40.

432. Kopal, Z. (1958) Prospecting the Moon without Rockets. New Sci 4:1052-1053.

433. Kopal, Z. (1958) Volcano on the Moon? New Sci 4:1362-1364.

434. Kotanyi, C., et al. (1983) Are There Jets in Spiral Galaxies? pp.97-98 in (eds) A. Ferrari, A.G. Pacholczyk. Astrophysical Jets. Boston, Riedel.

435. Kotelnikov, V.A., et al. (1985) Radar Study of Venus Surface by Venera-15 and -16 Spacecraft. Adv Space Res 5(8):5-16.

436. Koyama, K., et al. (1990) Is the 5-kpc Galactic Arm a Colony of X-Ray Pulsars? Nature 343:148-149.

437. Kozyrev, N.A. (1963) Volcanic Phenomena on the Moon. Nature 198:979.

438. Krasnopol'sky, V.A., Parshev, V.A. (1983) Photochemistry of the Venus Atmosphere. pp.431-458 in (eds) D.M. Hunten, et al., Venus. Tucson, Univ Arizona.

439. Krause, M., et al. (1987) Magnetic Field Structure in M81. pp.57-60 in (eds) Interstellar Magnetic Fields: Observations and Theory. NY, Springer-Verlag.

440. Kresakova, M. (1987) Associations Between Ancient Comets and Meteor Showers. Astron Astrophys 187:935-936.

441. Krimigis, S.M., et al. (1982) Low-Energy Hot Plasma and Particles in Saturn's Magnetosphere. Science 215:571-577.

442. Krimigis, S.M., et al. (1983) General Characteristics of Hot Plasma and Energetic Particles in the Saturnian Magnetosphere: Results from the Voyager Spacecraft. J Geophys Res 88A:8871-92.

443. Krimigis, S.M., et al. (1986) Magnetosphere of Uranus: Hot Plasma and Radiation Environment. Science 233:97-102.

444. Krimigis, S.M., et al. (1989) Hot Plasma and Energetic Particles in Neptune's Magnetosphere. Science 246:1483-9.

445. Krinov, E.L. (1949) Tunguska Meteorite. Pub Acad Sci (USSR):196.

446. Krolik, J.H. (1988) Emission Line Regions in Active Galactic Nuclei: A Unified View. pp.19-37 in (eds) H.R. Miller, P.J. Wiita, Active Galactic Nuclei. NY, Springer-Verlag.

447. Krolik, J.H. (1988) X-Ray Heated Winds in Seyfert Galaxies. Adv Space Res 8(2):53-59.

448. Kronberg, P.P. (1985) Discovery of an Entire Population of Variable Radio Sources in the Nucleus of M82. pp.79-80 in (ed) J.E. Dyson, Active Galactic Nuclei. GB, Manchester.

449. Kronberg, P.P. (1987) Magnetic Fields and Faraday-Active Clouds Out to the Distances of Quasars. pp.86-94 (eds) R. Beck, R. Grave. Interstellar Magnetic Fields—Observation and Theory. NY, Springer-Verlag.

450. Kronberg, P.P., Kim, K.-T. (1990) Detailed Measurement in the Halo in the Coma Cluster of Galaxies, and its Magnetic Field. Geophys Astrophys Fluid Dynamics 50:7-22.

451. Ksanfomality, L.V. (1985) Volcanism on Venus: Connecting Link? Adv Space Res 5(9):91-98.

452. Ksanfomality, L.V., et al. (1983) Electrical Activity of the Atmosphere of Venus. pp.565-603 in (eds) D.M. Hunten, et al., Venus. Tucson, Univ Arizona.

453. Kulik, L.A. (1927) History of the Bolide of June 30, 1908 [In Russian]. Doklady Akademiaa Nauk SSSR (A23):393-398 [and in English] Pop Astron (1935) 43:499-504.

454. Kulik, L.A. (1927) Problem of the Impact Area of the Tunguska Meteorite of 1908 [in Russian]. Doklady Akademiaa Nauk SSSR (A23):399-402.

455. Kulik, L.A. (1933) Preliminary Results of Meteorite Expeditions in the Decade 1921-1931 [in Russian]. Papers Lomonosov Institute Akademiaa Nauk SSSR (2):73-81 [and in English] Pop Astron (1936) 44:215-20.

456. Kulik, L.A. (1937) Question of the Meteorite of June 30,1908, in Central Siberia. Meteors and Meteorites:559-562.

457. Kulik, L.A. (1939) Data on the Tunguska Meteorite Obtained by 1939 [in Russian]. Doklady Akad Nauk SSSR (8):520-524.

458. Kumar, S., et al. (1985) Detection of Hot Hydrogen in the Upper Atmosphere of Saturn: Voyager UVS Observations. Bull Amer Astron Soc 17(3):3.4.

459. Kundt, W. (1987) Magnetic Fields in Supernovae and Supernova Shells. pp.185-192 in (eds) R. Beck, R. Grave. Interstellar Magnetic Fields—Observation and Theory. NY, Springer-Verlag.

460. Kurki-Suonio, H. (1991) Primordial Nucleosynthesis with Horizon-Scale Curvature Fluctuations. Phy Rev D Fields 43:1087-1105.

461. Kuz'min, A.D. (1983) Radio Astronomical Studies of Venus. pp.36-44 in (eds) D.M. Hunten, et al., Venus. Tucson, Univ Arizona.

462. Lada, C.J. (1988) On the Importance of Outflows for Molecular Clouds and Star Formation. pp.5-24 in (eds) R.E. Purditz, M. Fich. Galactic and Extraglactic Star Formation. Boston, Kluwer.

463. Lada, C.J., Shu, F.H. (1990) Formation of Sun-like Stars. Science 248:564-572.

464. Lafon, J.-P., et al. (1981) On the Electrostatic Potential and Charge of Cosmic Grains. 1. Theoretical Background and Preliminary Results. Astron Astrophys 95:295-303.

465. Lal, D. (1988) Temporal Variations of Low-Energy Cosmic-Ray Protons on Decadal and Million Year Time-Scales: Implications of Their Origin. pp.337-346 in (eds) C.-G. Falthammar, et al., Plasma and the Universe. Boston, Kluwer Acad. [and] Astrophys Space Phys 144:337-346.

466. Lanzerotti, L.J., et al. (1987) Experimental Study of Erosion of Methane Ice by Energetic Ions and Some Considerations for Astrophysics. Astrophys J 313(2):910-19.

467. Larson, S. (1988) Comet Halley: Gas and Dust Jets in a Spin. Nature 332:681-682.

468. Larson, S., et al. (1987) Comet P/Halley Near-Nucleus Phenomena in 1986. Astron Astrophys 187:639-644.

469. Latham, G., et al. (1971) Moonquakes. Science 174:687-692.

470. Laurent, B. E., et al. (1988) On the Dynamics of the Metagalaxy. pp.639-658 in (eds) C.-G. Falthammer, et al., Plasma and the Universe. Boston, Kluwer [and] Astrophys Space Phys 144:639-658.

471. Lawrence, E.N. (1955) Raining Fish. Weather 10:345-6.

472. Leblanc, Y. (1990) Radio Sources in the Magnetospheres of Jupiter, Saturn and Uranus After Voyager Mission. Adv Space Sci 10(1):39-48.

473. Lecacheux, A., et al. (1986) Magnetospheric Low Frequency Radio Emissions at Uranus. Bull Amer Astron Soc 18:765-6.

474. Levin, B.Y., Bronshten, V.A. (1986) Tunguska Event and the Meteors with Terminal Flares. Meteorites 21(2):199-215.

475. Levy, R.L., et al. (1970) Organic Analysis of the Pueblito De Allende Meteorite. Nature 227:148-150.

476. Lewis, J.S. (1980) Lightning on Jupiter: Rate, Energetics and Effects. Science 210:1351-2.

477. Lewis, J.S. (1988) Origin and Composition of Mercury. pp.651-666 in (eds) F. Vilas, et al. Mercury. Tucson, Univ Ariz.

478. Libby, L.M., Libby, W.F. (1975) Comparison of Magnetosphere and Radio Emissions of Jupiter with Earth. pp.1546-1551 in Conf Pap Int Cosmic Ray Conf 14th vol 4. Garchine, Germany. Max-Planck Inst Extraterr Phys.

479. Lightman, A.P. (1988) Relativistic Plasmas. Adv Space Res 8(2/3):(2)547-(2)554.

480. Lilly, P.A. (1981) Shock Metamorphism in the Vrederfort Collar: Evidence for Internal Shock Sources. J Gepophys Res 86:10689-10700.

481. Limaye, S.S. (1985) Venus Atmospheric Circulation: Observations and Implications of the Thermal Structure. Adv Space Res 5(9):51-62.

482. Lindley, D. (1989) Supernova Springs New Surprise. Nature 337:595.

483. Lindley, D. (1990) An Excess of Perfection. Nature 343:207.

484. Lindzen, R.S., Teitelbaum, H. (1984) Venus Zonal Wind Above the Cloud Layer. Icarus 57:356-361.

485. Lingenfelter, R.E., et al. (1968) Lunar Rivers. Science 161:266-269.

486. Lizano, S. (1989) Magnetic Fields and Star Formation in Molecular Clouds. Rev Mexicana Astron Astrofis 18:11-21.

487. Lovelace, R.V.E. (1976) Dynamo Model of Double Radio Sources. Nature 262:649-652.

488. Lucchitta, B.K., et al. (1981) Did Ice Streams Carve Martian Outflow Channels? Nature 290:759-763.

489. Lundstedt, H., Magnusson, P. (1987) Two Disconnection Events in Comet P/Halley and Possible Solar Causes. Astron Astrophys 187:261-263.

490. Lyuti, V.M. (1977) Optical Variability of the Nuclei of Seyfert Galaxies. II. UBV and H-[Alpha] Photometry. Soviet Astron 21:655-664.

491. Maciel, W.J. (1989) Galactic Distribution, Radial Velocities and Masses of PN. pp.73-82 in (ed) S. Torres-Peimbert. Planetary Nebulae. Boston, Kluwer.

492. Maclellan, C.G., et al. (1983) Low-Energy Particles at the Bow Shock, Magnetopause and Outer Magnetosphere of Saturn. J Geophys Res 88A:8817-30.

493. Mahajan, K.K., Mayr, H.G. (1990) Mars Ionopause during Solar Minimum: A Lesson from Venus. J Geophys Res 95A:8265-8270.

494. Marcaide, J.M., et al. (1985) Simultaneous Dual-Wavelength VLBI Observations of the Compact Radio Source Near the Galactic Center. pp.50-53 in (ed) J.E. Dyson, Active Galactic Nuclei. GB, Manchester Univ.

495. Mark, K. (1995) Meteorite Craters. Tucson, Univ Ariz.

496. Marouf, E.A., et al. (1986) Profiling Saturn's Rings by Radio Occultation. Icarus 68:120-66

497. Marscher, A.P. (1987) Synchro-Compton Emission from Superluminal Sources. pp.280-300 in (eds) J.A. Zensus, T.J. Pearson, Superluminal Radio Sources. NY, Cambridge Univ.

498. Martin, P.G. (1985) Optical and Infra-Red Polarization of Active Galactic Nuclei. pp.194-214 in (ed) J.E. Dyson, Active Galactic Nuclei. GB, Manchester Univ.

499. Mayr, H.G., et al. (1985) Conjecture on Superrotation in Planetary Atmospheres: A Diffusion Model with Mixing Length Theory. Adv Space Res 5(9):63-68.

500. Mazets, E.P. (1988) Gamma-Ray Bursts: Current Status. Adv Space Res 8(2):669-677.

501. Mazur, P., et al. (1978) Biological Implications of the Vikings Mission to Mars. Space Sci Rev 22:3-34.

502. McClintock, J. (1988) X-Ray Properties of Galactic Black Holes. Adv Space Res 8(2/3):(2)191-(2)195.

503. McDonald, F.B., et al. (1979) Energetic Protons in the Jovian Magnetosphere. J Geophys Res 84A:2579-96.

504. McFadden, L.A., et al. (1987) Activity of Comet P/Halley 23-25 March, 1986: IUE Observations. Astron Astrophys 187:333-338.

505. McGill, G.E., (1982) International Conference on Venus Geology and Geophysics. Nature 296:14.

506. McGill, G.E., et al. (1983) Topography, Surface Properties and Tectonic Evolution. pp.69-130 in (eds) D.M. Hunten, et al., Venus. Tucson, Univ Ariz.

507. McGovern, W.E., Burk, S.D. (1972) Upper Atmospheric Thermal Structure of Jupiter with Convective Heat Transfer. J Atmos Sci 29(1):179-89.

508. McNaughton, N.J. (1981) Deuterium/Hydrogen Ratios in Unequilbrated Ordinary Chondrites. Nature 294:639-41.

509. McNutt, R.L., et al. (1981) Positive Ion Observations in the Middle Magnetosphere of Jupiter. J Geophys Res 86A:8319-42.

510. Meaburn, J., et al. (1985) Occultation of the Inner Seyfert Nucleus of NGC 4151? pp.184-188 in (ed) J.E. Dyson, Active Galactic Nuclei. GB, Manchester Univ.

511. Mead, K.N., et al. (1990) Molecular Clouds in the Outer Galaxy. IV. Studies of Star Formation. Astrophys J 354:492-503.

512. Meek, A. (1918) Shower of Sand Eels. Nature 102:46.

513. Meinschein, W.G. (1959) Origin of Petroleum. Bull Amer Assoc Petrol Geol 43:925-943.

514. Melosh, H.J., McKinnon W.B. (1988) Tectonics of Mercury. pp.374-400 in (eds) F. Vilas, et al. Mercury. Tucson, Univ Ariz.

515. Melia, F. (1988) Gamma-Ray Burst Reprocessing. Adv Space Res 8(2):641-652.

516. Mendis, D.A. (1987) A Cometary Aurora. Earth, Moon & Planets 39(1):17-20.

517. Merril, G.P. (1919) Cumberland Falls Meteorite. Science 50:90.

518. Metz, W.D. (1974) Mercury: More Surprise in the Second Assessment. Science 185:132.

519. Metz, W.D. (1974) Update on Mars: Clues About the Early Solar System. Science 183:187-189.

520. Metz, W.D. (1975) Quasars Flare Sharply: Explaining the Energy Gets Harder. Science 189:129.

521. Metz, W.D. (1978) HEAO Records Gamma-Ray Burst. Science 199:870.

522. Metzger, A.E., et al. (1983) Detection of X-Rays from Jupiter. J Geophys Res 88A:7731-7795.

523. Michaels, P.J., et al. (1982) Observations of a Comet on a Collision Course with the Sun. Science 215:1097-1102.

524. Middlehurst, B.M. (1966) Transient Changes in the Moon. The Observatory 86:239-242.

525. Mignard, F. (1984) Effects of Radiation Forces on Dust Particles. pp.333-366 in (eds) R. Greenberg, A. Brahic, Planetary Rings. Tucson, Univ Arizona.

526. Milgrom, M. (1989) Alternatives to Dark Matter. Comments Astrophys 13:215-230.

527. Mills, A.A. (1970) Transient Lunar Phenomena and Electrostatic Glow Discharges. Nature 225:929-930.

528. Milton, D.J. (1974) Carbon Dioxide Hydrate and Flood on Mars. Science 183:654-655.

529. Miner, E.D. (1995) Uranus: The Planet, Rings and Satellites. NY, Ellis Horwood/Wiley.

530. Miralda-Escude', J., Ostriker, J.P. (1990) What Produces the Ionizing Background at Large Redshift? Astrophys J 350:1-22.

531. Moore, P. (1977) Linne Controversy: A Look Into the Past. Brit Astron Assoc J 87:363-68.

532. Morfill, G.E. (1982) Formation of Spokes in Saturn's Rings. pp.551-568 in (ed) A. Brahic, Anneaux des Planetes, Planetary Rings. Toulouse, France, Cepadues-Editions.

533. Morfill, G.E., et al. (1988) Thermal Cycling and Fluctuations in the Protoplanetary Nebula. Icarus 76:391-403.

534. Morgan, J.W., et al. (1971) Glazed Lunar Rocks: Origin by Impact. Science 172:556-557.

535. Morris, S.L., et al. (1985) Velocity Field in the Seyfert Galaxy NGC 5643. pp.178-183 in (ed) J.E. Dyson, Active Galactic Nuclei. GB, Manchester Univ.

536. Mukhin, L.M. (1983) Problem of Rare Gases in the Venus Atmosphere. pp.1037-1044 in (eds) D.M. Hunten, et al., Venus. Tucson, Univ Arizona.

537. Muller, P.M., Sjorgen, W.L. (1968) Mascons: Lunar Mass Concentrations. Science 161:680-684.

538. Murphy, B.W., et al. (1988) Evolution of Active Galactic Nuclei: A Multi-Mass Model. pp.421-423 in (eds) H.R. Miller, P.J. Wiita, Active Galactic Nuclei. NY, Springer-Verlag.

539. Mutel, R.L., Philips, R.B. (1987) Superluminal Motion in BL Lac: Evidence for Deceleration in Two Events. pp.60-66 in (eds) J.A. Zensus, T.J. Pearson. Superluminal Radio Sources. NY, Cambridge Univ.

540. Nagy, A.F., Brace, L.H. (1982) International Conference on Venus. Structure and Dynamics of the Ionosphere. Nature 296:19.

541. Nagy, A.F., Cravens, T.E. (1985) Recent Advances in Model Calculations of the Venus Ionosphere. Adv Space Res 5(9):135-143.

542. Nagy, A.F., et al. (1983) Basic Theory and Model Calculations of the Venus Ionosphere. I. pp.841-872 in (eds) D.M. Hunten, et al., Venus. Tucson, Univ Arizona.

543. Nagy, A.F., et al. (1986) Is Jupiter's Ionosphere a Significant Plasma Source for Its Magnetosphere? J Geophys Res 91A:351-4.

544. Nagy, B., et al. (1961) Mass Spectroscopic Analysis of the Orgueil Meteorite: Evidence for Biogenic Hydrocarbons. Ann N Y Acad Sci 93:27-35.

545. Napier, W., et al. (1988) Are Redshifts Really Quantizated. pp.191-194. in (eds) F. Bertola, et al., New Ideas in Astronomy. NY, Cambridge Univ.

546. Narlikar, J.V. (1988) Noncosmological Redshifts, Theoretical Alternatives. pp.243-256. in (eds) F. Bertola, et al., New Ideas in Astronomy. NY, Cambridge Univ.

547. Narlikar, J.V., Apparao, K.M.V. (1975) White Holes and High Energy Astrophysics. Astrophys Space Sci 35:321-336.

548. National Research Council, Astronomy Survey Committee (1972-73) Astronomy and Astrophysics for the 1970's. Vol.1. Wash, DC, Natl Acad Sci.

549. Neckel, T., Munch, G. (1987) Photometry of Comet P/Halley at Near Post-Perihelion Phases. Astron Astrophys 187:581-584.

550. Nelson, A.H. (1987) Galaxy Magnetic Fields and Hidden Matter in Galaxies. pp.142-145 in (eds) R. Beck, R. Grave. Interstellar Magnetic Fields—Observations and Theory. NY, Springer-Verlag.

551. Ness, N.F., et al. (1974) Magnetic Field Observations Near Mercury. Science 185:151-160.

552. Ness, N.F., et al. (1986) Magnetic Field of Uranus: Voyager Results. Bull Amer Astron Soc 18(3):765.

553. Ness, N.F., et al. (1989) Magnetic Fields at Neptune. Science 246:1473-1478.

554. Neugebauer, G., et al. (1984) Early Results from the Infrared Astronomical Satellite. Science 224:14-21.

555. Newman, M., et al. (1984) Zonal Winds in the Middle Atmosphere of Venus from Pioneer Venus Radio Occultation Data. J Atmosph Sci 41:1901-1913.

556. Niedner, M.B., Jr., Schwingenschuh, K. (1987) Plasma-Tail Activity at the Time of the Vega Encounters. Astron Astrophys 187:103-108.

557. Nininger, H.H. (1963) Meteorite with Unique Features. Science 139:345-347.

558. Ogilvie, K.W., et al. (1977) Observations of the Planet Mercury by the Plasma Electron Experiment: Mariner 10. J Geophys Res 82:1807-24.

559. O'Keefe, J.A. (1970) Tektite Glass in Apollo 12 Sample. Science 168:1209-1210.

560. O'Keefe, J.A. (1976) Tektites and Their Origin. NY, Elsevier.

561. O'Keefe, J.A. (1985) The Coming Revolution in Planetology. Eos 66:89-90.

562. O'Keefe, J.A. (1985) Terminal Cretaceous Event: Circumterrestrial Rings of Tektite Glass Particles? Cretaceous Res 6:261-269.

563. Olavesen, A.H., Wickramasinghe, N. (1978) Cosmo-Chemistry and Evolution. Nature 275:694.

564. Oliver, C.P. (1928) Great Siberian Meteorite. Sci Amer 139(1):42-44.

565. Opp, A.G. (1980) Scientific Results from the Pioneer Saturn Encounter Summary. Science 207:401-403.

566. Orr, M.J.L., Browne, I.W.A. (1982) Relativistic Beaming and Quasar Statistics. Mon Not Roy Astron Soc 200:1067-1080.

567. Ostriker, J.P., et al. (1986) Cosmological Effects of Superconducting Strings. Phys Lett 180B:231-239.

568. Owen, T., Terrile, R.J. (1981) Colors on Jupiter. J Geophys Res 86A:8797-8814.

569. Paczynski, B. (1986) Gamma-Ray Bursters at Cosmological Distances. Astrophys J 308:L43-L46.

570. Pagel, B.E.J. (1985) Summary. pp.373-374 in (ed) J.E. Dyson, Active Galactic Nuclei. GB, Manchester Univ.

571. Parker, E.N. (1983) Magnetic Fields in the Cosmos. Sci Amer 249(2):44-65.

572. Parkin, D.W., et al. (1962) Metallic Cosmic Dust with Amorphous Attachments. Nature 193:639-642.

573. Parravano, A. (1989) Self-regulated Star Formation Rate as a Function of Global Galactic Parameters. Astrophys J 347:812-816.

574. Pascoli, G. (1989) Some Hypothesized Observational Aspects of Magnetic Fields in Protoplanetary Nebulae. p.455 in (ed) S. Torres-Peimbert. Planetary Nebulae. Boston, Kluwer.

575. Pauliny-Toth, I.I.K. (1987) Structural Variations in the Quasar 3C454.3. pp.55-59 in (eds) J.A. Zensus, T.J. Pearson, Superluminal Radio Sources. NY, Cambridge Univ.

576. Paviov, A.V. (1985) Model of Composition of the Mars Ionosphere in the Photochemical Equilibrium Region. Cosmic Res 23(2):236-42.

577. Peale, S. (1976) Oribtal Resonances in the Solar System. Ann Rev Astron Astrophys 14:215-246.

578. Peale, S.J. (1988) Rotational Dynamics of Mercury and the State of its Core. pp.461-494 in (eds) F. Vilas, et al. Mercury. Tucson, Univ Ariz.

579. Pearson, T.J., Zensus, J.A. (1987) Introduction. pp.1-11 in (eds) J.A. Zensus, T.J. Pearson, Superluminal Radio Sources. NY, Cambridge Univ.

580. Pearson, T.J., et al. (1987) Quest for Superluminal Sources. pp.94-103 in (eds) J.A. Zensus, T.J. Pearson, Superluminal Radio Sources. NY, Cambridge Univ.

581. Pecker, J.-C. (1988) Difficulties of Standard Cosmologies. pp.295-312 in (eds) F. Bertola, et al., New Ideas in Astronomy. NY, Cambridge Univ.

582. Pedersen, B.M., et al. (1981) Low-Frequency Plasma Waves Near Saturn. Nature 292:714-716.

583. Pedler, A., et al. (1985) MERLIN Observations of the Radio Structure of Seyfert Nuclei. pp.87-92 in (ed) J.E. Dyson, Active Galactic Nuclei. GB, Manchester Univ.

584. Perez-de-Tejada, H. (1991) Momentum Transport at the Mars Magnetosphere. J Geophys Res 96A:11155-11165.

585. Perola, G.C. (1983) Concluding Remarks: A Progress Report on Our Understanding of Jets. pp.315-320 in (eds) A. Ferrari, A.G. Pacholczyk, Astrophysical Jets. Boston, Riedel.

586. Perry, S.J. (1884) Extraordinary Darkness at Midday [and] Black Rain. Nature 30:6 & 32.

587. Perryman, M.P.C. (1985) MR 2251-178: Gravitational Interaction Associated with Quasar Activity. pp.189-193 in (ed) J.E. Dyson, Active Galactic Nuclei. GB, Manchester.

588. Peterson, B.M., et al. (1991) Steps Towards Determination of the Size and Structure of the Broad-Line Region in Active Galactic Nuclei. II. An Intensive Study of NGC 5548 at Optical Wavelengths. Astrophys J 368:119-137.

589. Peterson, B., Ferland, G.J. (1986) Accretion Event in the Seyfert Galaxy, NGC 5548. Nature 324:345-6.

590. Petrosian, V., Caditz, D. (1989) Quasar and AGN Evolution. pp.59-61 in (eds) D.E. Osterbrock, J.S. Miller. Active Galactic Nuclei. Boston, Kluwer.

591. Phillips, J.L., et al. (1986) Venus Ultraviolet Aurora: Observations at 130.4 nm. Geophys Res Lett 13:1047-1050.

592. Phillips, R.J., Malin, M.C. (1983) The Interior of Venus and Tectonic Implications. pp.159-214 in (eds) D.M. Hunten, et al., Venus. Tucson, Univ Arizona.

593. Phinney, E.S. (1983) Black Hole-Driven Hydromagnetic Flows—Flywheels vs. Fuel. pp.201-213 in (eds) A. Ferrari, A.G. Pacholczyk, Astrophysical Jets. Boston, Riedel.

594. Phinney, E.S. (1987) How Fast Can a Blob Go? pp.301-305 in (eds) J.A. Zensus, T.J. Pearson, Superluminal Radio Sources. NY, Cambridge Univ.

595. Pielke, R.A. (1975) Ice Fall from a Clear Sky in Fort Pierce, Fla. Weatherwise 28:156-160.

596. Pike, R.J. (1985) Some Morphologic Systematics of Complex Impact Structures. Meteorites 20(1):49-68.

597. Pilcher, C.B. (1980) Images of Jupiter's Sulfur Ring. Science 207:181-3.

598. Pirraglia, J.A. (1984) Meridional Energy Balance of Jupiter. Icarus 59:169-176.

599. Pither, C.M. (1963) Origin of the Cytherean Cusp Caps. British Astron Assoc J 73:197-99.

600. Podgorny, I.M. (1983) Laboratory Simulation of the Interaction Between the Solar Wind and Venus. pp.994-1002 in (eds) D.M. Hunten, et al., Venus. Tucson, Univ Arizona.

601. Pool, R. (1990) Case Continues for Metallic Hydrogen. Science 247:1545-1546.

602. Porcas, R.W. (1983) Recent Observations of Superluminal Sources. pp.47-49 in (eds) A. Ferrari, A.G. Pacholczyk, Astrophysical Jets. Boston, Reidel.

603. Porcas, R.W. (1985) Radio Observations of Active Galactic Nuclei. pp.22-49 in (ed) J.E. Dyson, Active Galactic Nuclei. GB, Manchester Univ.

604. Porcas, R.W. (1987) Summary of Known Superluminal Sources. pp.12-25 in (eds) J.A. Zensus, T.J. Peterson. Superluminal Radio Sources, Cambridge Univ.

605. Pottasch, S.R. (1984) Planetary Nebulae. Boston, D Reidel.

606. Pronik, I., Metik, L. (1988) Peculiarities of the NGC 1275 Circumnuclear Region. pp.119-122 in (eds) F. Bertola. New Ideas in Astonomy. NY, Cambridge Univ.

607. Pudritz, R.E. (1988) Origin of Bipolar Outflows. pp.135-158. in (eds) R.E. Purditz, M. Fich. Galactic and Extraglactic Star Formation. Boston, Kluwer.

608. Pudritz, R.E., Silk, J. (1989) Origin of Magnetic Fields and Primordial Stars in Protogalaxies. Astrophys J 342:650-659.

609. Pyle, K.R., et al. (1983) Pioneer 11 Observations of Trapped Particle Absorption by the Jovian Ring and the Satellites 1979, J1, J2, and J3. J Geophys Res 88A:45-8.

610. Raeder, J., et al. (1987) Macroscopic Perturbations of the IMF by P/Halley as Seen by the Giotto Magnetometer. Astron Astrophys 187:61-64.

611. Rasmussen, K.L., et al. (1984) Nitrate in the Greenland Ice Sheet in the Years Following the 1908 Tunguska Event. Icarus 58:101-108.

612. Readhead, A.C.S., et al. (1978) A Jet in the Nucleus of NGC 6251. Nature 272:131-134.

613. Rees, M.J. (1990) "Dead Quasars" in Nearby Galaxies? Science 247:817-823.

614. Reich, W., et al. (1987) Observations of Linear Polarization in the Galactic Centre Region. pp.146-149 in (eds) R. Beck, R. Grave. Interstellar Magnetic Fields—Observations and Theory. NY, Springer-Verlag.

615. Reif, K. (1987) Radio Continuum Brightness Minimum Near Polaris. A Hole in the Interstellar Magnetic Field? pp.119-122 in (eds) R. Beck, R. Grave. Interstellar Magnetic Fields—Observations and Theory. NY, Springer-Verlag.

616. Reitsema, H.J., et al. (1989) Active Polar Region on the Nucleus of Comet Halley. Science 243:198-200.

617. Remillard, R.A., et al (1991) A Rapid Energetic X-ray Flare in the Quasar PKS 0558-504. Nature 350:589-91.

618. Rettig, T.W., et al. (1987) Observations of the Coma of Comet P/Halley and the Outburst of 1986 March 24-25 (UT). Astron Astrophys 187:249-255.

619. Robson, E.I., et al. (1985) Flare in the Infra-red to Radio Continuum of 3C 273. pp.147-152 in (ed) J.E. Dyson. Active Galactic Nuclei, GB, Manchester Univ.

620. Roederer, J.G., et al. (1977) Jupiter's Internal Magnetic Field Geometry Relevant to Particle Trapping. J Geophys Res 82:5187-5194.

621. Rubashev, B.M. (1964) Problems of Solar Activity. Moscow-Leningrad, Nauk. Trans NASA TT-F244. Wash, DC, USGPO.

622. Ruderman, M., et al. (1989) On the Origin of Pulsed Emission from the Young Supernova Remnant SN1987A. Astrophys J 346:L77-L80.

623. Rudnick, L. (1987) Different Perspective on Superluminal Sources. pp.217-232 in (eds) J.A. Zensus, T.J. Pearson, Superluminal Radio Sources. NY, Cambridge Univ.

624. Rudnicki, K. (1988) Alternatives to Present Day Cosmological Principles. pp.313-317 in (eds) F. Bertola, et al., New Ideas in Astronomy. NY, Cambridge Univ.

625. Ruffini, R., Wilson, J.R. (1975) Relativistic Magnetohydrodynamical Effects of Plasma Accreting Into a Black Hole. Phys Rev D12:2959-2962.

626. Runcorn, K. (1982) Moon's Deceptive Tranquility. New Sci 96:174-180.

627. Russell, C.T., et al. (1990) Upstream Waves at Mars: Phobos Observations. Geophys Res Lett 17:897-900.

628. Russell, C.T., et al. (1983) Unusual Interplanetary Event: Encounter with a Comet? Nature 305:612.

629. Russell, C.T., et al. (1984) Possible Observation of a Cometary Bow Shock. Geophys Res Lett 11:1022.

630. Russell, C.T., et al. (1985) Mass-Loading and the Formation of the Venus Tail. Adv Space Res 5(9):177-184.

631. Russell, C.T., Vaisberg, O. (1983) Interaction of the Solar Wind with Venus. pp.873-940 in (eds) D.M. Hunten, et al., Venus. Tucson, Univ Arizona.

632. Ruzmaikin, A. (1987) Magnetic Fields of Galaxies. pp.16-22 in (eds) R. Beck, R. Grave, Interstellar Magnetic Fields—Observations and Theory. NY, Springer-Verlag.

633. Sadeh, D. (1972) Possible Sidereal Period for the Seismic Lunar Activity. Nature 240:139-140.

634. Saito, T., et al. (1987) Possible Models on Disturbances of the Plasma Tail of Comet Halley During the 1985-1986 Apparition. Astron Astrophys 187:201-208.

635. Sandel, B.R., Broadfoot, A.L. (1981) Morphology of Saturn's Aurora. Nature 292:679-82.

636. Sandel, B.R., et al. (1982) Extreme Ultraviolet Observations from the Voyager 2 Encounter with Saturn. Science 215:548-553.

637. Sanders, R. (1988) Alternatives to Missing Mass. pp.279-294 in (eds) F. Bertola, et al., New Ideas In Astronomy. NY, Cambridge Univ.

638. Sanders, W., Frenk, C. (1991) Density Field of the Local Universe. Nature 349:32-38.

639. Saslaw, W. (1988) Ejection Mechanisms in Galaxies. pp.201-212 in (eds) F. Bertola, et al. New Ideas In Astronomy. NY, Cambridge Univ.

640. Scarf, F.L. (1985) Lightning on Venus. Adv Space Res 5(8):31-36.

641. Scarf, F.L., et al. (1985) Current-Driven Plasma Instabilities and Auroral-Type Particle Acceleration at Venus. Adv Space Res 5(9):185-91.

642. Scarrott, S.M., et al. (1987) Optical Polarisation Studies of Star Formation Regions. pp.161-165 in (eds) R. Beck, R. Grave. Interstellar Magnetic Fields—Observations and Theory. NY, Spring-Verlag.

643. Schaffer, D.B., Marscher, A.P. (1987) 4C39.25: Superluminal Motion Between Stationary Components. pp.67-71 in (eds) J.A. Zensus, T.J. Pearson, Superluminal Radio Sources. NY, Cambridge Univ.

644. Scheuer, P.A.G. (1987) Tests of Beaming Models. pp.104-113 in (eds) J.A. Zensus, T.J. Pearson, Superluminal Radio Sources. NY, Cambridge Univ.

645. Schlickheiser, R. (1985) Particle Acceleration in Active Galactic Nuclei. pp.355-360 in (ed) J.E. Dyson, Active Galactic Nuclei. GB, Manchester Univ.

646. Schneider, S.E., et al. (1983) Discovery of a Large Intergalactic H I Cloud in the M96 Group. Astrophys J 273:L1-L5.

647. Schnur, G., Kreitschmann, J. (1988) NGC 1808: Jet Activity and Evidence for Non-Doppler Redshifts. pp.149-154 in (eds) F. Bertola, et al., New Ideas in Astronomy. NY, Cambridge.

648. Schove, D.J. (ed) (1983) Sunspot Cycles. Stroudsberg, Pa, Hutchinson Ross.

649. Schramm, D.N., Truran, J.W. (1990) New Physics from Supernova 1987A. Phys Rpts 189(2):89-126.

650. Schubert, G. (1983) General Circulation and the Dynamical State of the Venus Atmosphere. pp.681-765 in (eds) D.M. Hunten, et al., Venus. Tucson, Univ Arizona.

651. Schulman, L.S., Seiden, P.E. (1986) Hierarchal Structure in the Distribution of Galaxies. Astrophys J 311:1-5.

652. Schulman, L.S., Seiden, P.E. (1986) Percolation and Galaxies. Science 233:425-431.

653. Schwarz, G., et al. (1987) Detailed Analysis of a Surface Feature on Comet P/Halley. Astron Astrophys 187:847-851.

654. Scoville, N., Norman, C. (1989) Evolution of Starburst Galaxies to Active Galactic Nuclei. pp.65-68 in (eds) D.E. Osterbrock, J.S. Miller. Active Galactic Nuclei. Boston, Kluwer.

655. Seiff, A. (1983) Thermal Structure of the Atmosphere of Venus. pp.215-279 in D.M. Hunten, et al., Venus. Tucson, Univ Arizona.

656. Sekanina, Z. (1983) Tunguska Event: No Cometary Signature in Evidence. Astron J 88:1382-1414.

657. Sekanina, Z. (1987) Dust Environment of Comet P/Halley: A Review. Astron Astrophys 187:789-795.

658. Sekanina, Z., et al. (1987) Sunward Spike of Haley's Comet. Astron Astrophys 187:645-49.

659. Sentman, D.D., Goertz, C.K. (1978) Whistler Mode Noise in Jupiter's Inner Magnetosphere. J Geophys Res 83A:3151-65.

660. Shaffer, D.B., Marscher, A.P. (1987) 4C39.25: Superluminal Motion Between Stationary Components. pp.67-71 in (eds) J.A. Zensus, T.J. Pearson. Superluminal radio Sources. NY, Cambridge Univ.

661. Shapiro, S.L., Teukolsky, S.A. (1991) Formation of Naked Singularities: The Violation of Cosmic Censorship. Phys Rev Lett 66:994-997.

662. Sharp, N. (1988) Imaging of the NGC 5296/7 System. pp.145-148 in (eds) F. Bertola, et al. New Ideas in Astronomy. NY, Cambridge Univ.

663. Sharp, R.P., Malin, M.C. (1975) Channels on Mars. Geol Soc Amer Bull 85:593-609.

664. Shemansky, D.E., Ajello, J.M. (1983) Saturn Spectrum in the EUV-Electron Excited Hydrogen. J Geophys Res 88A:459-64.

665. Shemansky, D.E., Smith, G.R. (1986) Implication for the Presence of a Magnetosphere on Uranus in the Relationship of EUV and Radio Emission. Geophys Res Lett 13(1):2-5.

666. Shlosman, I., Sikora, M. (1988) Leakage of UHE Photons from AGNs: Production of X-ray and Gamma-ray Halos within 10-30 Kpc. pp.283-285 in (eds) H.R. Miller, P.J. Wiita. Active Galactic Nuclei. NY, Springer-Verlag.

667. Shore, S. (1987) Star Clusters. pp.645-657 in (ed) R.A Meyer, Encyclopedia of Astronomy and Astrophysics. NY, Academic.

668. Shorthill, R.W., Saari, J.M. (1965) Nonuniform Cooling of the Eclipsed Moon. Science 150:210-212.

669. Shu, F.H. (1984) Waves in Planetary Rings. pp.513-561 in (eds) R. Greenberg, A. Brahic, Planetary Rings. Tucson, Univ Arizona.

670. Sikora, M., et al. (1989) Relativistic Neutrons in Active Galactic Nuclei. Astrophys J 341:L33-L35.

671. Simon, R.S., et al. (1987) Superluminal Motion Towards a Stationary Component in Quasar 3C395. pp.72-75 in (eds) J.A. Zensus, T.J. Pearson, Superluminal Radio Sources. NY, Cambridge Univ.

672. Simpson, J.A., et al. (1980) Saturnian Trapped Radiation and Its Absorption by Satellites and Rings: The First Results from Pioneer 11. Science 207:411-415.

673. Sittler, E.C., et al. (1983) Survey of Low-Energy Plasma Electrons in Saturn's Magnetosphere: Voyagers 1 and 2. J Geophys Res 88A:8847-70.

674. Skinner, T.E., et al. (1984) IUE Observations of Longitudinal and Temporal Variations in the Jovian Auroral Emission. Astrophys J 278:441-8.

675. Skinner, T.E., Moos, H.W. (1984) Comparison of the Jovian North and South Pole Aurorae Using the IUE Observatory. Geophys Res Lett 11:1107-1110.

676. Smith, B.A. (1984) Future Observation of Planetary Rings from Groundbased Observations and Earth-Orbiting Satellites. pp.704-712 in (eds) R. Greenberg, A. Brahic, Planetary Rings. Tucson, Univ Arizona.

677. Smith, B.A., et al. (1981) Encounter with Saturn: Voyager 1 Imaging Science Results. Science 212:163-191.

678. Smith, B.A., et al. (1987) Rejection of a Proposed 7.4-Day Rotation Period of the Comet Halley Nucleus. Nature 326:573-574.

679. Smith, B.A., et al. (1989) Voyager 2 at Neptune: Imaging Science Results. Science 246:1422-1449.

680. Smith, E.J., et al. (1978) Compression of Jupiter's Magnetosphere by the Solar Wind. J Geophys Res 83A:4733-42.

681. Smith, E.J., Tsurutani, B.T. (1983) Saturn's Magnetosphere: Observations of Ion Cyclotron Waves Near the Dione L Shell. J Geophys Res 88A:7831-6.

682. Smith, M.G. (1985) Infra-Red Observations of the Activity in Galactic Nuclei. pp.103-142 in (ed) J.E. Dyson, Active Galactic Nuclei. GB, Manchester Univ.

683. Smith, M.G. (1989) QSO Luminosity Functions and Evolution. pp.1-24 in (eds) D.E. Osterbrock, J.S. Miller. Active Galactic Nuclei. Boston, Kluwer Acad.

684. Smith, P.H., et al. (1986) Jupiter's Stratospheric Haze: Polar Versus Equatorial. Bull Amer Astron Soc 18(3):772.

685. Smith, P.H., Tomasko, M.G. (1984) Photometry and Polarimetry of Jupiter at Large Phase Angles. II. Polarimetry of the South Tropical Zone, South Equatorial Belt and the Polar Regions from the Pioneer 10 and 11 Missions. Icarus 58:35-73.

686. Smooth, F., et al. (1990) COBE Differential Microwave Radiometers: Instrument Design and Implementations. Astrophys J 360:685-695.

687. Sofue, Y. (1987) Global Structure of Magnetic Fields in Spiral Galaxies. pp.30-37 in (eds) Interstellar Magnetic Fields—Observations and Theory. NY, Springer-Verlag.

688. Sofue, Y., Reich, W. (1987) Magnetic Fields in the Galactic Center. pp.150-152 in (eds) Interstellar Magnetic Fields—Observations and Theory. NY, Springer-Verlag.

689. Song, G.X. (1982) On the Role of Mother-Planet Magnetic Field in Spiral Structure Explanation of Formation of Spokes. pp.569-574 in (ed) A. Brahic, Anneaux des Planetes. Planetary Rings. Toulouse, France, Cepadues-Editions.

690. Sromovsky, L.A., et al. (1985) New Atmospheric Temperature Results from the Pioneer Venus Entry Probes. Adv Space Res 5(9):37-40.

691. Stern, S.A., et al. (1988) Why is Pluto Bright? Implications of the Albedo and Light-curve Brehavior of Pluto. Icarus 75:485-498.

692. Stewart, A.I.F. (1987) Pioneer Venus Measurements of H, O, and C Production in Comet P/Halley Near Perihelion. Astron Astrophys 187:369-374.

693. Stewart, G.R. (1988) A Violet Birth for Mercury. Nature 335:496-497.

694. Stoker, C.R., et al. (1986) Convective Clouds on Uranus. Bull Amer Astron Soc 18:757-8.

695. Stone, E.C., et al. (1989) Energetic Charged Particles in Neptune's Magnetosphere. Science 246:1489-1494.

696. Stone, E.C., Miner, E.D. (1989) Voyager 2 Encounter with the Neptunian System. Science 246:1417-1421.

697. Stoneley, J. (1977) Cauldron of Hell: Tunguska. NY, Simon & Schuster.

698. Storeb, P.B. (1962) Rainfall Initiation by Meteor Particles. Nature 194:524-527.

699. Strom, K.M., et al. (1989) Circumstellar Material Associated with Solar-Type Pre-Main-Sequence Stars: A Possible Constraint on the Time-Scale for Planet Building. Astronomical J 97:1451-1470.

700. Strom, R.G., Neukum, G. (1988) Cratering Record on Mercury and the Origin of Impacting Objects. pp.336-374 in (eds) F. Vilas, et al. Mercury. Tucson, Univ Ariz.

701. Strom, S.E., et al. (1988) Energetic Winds and Circumstellar Disks Associated with Low Mass Young Stellar Objects. pp.53-88 in (eds) R.E. Purditz, M. Fich. Galactic and Extraglactic Star Formation. Boston, Kluwer Acad.

702. Sulentic, J. (1988) Tests of the Discordant Redshift Hypothesis. pp.123-144 in (eds) F. Bertola, et al. New Ideas in Astronomy. NY, Cambridge Univ.

703. Suslov, I.M. (1927) In Search of the Great Meteorite of 1908 [in Russian]. Mirovedenie 16(1):13-18.

704. Sykes, R.V., et al. (1987) IRAS Serendipitous Survey Observations of Pluto and Charon. Science 237:1336-1340.

705. Tariq, G.F., et al. (1985) Electrodynamic Interaction of Ganymede with the Jovian Magnetosphere and the Radial Spread of Wake-Associated Disturbances. J Geophys Res 90A:3995-4009.

706. Taylor, F.W., et al. (1979) Polar Clearing in the Venus Clouds Observed from the Pioneer Orbiter. Nature 279:612-4.

707. Taylor, F.W., et al. (1983) Thermal Balance of the Middle and Upper Atmosphere of Venus. pp.650-680 in (eds) D.M. Hunten, et al., Venus. Tucson, Univ Ariz.

708. Taylor, F.W., et al. (1985) Temperature Structure and Dynamics of the Middle Atmosphere of Venus. Adv Space Res 5(9):5-23.

709. Taylor, H.A., et al. (1985) In Situ Results on the Variation of Neutral Hydrogen at Venus. Adv Space Res 5(9):125-128.

710. Taylor, W.W.L., et al. (1979) Evidence for Lightning on Venus. Nature 279:614.

711. Thomas, P.G., et al. (1988) Tectonic History of Mercury. pp.401-428 in (eds) F. Vilas, et al. Mercury. Tucson, Univ Ariz.

712. Thomasko, M.G., et al. (1984) Clouds and Aerosols in Saturn's Atmosphere. pp.150-194 in (eds) T.Geherels, M.S. Matthews. Saturn. Tucson, Univ Arizona.

713. Tifft, W. (1988) Quantization and Time Dependence in the Redshift. pp.173-190 in (eds) F. Bertola, et al., New Ideas in Astronomy. NY, Cambridge Univ.

714. Treves, A. (1988) Recent Observations of BL Lac Object PKS 2155-30. pp.195-200 in (eds) F. Bertola et al., New Ideas in Astronomy. NY, Cambridge Univ.

715. Trimble, V. (1988) New- and Old-Ideas on the Universal J(M) Relationship. pp.239-242 in (eds) F. Bertola, et al., New Ideas in Astronomy. NY, Cambridge Univ.

716. Trouvelot, L. (1878) Moon's Zodiacal Light. Amer J Sci 3(15):88-89.

717. Turco, R.P., et al. (1982) Analysis of the Physical, Chemical, Optical and Historical Impacts of the 1908 Tunguska Meteor Fall. Icarus 50:1-52.

718. Tyler, G.L., et al. (1989) Voyager Radio Science Observations of Neptune and Triton. Science 246:1466-1473.

719. Unger, S.W., et al. (1985) Young Supernovae in the Star Burst Galaxy M82. pp.73-78 in (ed) J.E. Dyson, Active Galactic Nuclei. GB, Manchester Univ.

720. Unson, J.M., et al. (1990) Central Galaxy in Abell 2029: An Old Supergaint. Science 250:539-541.

721. Unwin, S.C. (1987) Superluminal Motion in the Quasar 3C279. pp.34-39 in (eds) J.A. Zensus, T.J. Pearson, Superluminal Radio Sources. NY, Cambridge Univ.

722. Urey, H.C. (1966) Biological Materials in Meteorites: A Review. Science 151:157-166.

723. Urey, H.C. (1967) Water on the Moon. Nature 216:1094-1095.

724. Valentijn, E.A. (1990) Opaque Spiral Galaxies. Nature 346:153-155.

725. Van Allen, J.A., et al. (1980) Saturn's Magnetosphere, Rings, and Inner Satellites. Science 207:415-421.

726. Van Biesbroeck, G. (1924) Luminous Extension at the Terminator of Mars. Pop Astron 32:589-591.

727. van den Bergh, S. (1990) Supernova Rates and Bursts of Star Formation. Astron Astrophys 231:L27-L28.

728. van den Bergh, S., et al. (1990) Supernova Rates and Galaxy Inclinations. Astrophys J 359:277-279.

729. van Diggelen, J. (1976) Is the Earth Expanding? Nature 262:675-676.

730. Van Flandern, T.C. (1976) Former Major Planet of the Solar System. Eos 57:280.

731. Viotti, R. (1987) Formation and Structure of Nebulae Around Symbiotic Objects Based on Radio to X-ray Observations (A Review). pp.163-170 in (ed) A. Preite Martinez, Planetary and Protoplanetary Nebulae: from IRAS to ISO. Boston, Reidel.

732. Vogel, S.N., et al. (1988) Star Formation in Giant Molecular Associations Synchronized by a Spiral Density Wave. Nature 334:402-406.

733. Volk, H.J., Morfill, G.E. (1991) Physical Processes in the Protoplanetary Disk. Space Sci Rev 56:65-73.

734. von Zahn, U., et al. (1983) Composition of the Venus Atmosphere. pp.299-430 in (eds) D.M. Hunten, et al., Venus. Tucson, Univ Arizona.

735. Waldrop, M.M. (1989) Collision and Cannibalism Shape Galaxies. Science 243:607-608.

736. Waldrop, M.M. (1989) Feeding the Monster in the Middle. Science 243:478.

737. Waldrop, M.M. (1989) High-Energy Summer for Astrophysics. Science 245:129.

738. Waldrop, M.M. (1989) Supernova 1987A Pulsar: Found? Science 243:892.

739. Walker, R.C., et al. (1987) 3C120. pp.48-54 in (eds) J.A. Zensus, T.J. Pearson, Superluminal Radio Sources. NY, Cambridge Univ.

740. Wampler, J., Burke, W. (1988) Cosmological Models with Non-Zero Lambda. pp.317-326 in (eds) F. Bertola, et al., New Ideas In Astronomy. NY, Cambridge Univ.

741. Wang, D.Y., Peng, Q.H. (1986) Monopole Model for Annihilation Line Emission from the Galactic Center. Adv Space Res 6(4):177-179.

742. Watanabe, J., et al. (1987) Outburst of Comet P/Halley on December 12, 1985. Astron Astrophys 187:229-332.

743. Wark, D. (1988) Meteorites: News from the Early Solar System. Nature 331:387.

744. Warren-Smith, R.F., et al. (1987) Observation of Magnetic Field Structure Around Newly Formed Stars. pp.167-170 in (eds) R. Beck, R. Grave. Interstellar Magnetic Fields: Observations and Theory. NY, Springer-Verlag.

745. Warwick, J.W., et al. (1982) Saturn Electrostatic Discharges. pp.299-306 in (ed) A. Brahic, Anneaux des Planetes. Planetary Rings. Toulouse, France, Cepadues-Editions.

746. Warwick, J.W., et al. (1989) Voyager Planetary Radio Astronomy at Neptune. Science 246:1498-1501.

747. Watanabe, J., et al. (1987) Outburst of Comet P/Halley on December 12, 1985. Astron Astrophys 187:229-232.

748. Wehrle, A.E. (1989) Superluminal Motion in CTA 102. pp.529-530 in (eds) D.E. Osterbrock, J.S. Miller. Active Galactic Nuclei. Boston, Kluwer.

749. Welch, W.J., et al. (1985) Gas Jets Associated with Star Formation. Science 228:1389-95.

750. Weldon, R.J., et al. (1982) Shock-Induced Color Changes in Nontronite: Implications for the Martian Fines. J Geophys Res 87B:10102-14.

751. Wells, R.A. (1979) Geophysics of Mars. NY, Elsevier.

752. Wesson, P.S. (1978) Cosmology and Geophysics. NY, Oxford.

753. Wesson, P.S. (1979) Self-Similarity and the Angular Momenta of Astronomical Systems: A Basic Rule in Astronomy. Astron Astrophys 80:296-300.

754. Wesson, P.S. (1981) Clue to the Unification of Gravitation and Particle Physics. Phys Rev 23D:1730-1734.

755. Whipple, E.C. (1981) Potentials of Surfaces in Space. Rep Progr Phys 44:1197-1250.

756. Whipple, F.J.W. (1930) Great Siberian Meteor and the Waves It Produced. Q J Roy Meteorol Soc 16:287-304.

757. Whipple, F.L. (1987) Cometary Nucleus: Current Concepts. Astron Astrophys 187:852-58.

758. Whitten, R.C., et al. (1975) Possible Ozone Depletions Following Nuclear Explosions. Nature 257:38-39.

759. Whitworth, A.P. (1989) Star Formation: When Protostars Cease Trading. Nature 342:126.

760. Wickramasinghe, C. (1977) Where Life Begins. New Sci 74:119-121.

761. Wielebinski, R. (1987) Magnetic Fields in the Galaxy. pp.96-99 in (eds) R. Beck, R. Grave. Interstellar Magnetic Fields—Observations and Theory. NY, Springer-Verlag.

762. Wiita, P.J., Rosen, A. (1989) Interactions of Jets with Interstellar and Intergalactic Media. pp.467-469 in (eds) D.E. Osterbrock, J.S. Miller. Active Galactic Nuclei. Boston, Kluwer.

763. Willson, L.A. (1990) Why a Meeting on Angular Momentum and Mass Loss for Hot Stars? pp.1-5 in (eds) L.A. Willson, R. Stalio, Angular Momentum and Mass Loss for Hot Stars. Boston, Kluwer.

764. Wilson, L.J. (1937) Apparent Flashes Seen on Mars. Pop Astron 45:430-432.

765. Wilson, A.S (1988) Extended Ionized Nebulosities in Seyfert Galaxies. Adv Space Res 8(2/3):(2)27-(2)37.

766. Witzel, A. (1987) Superluminal Motion and Other Indications of Bulk Relativistic Motion in a Complete Sample of Radio Sources from the S5 Survey. pp.83-93 in (eds) J.A. Zensus, T.J. Pearson, Superluminal Radio Sources. NY, Cambridge.

767. Wolf, R.S., et al. (1985) Cometary Rays: Magnetically Channeled Outflow. Bull Amer Astron Soc 17(3):1.21.

768. Wolfe, J.H., et al. (1980) Preliminary Results on the Plasma Environment of Saturn from the Pioneer II Plasma Analyzer Experiment. Science 207:403-407.

769. Woltjer, L. (1988) Quasar Redshifts. pp.219-222 in (eds) F. Bertola, et al., New Ideas in Astronomy. NY, Cambridge.

770. Womack, M., et al. (1991) N2H+ in Orion: Chemical Clues to the Dynamics of the Quiescent Gas. Astrophys J 370:L99-L102.

771. Worrall, D.M. (1987) Superluminal Radio Sources: What Does X-Ray Emission Tell Us? pp.251-259 in (eds) J.A. Zensus, T.J. Pearson, Superluminal Radio Sources. NY, Cambridge.

772. Wouterloot, J.G.A., et al. (1990) IRAS Sources Beyond the Solar Circle: II. Distribution in the Galactic Warp. Astron Astrophys 230:21-36.

773. Yanny, B. (1990) Emission-Line Objects Near QSO Absorbers. II. Analysis and Implications of Star-Forming Objects at Moderate Redshift. Astrophys J 351:396-405.

774. Yanny, B., et al. (1990) Emission-Line Objects Near QSO Absorbers. I. Narrow-Band Imaging and Cadidate List. Astrophys J 351:377-395.

775. Yusef-Zadeh, F., et al. (1984) Large Highly Organized Radio Structures Near the Galactic Center. Nature 310:557-561.

776. Yusef-Zadeh, F. (1989) Summary of IAU Symposium No. 136 on the Galactic Center. Comments Astrophys 13:273-294.

777. Zamorani, G. (1985) X-Ray Absorption and Variability in Quasars. pp.233-237 in (ed) J.E.D. Dyson, Active Galactic Nuclei. GB, Manchester Univ.

778. Zarka, P., Pedersen, B.M. (1982) Statistical Study of Saturn Electrostatic Discharges Observed by the Voyager Planetary Radio Astronomy Experiment. pp.307-317 in (ed) A. Brahic, Anneaux des Planetes. Planetary Rings. Toulouse, France, Cepadues-Editions.

779. Zarka, P., Lecacheux, A. (1987) Beaming of Uranian Nightside Kilometric Radio Emission and Inferred Source Location. J Geophys Res 92A:15177-15187.

780. Zensus, J.A. (1987) 3C273: Archetype of Superluminal Sources. pp.26-32 in (eds) J.A. Zensus, T.J. Pearson, Superluminal Radio Sources. NY, Cambridge Univ.

781. Zigel, F. (1961) Nuclear Explosion Over the Tiaga: Study of the Tunguska Meteorite [in Russian]. Znaniye-Sila (12):24-27 [and in English] Joint Pub Res Service, JPRS-13480 (April 1962) Wash, DC.

782. Zolotov, A.V. (1967) Problems of the Dust Structure of the Tungus Cosmic Body [in Russian]. pp.173-187 in (ed) M.V. Tronov, The Problem of the Tungus Meteorite. Tomsk Univ.

783. Zolotov, A.V. (1967) Relationship of the Geomagnetic Effect Caused by an Atomic Explosion in the Air with the Blast [in Russian]. pp.162-168 in (ed) M.V. Tronov. Problem of the Tungus Meteorite. Tomsk Univ.

784. Zotkin, I.T., Tsikulin, M.A. (1966) Scale Reproduction of the Explosion of the Tungus Meteorite [in Russian]. Doklady Akademiaa Nauk SSSR 167:59-62.

Tome Five

1. Abe, K., Kanamori, H.J. (1975) Temporal Variation of the Activity of Intermediate and Deep Focus Earthquakes. J Geophys Res 84:3589-3595.

2. Abraham, H.J.M. (1986) Features of the Chandlerian Nutation. Proc Astron Soc Austral 6(4):416-424.

3. Abrami, G. (1972) Correlations Between Lunar Phases and Rhythmicities in Plant Growth Under Field Conditions. Canad J Bot 50:2157-2166.

4.   Abrami. G., Piccardi, G. (1972) Seed Germination as a Biological Test for the Study of Fluctuating Phenomena. p.135 in (eds) S.W. Tromp, J.J. Bouma. Biometerological Aspects of Plants and Animals in Human Life. Amsterdam, Swets & Zeitlinger.

5.   Adderley, E.E., Bowen, E.G. (1962) Lunar Component in Precipitation Data. Science 137:749-750.

6.   Ahluwalia, D.V., Sirag, S.P. (1979) Gravitational Magnetism. Nature 278:535-538.

7.   Ahluwalia, D.V., Wu, T.Y. (1978) On the Magnetic Field of Cosmological Bodies. Lettere al Nuovo Cimento 23:406-408.

8.   Akasofu, S.-I. (1979) Magnetospheric Substorms and Solar Flares. Solar Phys 64:333-348.

9.   Akasofu, S.-I. (1988) Electric-Current Description of Solar Flares. pp.303-310 in (eds) C.-G. Falthhammar et al. Plasma and the Universe. Boston, Kluwer Acad Pubs [and] Astrophys Space Phys 144:303-310.

10.   Alabovskii, Yu. I., Babenko, A.N. (1977) Mortality from Vascular Diseases of the Brain in Years with Different Levels of Magnetic Activity. pp.213-217 in (eds) M.N. Gnevyshev, A.I. Oi'. Effects of Solar Activity on the Atmosphere and Biosphere. Jerusalem, Israel Program for Sci Trans.

11.   Alcock, N., Quittner, J. (1978) Prediction of Civil Violence to the Year 2001. J Interdiscipl Cycle Res 9:307-324.

12.   Alfven, H. (1981) Cosmic Plasma. Boston, D Reidel.

13.   Alfven, H. (1984) Magnetospheric Research and the History of the Solar System. Eos 65:769-770.

14.   Allais, M.F.C. (1959) Should the Laws of Gravitation Be Reconsidered? Part I. Aerospace Eng 18(9):46-52.

15.   Allais, M.F.C. (1959) Should the Laws of Gravitation Be Reconsidered? Part II. Aerospace Eng 18(10):51-55.

16.   Allais, M.F.C. (1959) Should the Laws of Gravitation Be Reconsidered? Commentary Note. Aerospace Eng 18(11):55

17.   Allan, A.H. (1964) Interfering Radio Signals on 18 Kc/s Received in New Zealand. Nature 201:1016-1017.

18.   Alldredge, L.R. (1975) Hypothesis for the Source of Impulses in Geomagnetic Secular Variations. J Geophys Res 80:1571-1578.

19.   Alldredge, L.R. (1977) Geomagnetic Variations with Periods from 13 to 30 Years. J Geomag Geoelect 29:123-135.

20.   Alter, D. (1929) Critical Test of the Planetary Hypothesis of Sun Spots. Mon Weather Rev 57:143-146.

21.   Angell, J.L., Korshover, J. (1974) Quasi-Biennial and Long-Term Fluctuations in the Centers of Action. Mon Weather Rev 102:669-680.

22.   Anonymous (1915) Reaction of the Planets Upon the Sun. Pop Astron 23:324-325.

23.   Anonymous (1947) Effect of the Moon on Radio Wave Propagation. Nature 159:396.

24.   Anonymous (1969) Moral Support for Einstein. Nature 222:1117-1118.

25.   Anonymous (1971) Chandler Wobble, Earthquake Correlations. Nature 229:227.

26.   Anonymous (1974) Diminishing Gravity is No Joke. New Sci 63:711.

27.   Anonymous (1977) Solar Flares: Link to Thunderstorms. Sci News 111:389.

28.   Anonymous (1977) Super Lightning Detected by Satellite. SciNews 112:15.

29. Anonymous (1981) Highly Destructive Lightning Strikes Oklahoma. Geo J 3:150.

30. Anonymous (1981) NOAA Finds Lightning Superbolts in Oklahoma Storms. Amer Meteorol Soc Bull 62:1060.

31. Anonymous (1981) Patterns of Thunderbolts. New Sci 92:102.

32. Anonymous (1982) Sun and Moon Affect the Weather Out West. New Sci 96:20.

33. Anonymous (1989) Blame it on the Moon; Australian 'Solar Varves' Turn out to be Mostly Lunar. Sci Amer 260(2):18.

34. Anonymous (1990) Serious Shortfall of Solar Neutrinos. Sci News 138:141.

35. Arabaji, W.I. (1976) On the Problem of Ball Lightning. J Geophys Res 81:6455.

36. Astro News (1979) Currents of Sun Linked to Solar Cycle. Astron 7(7):58-59.

37. Atkinson, L.P., et al. (1989) Hydrographic Variability of Southeastern United States Shelf and Slope Waters During the Genesis of Atlantic Lows Experiment: Winter 1986. J Geophys Res 94C:10699-10713.

38. Attolini, M.R., et al. (1988) On the Existence of the 11-Year Cycle in Solar Activity Before the Maunder Minimum. J Geophys Res 93A:12729-12734.

39. Bachmann, E. (1965) [Who had Heaven and Earth Measured?] Wer hat Himmel und Erde gemessen? Zurich, Buechergilde Gutenberg.

40. Bagby, J.P. (1975) Sunspot Cycle Periodicities. Nature 253:482.

41. Bagby, J.P. (1990) Review of Some Recent Anomalous Gravity Experiments. Cycles 41:134-139.

42. Bahcall, J.N. (1987) Neutrinos and Sunspots. Nature 330:318.

43. Bahcall, J.N. (1989) Neutrino Astrophysics. Cambridge Univ.

44. Bahcall, J.N. (1990) Solar-Neutrino Problem. Sci Amer 262(5):54-61.

45. Bai, T., Sturrock, P.A. (1987) 152-Day Periodicity of the Solar Flare Occurrence Rate. Nature 327:601-604.

46. Bakun, W.H. (1987) Future Earthquakes. Rev Geophys 25:1135-1138.

47. Baldwin, B., et al. (1976) Stratospheric Aerosols and Climatic Changes. Nature 263:551-54.

48. Bandeen, W.R., Maran, S.P. (1974) Possible Relationships Between Solar Activity and Meteorological Phenomena. Proc Symp NASA-Goddard Space Fl Ctr, 7-8 Nov 1973, NASA Rpt X-901-74-156.

49. Barnett, T.P. (1989) Solar-Ocean Relation: Fact or Fiction? Geophys Lett 16:803-806.

50. Barnothy, J., Forro, M. (1948) Lethal Effect of Cosmic Ray Showers on the Progeny of Animals. Experientia 4:1-7.

51. Bartenwerfer, D. (1973) Differential Rotation, Magnetic Fields and the Solar Neutrino Flux. Astron Astrophys 25:455-456.

52. Baumer, H., Eichmeier, J. (1978) Elimination of Meterotropic Coagulation on Gelatin Films by Means of a Faraday Cage. Intl J Biometerol 22:227-232.

53. Bayal, M.H. (1969) About the influence of Solar Activity on Atmospheric Precipitation. [In Russian] Trudy Kazh, Nauchno-issledov Gidromet Instit 38:22-24.

54. Baxenall, J. (1925) Meterological Periodicities of the Order of a Few Years. Q J Roy Meteor Soc 51:371-392.

55. Becker, U. (1955) Relation Between Bismuthoxychloride Precipitation (Piccardi's Inorganic Test) and Solar Activity. Arch Meteorol Geophys Bioklim B6:511-516.

56.   Beer, J., et al. (1990) Use of 10Be in Polar Ice to Trace the 11-Year Cycle of Solar Activity. Nature 347:164-166.

57.   Belov, S.V. (1986) Periodicity of Present and Past Volcanism of the Earth. Doklady Akad Nauk SSSR 291:421-425.

58.   Bender, M., et al. (1985) Isotopic Composition of Atmospheric O2 in Ice Linked with Deglaciated Global Primary Productivity. Nature 318:349-352.

59.   Ben-Menahem, A., Israel, M. (1970) Effects of Major Seismic Events on the Rotation of the Earth. J Geophys 19:367-393..

60.   Best, S. (1978) Lunar Influence on Plant Growth: A Review of the Evidence. Phenomena (May-Aug):18-21.

61.   Bieber, J.W., et al. (1990) Variation of the Solar Neutrino Flux with the Sun's Activity. Nature 348:407-411.

62.   Bigg, E.K. (1967) Influence of the Planet Mercury on Sunspots. Astron J 72:463-466.

63.   Bitvinskas, T.T.(1977) Relationship Between Solar Activity, Climate and the Growth of Tree Stands. pp.53-57 in (eds) M.N. Gnevyshev, A.I. Oi'. Effects of Solar Activity on the Atmosphere and Biosphere. Jersulem, Israel Prog Sci Trans.

64.   Blanco, C., Catalano, S. (1975) Correlation of the Air Temperature at Catania and the Sunspot Cycle. J Atmos Terr Phys 37:185-187.

65.   Blanton, J.O., et al. (1989) Wind Stress and Heat Fluxes Observed During Winter and Spring 1986. J Geophys Res 94C:10686-10698.

66.   Bolotinskaya, M., Sleptzov, B. (1964) About the Influence of Solar Activity on Long-Term Changes of the Frequency of Atmospheric Circulation Forms. Problemy Artiki i Antartki 18:48-56.

67.   Bortels, H. (1954) [Two Simple Meteorological Models for Understanding Reactions in Relationship to Low Pressure and Solar Activity]. Zwei Einfache Model-Meteoroligischer Reaktionen in Ihren Beziehungen zu Luftdruckanderungen and zur Solaraktiviat. Arch Meteor Geophys Bioklim 5:234-257.

68.   Bortels, H., et al. (1963) [Variations in the Limit of the Minimum Level for Freezing Water Correlated with Changes in Atmospheric Pressure and Solar Activity] Wechselende Gefrierbereitschaft Kleiner Wassermengen im Vergleich mit Anderungen Des Luftdrucks und der Sonnenaktivitat. pp.1053-1063 in (eds) S.W. Tromp, W.H. Weihe, Biometeorology. Vol.2. Amsterdam, Swets & Zeitlinger.

69.   Bossolasco, M., et al. (1972) Solar Flare Control of Thunderstorm Activity. in (ed) G. Alverti. Studi in Onore. Italy, Rpt Univ Navali di Napoli.

70.   Bowhill, S.A. (1970) Ionosphere Effects in Solar Eclipses. pp.3-17 in (ed) M. Anastassiades. Solar Eclipses and the Ionosphere. NY, Plenum.

71.   Bown, W. (1991) Solar Flares Send Navigators Off in the Wrong Direction. New Sci 130(22 June):15.

72.   Bradley, D.W., et al. (1962) Lunar Synodical Period and Widespread Precipitation. Science 137:748-749.

73.   Bray, R.J., Loughhead, R.E. (1979) Sunspots. NY, Dover Pubs..

74.   Brier, G.W. (1965) Lunar Tides, Precipitation Variations and Rainfall Calendaricities. Trans N Y Acad Sci 27:676-688.

75.   Brier, G.W., Bradley, D.A. (1964) Lunar Synodical Period and Precipitation in the United States. J Atmos Sci 21:386-395.

76. Brosche, P. (1980) Mass-Angular Momentum-Diagram of Astronomical Objects. pp.375-382 in P.G. Bergmann, V. De Sabatta (eds) Cosmology and Gravitation: Spin, Torsion, Rotation and Supergravity. NY, Plenum.

77. Brown, E.W. (1900) Possible Explanation of the Sunspot Period. Mon Not Roy Aston Soc: 60.

78. Brown, F.A., Chow, C.S. (1973) Interorganismic and Environmental Influences Through Extremely Weak Electromagnetic Fields. Biol Bull (Woods Hole, Mass) 144:437-461.

79. Brown, F.A., Chow, C.S. (1973) Lunar-Correlated Variations in Water Uptake by Bean Seeds. Biol Bull (Woods Hole, Mass) 145:265-278.

80. Brown, F.A., Park, Y.A. (1965) Phase-Shifting, A Lunar Rhythm in Planarians by Altering the Horizontal Magnetic Vector. Biol Bull (Woods Hole, Mass) 129:78-86.

81. Brown, G.M. (1974) New Solar-Terrestrial Relationship. Nature 251:592-594.

82. Browne, P.F. (1987) Magnetic Vortex Tubes and Charge Acceleration. pp.211-221 in (eds) R. Beck, R. Grave. Interstellar Magnetic Fields—Observation and Theory. NY, Springer Verlag..

83. Budyko, M.I. (1968) On the Causes of Climatic Variations. Sverig Meteor Och Hydrol Instit Meddelanden, Stockholm Series B 28:6-13.

84. Budyko, M.I. (1969) Effect of Solar Variations on the Climate of the Earth. Tellus 21:611-9

85. Bureau, R.A., Craine, L.B. (1970) Sunspots and Planetary Orbits. Nature 228;984.

86. Burlaga, L.F., et al. (1989) Large-Scale Fluctuations in the Solar Wind at 1 AU:1978-1982. J Geophys Res 94A:177-184.

87. Burr, H.S. (1945) Diurnal Potentials in the Maple Tree. Yale J Biol Med 17:727.

88. Burr, H.S. (1972) Blueprint for Immortality: The Electric Patterns of Life. London, Neville Spearman.

89. Callis, L.B., Natarajan, M. (1986) Ozone and Nitrogen Dioxide Changes in the Stratosphere During 1979-1984. Nature 323:772-777.

90. Callis, L.B., et al. (1979) Ozone and Temperature Trends Associated with the 11-Year Solar Cycle. Science 204:1303-6.

91. Campbell, B.O. (1964) Solar and Lunar Periodicities in Oxygen Consumption by the Mealworm, *Tenebrio Molitor*. Northwestern Univ, Thesis.

92. Campbell, W.H. (1983) Possible Tidal Modulation of the Indian Monsoon Onset. PhD Thesis. Madison, Univ Wisconsin.

93. Campbell, W.H., et al. (1983) Long-Period Tidal Forcing of Indian Monsoon Rainfall: An Hypothesis. J Climate Appl Meteorol 22:289-296.

94. Capel-Boute, C. (1962) [Chemical and Clinical Tests in the Study of Environmental Geophysical Factors] Tests Chimiques et Tests Cliniques dans L'etude des Facteurs Geophysiques de L'ambiance. pp.239-261 in IX Conegno della Salute, Ereditarieta-Ambiente Alimentazione, Ferrara.

95. Carpenter, T.H., et al. (1972) Observed Relationships Between Tidal Cycles and Formation of Hurricanes and Tropical Storms. Mon Weather Rev 100:451-460.

96. Carrea, G. (1978) Sun-Weather Relationships at Oxford. J Atmos Terrestr Phys 40:179-185.

97. Carter, W.E., et al. (1984) Variations in the Rotation of the Earth. Science 224:956-961.

98. Casetti, G., et al. (1981) A Statistical Analysis in Time of the Eruptive Events on Mount Etna (Italy) from 1323 to 1980. Bull Volcanol 44:283-294.

99.   Cevolani, G., et al. (1986) Atmospheric Tides and Periodic Variations in the Precipitation Field. Nuova Cimento (Italy) 9C:729-60.

100. Challinor, R.A. (1971) Variations in the Rate of Rotation of the Earth. Science 172:1022-4.

101. Chapman, S. (1961) Earth's Magnetism. NY, J. Wiley.

102. Chen-Ju-Ying (1984) Tendency Prediction of Precipitation and Inundation in July in the Sechwan Basin, China. J Climatol 4(5):521-9.

103. Chernosky, E.J. (1966) Double Sunspot-Cycle Variation in Terrestrial Magnetic Activity, 1884-1963. J Geophys Res 71:965-974.

104. Chernyshev, V.B. (1977) Disturbance Level of the Geomagnetic Field and the Motor Activity of Insects. pp.247-258 in (eds) M.N. Gneyshev, A.I. Oï. Effects of Solar Activity on the Atmosphere and Biosphere. Jerusalem Prog Sci Trans.

105. Chinnery, M.A., Landers, T.E. (1975) Evidence for Earthquake Triggering Stress. Nature 258:490-493.

106. Chitre, S.M., et al. (1973) Solar Neutrinos and a Central Magnetic Field in the Sun. Astrophys Lett 14:37-40.

107. Chizhevsky, A.L. (1934) [Effect of Ionization in an Artificial Atmosphere Upon Healthy Organisms and Diseased Organisms] Action de L'Ionisation de L'Atmosphére Artificelle de L'Air sur les Organismes Sains et les Organismes Maladies. pp.661-673 in (ed) M. Piery, Traite de Climatologue Biologique et Medicale. Vol. 1. Paris, Masson.

108. Chizhevsky, A.L. (1938) [Epidemics and Changes in the Electromagnetism of the Outer Atmosphere] Les Épidemies et les Perturbations Electromagnetiques du Milieu Exterieur. Paris, Masson.

109. Chupp, E.L. (1990) Transient Particle Acceleration Associated with Solar Flares. Science 250:229-236.

110. Clark, D.H., Stephenson, F.R. (1978) Oriental Sun Spot Records: An Interpretation of the Pretelescopic Sunspot Records from the Orient. Q J Roy Astron Soc 19:387-410.

111. Claverie, A., et al. (1981) Rapid Rotation of the Solar Interior. Nature 293:443-445.

112. Clayton, D.O. et al. (1975) Solar Models of Low Neutrino Counting Rate: The Central Black Hole. Astrophys J 201:489-493.

113. Clayton, H.H. (1923) World Weather: Including a Discussion of the Influence of Variations of Solar Radiation on the Weather and Meteorology of the Sun. NY, MacMillan.

114. Cliver, E.W., et al. (1982) Injection Onsets of 2 GeV Protons, 1 MeV Electrons, and 100 KeV Electrons in Solar Cosmic Ray Flares. Astrophys J 260:362-370.

115. Coe, R.S., Prevot, M. (1989) Evidence Suggesting Extremely Rapid Field Variation During a Geomagnetic Reversal. Earth Planet Sci Lett 92:292-298.

116. Corliss, W. (comp) (1982) Lightning, Auroras, Nocturnal Lights and Related Luminous Phenomena. A Catalog of Geophysical Anomalies. Glen Arm, MD, Sourcebook Project.

117. Corliss, W. (comp) (1983) Earthquakes, Tides, Unidentified Sounds and Related Phenomena: A Catalog of Geophysical Anomalies. [and] (1989) Anomalies in Geology. Glen Arm, Md, Sourcebook Project.

118. Corliss, W. (comp) (1985) The Moon and the Planets: A Catalog of Astronomical Anomalies. Glen Arm, Md, Sourcebook Project.

119. Coroniti, S.C. (ed) (1965) Problems of Atmospheric and Space Electricity. Proc 3rd Intl Conf, Montreux Switz, 5-10 May 1963. NY, Elsevier.

120. Courtillot, V., LeMouel, J.L. (1984) Geomagnetic Secular Variation Impluses. Nature 311:709-716.

121. Craig, P.P., Watt, K. (1985) The Kondratmeff Cycle and War: How Close is the Connection? Cycles 36:43-46.

122. Creber, G.T. (1975) Effects of Gravity and the Earth's Rotation on the Growth of Wood. pp.75-87 in (eds) G.D. Rosenberg, S.K. Runcorn, Growth Rhythms, and the History of the Earth's Rotation. NY, J Wiley.

123. Crutzen, P.J., et al. (1975) Solar Proton Events: Stratospheric Sources of Nitric Oxide. Science 189:457-459.

124. Currie, G.B. (1980) Detection of the 11-Year Sunspot Cycle Signal in the Earth's Rotation. Geophys J Roy Astron Soc 61:131-140.

125. Currie, R.G. (1949) Period, Q and Amplitude of the Pole Tide. Geophys J Roy Astron Soc 43:73-86.

126. Currie, R.G. (1952) Spectrum of Sea Level from 4 to 40 Years. Geophys J Roy Astron Soc 46:513-520.

127. Currie, R.G. (1973) Geomagnetic Line Spectra—2 to 70 Years. Astrophys Space Sci 21:425-438.

128. Currie, R.G. (1974) Solar Cycle Signal in Surface Air Temperature. J Geophys Res 79:5657-5660.

129. Currie, R.G. (1975) Lunar Terms in the Geomagnetic Spectrum at Hermanus. J Atmos Terres Phys 32:439-46.

130. Currie, R.G. (1979) Distribution of Solar Cycle Signal in Surface Air Temperature Over North America. J Geophys Res 84:753-761.

131. Currie, R.G. (1980) Detection of the 11-Year Sunspot Cycle Signal in Earth Rotation. Geophys J Roy Astronom Soc 61:131-139.

132. Currie, R.G. (1981) Amplitude and Phase of the 11-Year Term in Sea Level: Europe. Geophys J Roy Astron Soc 67:547-556.

133. Currie, R.G. (1981) Evidence for 18.6-Year Signal in Temperature and Drought Conditions in North America Since A.D. 1800. J Geophys Res 86:11055-11064.

134. Currie, R.G. (1981) Solar Cycle Signal in Air Temperature in North America: Amplitude, Gradient, Phase and Distribution. J Atmos Sci 38:808-818.

135. Currie, R.G. (1981) Solar Cycle Signal in Earth Rotation: Nonstationary Behavior. Science 211:386-389.

136. Currie, R.G. (1982) Evidence for 18.6-Year Term in Air Pressure in Japan and Geophysical Implications. Geophys J Roy Astron Soc 69:321-327.

137. Currie, R.G. (1983) Detection of 18.6-Year Nodal Induced Drought in the Patagonian Andes. Geophys Res Lett 10:1089-1092.

138. Currie, R.G. (1984) Evidence for 18.6-Year Lunar Nodal Drought in Western North America During the Past Millennium. J Geophys Res 89:1295-1308.

139. Currie, R.G. (1984) On Bistable Phasing of 18.6-Year Induced Flood in India. Geophys Res Lett 11:50-53.

140. Currie, R.G. (1984) Periodic 18.6-Year and Cyclic 11-Year Induced Drought and Flood in Western North America. J Geophys Res 89:7215-7230.

141. Currie, R.G. (1987) Examples of 18.6-Year and 11-Year Terms in World Weather Records. The Implications. in (eds) M.R. Rampino, et al., Climate: History, Periodicity and Predictability. NY, Van Nostrand.

142. Currie, R.G. (1987) On Bistable Phasing of 18.6-Year Induced Drought and Flood in the Nile Records Since AD 650. J Climatol 7:373-389.

143. Currie, R.G. (1987) Periodic 18.6-Year Signal in Northeastern United States Precipitation Data. J Climatol 8:255-281.

144. Currie, R.G. (1988) Trinity Wave in Climate and Economics. in (eds) G.J. Erickson, C.R. Smith. Maximum-Entropy and Bayesian Methods in Applied Statistics. Boston, Kluwer.

145. Currie, R.G. (1988) Climatically Induced Cyclic Variations in United States Crop Production: Implications in Economic and Social Science. pp.181-242 in G.J. Erickson, C.R. Smith. Maximum-Entropy and Bayesian Methods in Applied Statistics. Vol.2. Boston, Kluwer Acad.

146. Currie, R.G., et al. (1985) Cyclic Variations in U.S. Corn Production. Paper presented in England, Europe and Federal Reserve Bank. Minneapolis, May.

147. Currie, R.G., Fairbridge, R.W. (1985) Periodic 18.6-Year and Cyclic 11-Year Induced Drought and Flood in Northeastern China and Some Global Implications. Q Sci Rev 4:109-134.

148. Davison, C. (1927) Clustering and Periodicity of Earthquakes. Nature 120:587-588.

149. Davison, C. (1934) Diurnal Periodicity of Earthquakes. J Geol 42:449-468.

150. Davydova, A.I. (1971) About Spectral Methods of Hydrological Data Investigations. Metrorologia i Gidrologia 3:72-79.

151. Dehsara, M., Cehak, K. (1970) A Global Survey on Periodicities in Annual Mean Temperatures and Precipitation Totals. Arch Met Geoph Biokl Ser B 18:253-275.

152. DeJager, C. (1988) Solar Flares Through Electric Current. pp.311-320 in (eds) C.-G. Falthammer, et al., Plasma and the Universe. Boston, Kluwer Acad Pubs [and] Astrophys Space Phys 144:311-320.

153. De Mendoca Dias, A.A. (1962) Volcano of Capelinhos (Azores), Solar Activity and the Earth Tides. Bull Volcanol 24:211-21.

154. De Maria, M., et al. (1989) Observations of Mesoscale Wave Disturbances During the Gensis of Atlantic Lows Experiment. Mon Weather Rev 117:826-842.

155. Dessler, A.J. (1975) Some Problems in Coupling Solar Activity to Meteorological Phenomena. Possible Relationships Between Solar Activity and Meteorological Phenomena. NASA Spec Pub SP-366.

156. Dewey, E.R. (1968) Economic and Sociological Phenomena Related to Solar Activity and Influences. Cycles 19:201-214.

157. Dewey, E.R. (1971) Cycle Synchronicities. J Interdiscipl Cycle Res 2(3):331-362 [and] (1970) Cycles 21:190-223.

158. Dewey, E.R. (1971) Study of Possible Cyclic Patterns in Human Aggressiveness Leading to National and Internal Conflicts. J Interdiscipl Cycle Res 12(1):17-21.

159. Dewey, E.R. (1985) What Forces Could Cause Cycles? Cycles 36(3):68-69.

160. Dewey, E.R. (1989) Key to Sunspot-Planetary Relationship. Cycles 40:163-166.

161. Dewey, E.R. (1989) A List of Rhythms Variously Determined and/or Alleged. Cycles 40:238-241.

162. Dey, L.M. (1882) Sci Amer 47:16.

163. Dezso, L., et al. (1984) Sunspot Motions and Magnetic Shears as Precursors of Flares. Adv Space Res 4(7):57-60.

164. Dicke, R.H. (1978) Is There a Chronometer Hidden Deep in the Sun? Nature 276:676-680.

165. Dickinson, R.E. (1975) Solar Variability and the Lower Atmosphere. Bull Amer Meteorol Soc 56:1240-48.

166. Dickman, S.R. (1986) Damping of the Chandler Wobble and the Pole Tide. pp.203-228 in (ed) A. Cazenave, Earth Rotation: Solved and Unsolved Problems. Boston, Riedel.

167. Dirks, R.A., et al. (1989) Genesis of Atlantic Lows Experiment (GALE): An Overview. Bull Amer Meteorol Soc 69:148-160.

168. Djurovic, D. (1986) Atmospheric Circulation, the Earth's Rotation and Solar Activity. pp.187-192 in (ed) A. Cazenave, Earth Rotation: Solved and Unsolved Problems. Boston, Riedel.

169. Djurovic, D., Paquet, P. (1989) A 120-day Oscillations in the Solar Activity and Geophysical Phenomena. Astron Astrophys 218:302-306.

170. Dobson, H.W., Hedeman, E.R. (1964) An Unexpected Effect in Solar Cosmic Ray Data Related to 29.5 Days. J Geophys Res 69:3965-3971.

171. Douglass, A.E. (1971) Climatic Cycles and Tree Growth. NY, Whelon and Wesley, Stechert-Hafner Service Agency.

172. Dubrov, A. P. (1978) Geomagnetic Field and Life: Geomagnetobiology. NY, Plenum.

173. Duffy, P.B., et al. (1977) Influence of Interplanetary Magnetic Polarity on the Area of Tropospheric Troughs. Eos 58:1220.

174. Dzurisin, D. (1980) Influence of Fortnightly Earth Tides at Kilauea Volcano, Hawaii. Geophys Res Lett 7:925-928.

175. Eddy, J.A. (1976) Maunder Minimum. Science 192:1189-1202.

176. Eddy, J.A. (1976) The Sun Since the Bronze Age. Amer Geophys Union Paper, Intl Symp Solar-Terrestrial Physics. Wash, DC, Amer Geophys U.

177. Eddy, J.A. (1977) Anomalous Solar Rotation in the Early 17th Century. Science 198:824-9.

178. Eddy, J.A. (1978) Historical and Arboreal Evidence for a Changing Sun. pp.11-34 in (ed) J.A. Eddy, The New Solar Physics. Boulder, Colo, Westview.

179. Eddy, J.A. (1980) Historical Record of Solar Activity. pp.119-134 in (eds) R.O. Pepin, et al., Ancient Sun: Fossil Record in the Earth, Moon and Meteorites. NY, Pergammon.

180. Eddy, J.A. (1980) Solar Influence on the Earth's Atmosphere. pp.125-130 in Atmospheric Sciences: National Objectives for the 1980's. Wash, DC, Natl Acad Sci.

181. Eichmeier, J., Buger, P. (1969) [The Influence of Lightning's Electromagnetic Effects Upon the Precipitation Reaction of Bismuth Chloride According to the Piccardi Test] Uber den Einfluss Elektromagneitscher Strahlung auf die Bismuthchlorid-Fallungsreaktion nach Piccardi. Intl J Biometeor 13:239-256.

182. Elton, C. (1977) Periodic Fluctuations in the Numbers of Animals, Their Causes and Effects. Brit J Exp Biol 2:1924.

183. Emetz, A.I., Korsun, A.A. (1979) On the Long-Period Variations in the Rate of the Earth's Rotation. pp.59-60 in (eds) D.D. McCarthy, J.D. Pilkington. Time and the Earth's Rotation. Boston, Reidel.

184. Eos (1985) Transactions. p.441. (Amer Geophys Union).

185. Epperson, D. (1989) The 9-Year Cycle. Cycles 40:338-341.

186. Epperson, D. (1990) Wheeler's Climate Curve and War. Cycles 41:265-282.

187. Eubanks, T.M., et al. (1988) Causes of Rapid Motions of the Earth's Pole. Nature 334:115-119.

188. Fairbridge, R.W. (1990) Solar and Lunar Cycles Embedded in the El Nino Periodicities. Cycles 41:66-73.

189. Fairbridge, R.W. (1984) Nile Floods as a Global Climatic/ Solar Proxy. pp.181-190 in (eds) N.A. Morner, W. Karlen, Climatic Changes on Yearly to Millennial Basis. Boston, Reidel.

190. Fairbridge, R.W. (1984) Planetary Periodicities and Terrestrial Climatic Stress. pp.509-520 in (eds) N.A. Morner, W. Karlen, Climatic Changes on a Yearly to Millennial Basis. Boston, Reidel.

191. Fairbridge, R.W., Hameed, S. (1983) Phase Coherence of Solar Cycle Minima Over Two 178-Year Periods. Astron J 88:867-69.

192. Fairbridge, R.W., Hillaire-Marcel, C.H. (1977) An 8000-Year Palaeoclimatic Record of the Double-Hale's 45-Yr Solar Cycle. Nature 268:413-416.

193. Fairbridge, R.W., Sanders, J.E. (1987) Sun's Orbit, AD 750-2050: Basis for Perspectives on Planetary Dynamics and Earth-Moon Linkage. pp.446-471 in (ed) M.R. Rampino, et al., Climate: History, Periodicity and Predictability. NY, Van Nostrand Reinhold.

194. Farman, J.C. (1985) Large Losses of Total Ozone in Antarctica Reveal Seasonal CLOx/NOx Interaction. Nature 315:207-210.

195. Ferencz, Cs., Tarcsai, Gy. (1971) Interaction of Gravitational and Electromagnetic Fields or Another Effect? Nature 233:404-406.

196. Ferulano, F. (1986) Parma Earthquake (Northern Italy) and Main Aftershock—Nov. 9th, 1983. [In Italian] Boll Geofis Teor Appl (Italy) 28(109):65-72.

197. Feynman, J., Crooker, N.U. (1978) Solar Wind at the Turn of the Century. Nature 275:626-627.

198. Fischer, W.H., et al. (1968) Laboratory Studies on Fluctuating Phenomena. Intl J Biometeor 12:15-19.

199. Fisk, L.A. (1980) Solar Modulation of Galactic Cosmic Rays. pp.103-118 in (eds) R.O. Pepin, et al., The Ancient Sun: Fossil Record in the Earth, Moon and Meteorites. NY, Pergammon.

200. Fleming, S.T., Aitken, M.J. (1974) Radiation Dosage Associated with Ball Lightning. Nature 252:220-221.

201. Franz, R.C., et al. (1990) Television Image of a Large Upward Electrical Discharge Above a Thunderstorm System. Science 249:48-51.

202. Fremerman, B.J. (1977) Solar and Economic Relationships: A Revised Report. Cycles 28:149-50.

203. Friedman, H., Becker, R.O. (1963) Geomagnetic Parameters and Psychiatric Hospital Admissions. Nature 200:626-628.

204. Gaffin, S. (1987) Phase Difference Between Sea Level and Magnetic Reversal Rate. Nature 329:816-819.

205. Gaizauskas, V. (1983) Relation of Solar Flares to the Evolution and Proper Motions of Magnetic Fields. Adv Space Res 2(11):11-30.

206. Gary, D.E., Hurford, G.J. (1990) Multifrequency Observations of a Solar Microwave Burst with Two-Dimensional Spatial Resolution. Astrophys J 361:290-299.

207. Gauquelin, F. (1972) Possible Planetary Effects in Heredity: Refutation of Former Demographical and Astronomical Objections. J Interdiscipl Cycle Res 3:373-380.

208. Gauquelin, M. (1971) Methodological Model Analysing Possible Extra-Terrestrial Effects on the Daily Cycle of Birth. J Interdiscipl Cycle Res 2:219-226.

209. Gauquelin, M. (1972) Possible Planetary Effects at the Time of Birth of `Successful' Professionals: An Experimental Control. J Interdiscipl Cycle Res 3:381-390.

210. Gauquelin, M., et al. (1979) Personality and Position of the Planets at Birth. An Empirical Study. Brit J Soc Clin Psych 18:71-75.

211. Gerety, E.J. (1978) Possible Sun-Weather Correlation. Nature 275:775.

212. Gerety, E.J., et al. (1977) Sunspots, Geomagnetic Indices and the Weather: A Cross-Spectral Analysis, Activity and Global Weather Data. J Atmosph Sci 34:673-678.

213. Gergely, T.E. (1983) Summary of Significant Solar-Initiated Events During STIP Interval XII. Adv Space Res 2(11):271-84.

214. Gering-Galaktionova, I.V., Kupriyanov, S.N. (1977) Do Changes in Solar Activity Affect Onocological Morbidity? pp.224-228 in (eds) M.N. Gnevyshev, A.I. Oi'. Effects of Solar Activity on the Earth's Atmosphere and Biosphere. Jerusalem, Israel Prog Sci Trans.

215. Giordano, A. (1957) Blood Sedimentation Rate (A Study of the Influence of Environment). Intl J Bioclimatol Biometerol 1(4).

216. Glembotskiy, Y.L., et al. (1962) Effects of Cosmic Flight Factors on the Incidence of Recessive Lethal Mutations in the X-Chromosome of Drosophilia Melanogaster. pp.243-254 in (ed) N.M. Sisakyan, Problems of Space Biology. Vol. 1, Moscow, USSR Acad Sci Pub [in Russian]. Trans. NASA Report TTF-174.

217. Gnevyshev, M.N., et al. (1977) Sudden Death from Cardiovascular Diseases and Solar Activity. p.209 in (eds) M.N. Gnevyshev, A.I. Oi', Effects of Solar Activity on the Earth's Atmosphere and Biosphere. Jersusalem, Israel Prog Sci Trans.

218. Gnevyshev, M.N., Novikova, K.F. (1972) Influence of Solar Activity on the Earth Biosphere. Part I. J Interdiscipl Cycle Res 3:99-104.

219. Gnevyshev, M.N., Oi', A.I. (1977) Effects of Solar Activity on the Earth's Atmosphere and Biosphere. Jersusalem, Israel Prog Sci Trans.

220. Goodman, M., et al. (1979) Frozen Mammoth Muscle: Preliminary Findings. Paleopathol Newslett 25:3-5.

221. Goodwin, G.L., Hobson, G.J. (1978) Atmospheric Gravity Waves Generated During a Solar Eclipse. Nature 275:109-111.

222. Gray, W.M. (1990) Strong Association Between Western African Rainfall and U.S. Landfall of Intense Hurricanes. Science 249:1251-1255.

223. Gribbin, J. (1971) Relation of Sunspot and Earthquake Activity. Science 173:558.

224. Gribbin, J. (1973) Planetary Alignments, Solar Activity and Climatic Change. Nature 246:453-454.

225. Gribbin, J. (1975) Climate, The Earth's Rotation and Solar Variations. pp.413-425 in (eds) G.D. Rosenburg, S.K. Runcorn, Growth Rhythms and the History of the Earth's Rotation. NY, J Wiley.

226. Gribbin, J. (1977) Long-Term Astronomical Cycles and the Earth's Climate. J Interdiscipl Cycle Res 8:193-195.

227. Gribbin, J. (1977) New Chinese Results Tie Up Sun Cycles and Earth Weather. New Sci 76(1082):703.

228. Gribbin, J. (1978) Climatic Change. London, Cambridge Univ.

229. Gribbin, J. (1981) The Sun, the Moon and the Weather. New Sci 90(1258):754-66.

230. Gribbin, J. (1982) Stand By for Bad Winters. New Sci 96(1329):220-223.

231. Gribbin, J., Plagemann, J. (1973) Discontinuous Change in Earth's Spin Rate Following Great Solar Storm of August 1972. Nature 243:26-27.

232. Gribbin, J. Plagemann, S. (1975) Response to Meeus. Icarus 26:268-269.

233. Gross, R. (1979) Correlation Between Sunspot Cycles and Influenza. Cycles 30(3):59-6

234. Gupta, J.C., Chapman, S. (1969) Lunar Daily Harmonic Geomagnetic Variation as Indicated by Spectral Analysis. J Atmos Terres Phys 31: 233-252.

235. Hagyard, M.J., et al. (1984) Role of Magnetic Field Shear in Solar Flares. Adv Space Res 4(7):71-80.

236. Hale, C.J. (1987) Palaeomagnetic Data Suggest Link Between the Archaean-Proterozoic Boundary and Inner-Core Nucleation. Nature 329:233-237.

237. Hall, E. (1953) Introduction to Electron Microscopy. NY, McGraw-Hill, p.299.

238. Hameed, S. (1984) Fourier Analysis of Nile Flood Level. Geophys Res Lett 11:843-845.

239. Hameed, S., Currie, R.G. (1985) An Analysis of Long-term Variations in the Flood Levels of the Nile River. pp.68-69 in Third Conf Climate Variations & Symp on Contemporary Climate:1850-2100. Boston, Mass, Amer Meterolo Soc.

240. Hameed, S., et al. (1983) An Analysis of Periodicities in the 1470 to 1974 Beijing Precipitation Record. Geophys Res Lett 10:436-439.

241. Hamilton, W.L. (1973) Tidal Cycles of Volcanic Eruptions: Fortnightly to 19 Yearly Periods. J Geophys Res 78:3363-3375.

242. Haminov, N.A. (1968) Solar Activity and Long-Range Oscillations of Atmospheric Processes. Izvestiya Vses Geograf Obsheh 2:103-108.

243. Hamza, V.M., Beck, A.E. (1972) Terrestrial Heat Flow, the Neutrino Problem, and a Possible Energy Source in the Core. Nature 240:343-344.

244. Hannula, H. (1987) In Search of the Cause of Cycles. Cycles 38:159-166.

245. Hanson, K., Cotton, G. (1983) Temperature Variation in the United States During June: A Search for Causes and Mechanisms. pp.591-602 in (ed) B.M. McCormac, Weather and Climate Responses to Solar Variations. Boulder, Co, Assoc Univ.

246. Hanson, K., et al. (1987) Precipitation and the Lunar Synodic Cycle: Phase Progression Across the United States. J Clim App Meteorolo 26:1358-1362.

247. Harrison, R.A., Simnett, G.M. (1984) Do All Flares Occur within a Hierarchy of Magnetic Loops? Adv Space Res 4(7):199-202.

248. Hartzell, S., Heaton, T.H. (1989) Fortnightly Tide and the Tidal Triggering of Earthquakes. Bull Seismolo Soc Amer 79:1282-1286.

249. Harvey, J.W. (1983) Flare Build Up: 21 May 1980. Adv Space Res 2(11): 31-37.

250. Hasegawa, I. (1980) Catalogue of Ancient and Naked-Eye Comets. Vistas Astron 24:59-102.

251. Hauenschild, C. (1960) Lunar Periodicity. Cold Spr Harbor Symp Quant Biol 25:491-497.

252. Haynes, G. (1991) Mammoths, Mastodons and Elephants. NY, Cambridge Univ.

253. Heath, D.F., et al. (1975) Relation of the Observed Far Ultraviolet Solar Irradiance to the Solar Magnetic Sector Structure. Solar Phys 45:79-82.

254. Heath, D.F., et al. (1977) Solar Proton Event: Influence on Stratospheric Ozone. Science 197:886-9.

255. Heath, D.F., et al. (1986) Solar Proton Event: Influence on Stratospheric Ozone. Science 206:1272-1276.

256. Heaton, T.H. (1975) Tidal Triggering of Earthquakes. Geophys J Roy Astron Soc 43:307-326.

257. Heaton, T.H.E., et al. (1986) Climatic Influence on the Isotopic Composition of Bone Nitrogen. Nature 322:822-823.

258. Henoux, J.-C. (1984) Study of Energy Release in Flares. Adv Space Res 4(7):227-237.

259. Henriksen, K., et al. (1977) Lunar Influence on the Occurrence of Aurora. J Geophys Res 82:2842-2846.

260. Herman, J.R., Goldberg, R.A. (1978) Initiation of Non-Tropical Thunderstorms by Solar Activity. J Atm Terr Phys 40:121-134.

261. Herman, J.R., Goldberg, R.A. (1978) Sun, Weather and Climate. NASA SP-426. Wash, DC, USGPO.

262. Heyvaerts, J. (1983) High Energy Particle Acceleration in Flares. Adv Space Res 2(11):187-192.

263. Hill, B.T., Jones, S.J. (1990) Newfoundland Ice Extent and the Solar Cycle from 1860-1988. J Geophys Res 95:5385-5394.

264. Hillhouse, J., Cox, A. (1976) Brunhes-Matuyama Polarity Transition. Earth Planet Sci Lett 29:51-64.

265. Hines, C.O., Halevy, I. (1975) Reality and Nature of a Sun-Weather Correlation. Nature 258:313.

266. Hines, C.O., Halevy, I. (1977) On the Reality and Nature of a Certain Sun-Weather Correlation. J Atmos Sci 34:382-404.

267. Hoffman, G.W., Hopper, V.D. (1968) Electric Field and Conductivity Measurements in the Stratosphere. pp. 475-494 in (eds) S.C. Coronti, J. Hughes. Planetary Electrodynamics. Vol. 2. NY, Gordon & Breach.

268. Hoffman, K.A. (1986) Transitional Field Behaviour from the Southern Hemisphere Lavas: Evidence for Two-stage Reversals of the Geodynamo. Nature 320:228-32.

269. Hoffman, K.A. (1988) Ancient Magnetic Reversals: Clues to the Geodynamo. Sci Amer 258:76-83.

270. Holweger, H., et al. (1983) Sunspot Cycle and Associated Variation of the Solar Spectral Irradiance. Nature 302:125-126.

271. Hood, A.W. (1984) Stability of Magnetic Fields Relevant to Two-Ribbon Flares. Adv Space Res 4(7):49-52.

272. Hope-Simpson, R.E. (1979) Influence of Season Upon Type A Influenza. In (ed) S.W. Tromp, Biometeorological Survey 1973-1978. Vol.1, Part A. London, Heyden & Son.

273. Hope-Simpson, R.E. (1978) Sunspots and Flu: A Correlation. Nature 275:86.

274. Hoyle, F., Wickramasinghe, C. (1990) Sunspots and Influenza. Nature 343:304.

275. Hoyt, D.V. (1979) Variations in Sunspot Structure and Climate. Climatic Change 2:79-92.

276. Hsui, A. T. (1987) Borehole Measurement of the Newtonian Gravitational Constant. Science 237:881-883.

277. Hughes, D.W. (1974) Super-Rotation of the Upper Atmosphere. Nature 249:405-406.

278. Hughes, D.W. (1977) Planetary Alignments Don't Cause Earthquakes. Nature 265:13.

279. Hvistendahl, B. (1973) Correlation Between Polio and Sunspots and Northern Lights. Cycles 24(3):89-92.

280. Igarashi, C., Wakita, H. (1990) Groundwater Radon Anomalies Associated with Earthquakes. Tectonophysics 180:237-254.

281. Iijima, S., Okazaki, S. (1971) Short Period Terms in the Rate of Rotation and in the Polar Motion of the Earth. Bull Astron Soc Japan 24:109-125.

282. Illing, R.M.E., Hundhausen, A.J. (1983) Possible Observation of a Disconnected Magnetic Structure in a Coronal Transient. J Geophys Res 88:10210.

283. Imbrie, J., Imbrie, K.P. (1986) Ice Ages: Solving the Mystery. Cambridge, MA, Harvard.

284. Ip, W.H. (1976) Chinese Records on the Correlation of Heliocentric Planetary Alignments and Earthquake Activities. Icarus 29:435.

285. Iwasaki, N. (1982) Effect of the Northward Interplanetary Magnetic Field Inside the Region Surrounded with the Auroral Oval. J Geomag Geoelectr 34:9-25.

286. Jaggar, T.A. (1920) Seismometric Investigation of the Hawaiian Lava Column. Bull Seismol Soc Amer 10:155-275.

287. Jaggar, T.A., et al. (1924) Lava Tide, Seasonal Tilt, and Volcanic Cycle. Mon Weather Rev 52:142-145.

288. Jakubcova, I., Pick, M. (1987) Correlations Between Solar Motion, Earthquakes, and Other Geophysical Phenomena. Annales Geophysicae 5B: 135-142.

289. Jefferys, W.H. (1967) Nongravitational Forces and Resonances in the Solar System. Astron J 72:872-875.

290. Jiayu, X., et al. (1983) Morphology and Velocity Field of the Large Flare on the Solar Disc on July 14, 1980. Adv Space Res 2(11):221-224.

291. Johnson, M.O. (1946) Correlation of Cycles in Weather, Solar Activity, Geomagnetic Values and Planetary Configurations. SF, Philips & Van Orden.

292. Johnston, M.J.S., Mauk, F.J. (1972) Earth Tides and the Triggering of Eruptions from Mt. Stromboli, Italy. Nature 239:266-267.

293. Jones, L.M., Reasenberg, P. (1989) Preliminary Assessment of the Recent Increase of Earthquake Activity in the Los Angeles Region. U S Geol Survey Open-File Rpt 89-162.

294. Joselyn, J.A. (1985) Automatic Detection of Geomagnetic-Storm Sudden Commencements. Adv Space Res 5(4):193-198.

295. Ju-Ying, C. (1984) Tendency Prediction of Precipitation and Inundation in July in the Sichuan Basin, China. J Climatol 4:521-529.

296. Kalman, B. (1984) Magnetic Field Structure Changes in the Vicinity of Solar Flares. Adv Space Res 4(7):81-85.

297. Kanamori, H. (1977) Energy Release in Great Earthquakes. J Geophys Res 82:2981-2987.

298. Kardas, S.J. (1976) Significant Differences in Certain Forms of Human Behavior on Days of High and Low Solar Activity. 7th Intl Interdiscipl Cycle Res Symp, 23 June—3 July. Bad Homburg, Germany.

299. Kerr, R.A. (1979) East Coast Mystery Booms: Mystery Gone But Booms Linger On. Science 203:256.

300. Kerr, R.A. (1986) Taking Shots at Ozone Hole Theories. Science 234:817-818.

301. Kerr, R.A. (1988) Sunspot-Weather Link Holding Up. Science 242:1124-1125.

302. Kerr, R.A. (1989) Big Picture of the Pacific's Undulations. Science 243:739-740.
303. Khalil, M.A.K., Rasmussen, R.A. (1986) Interannual Variability of Atmospheric Methane: Possible Effects of the El Nino-Southern Oscillation. Science 232:56-58.
304. Killeen, T.L. (1985) Thermosphere as a Sink of Magnetospheric Energy: A Review of Recent Observations of Dynamics. Adv Space Res 5(4):257-266.
305. Kilston, S., Knopoff, L. (1983) Lunar-Solar Periodicities of Large Earthquakes in Southern California. Nature 304:21-24.
306. King, J.W. (1973) Solar Radiation Changes and the Weather. Nature 245:443-446.
307. King, J.W. (1974) Weather and the Earth's Magnetic Field. Nature 247:131-134.
308. King, J.W. (1975) Sun-Weather Relationships. Astronautics Aeronautics 13(4):10-19.
309. King, J.W., et al. (1977) Large Amplitude Standing Planetary Waves Induced in the Troposphere by the Sun. J Atmos Terr Phys 39:1357.
310. Kingan, S.G., et al. (1966) Apparent Decrease in Galactic Radio Noise During a Total Solar Eclipse. Nature 211:950.
311. Klecker, B. (1983) Propagation of Energetic Particles in the Solar Wind. Adv Space Res 2(11):285-292.
312. Klein, F.W. (1976) Earthquake Swarms and the Semidiurnal Solid Earth Tide. Geophys J Roy Astron Soc 45:245-295.
313. Klein, F.W. (1984) Eruption Forecasting at Kilauea Volcano Hawaii. J Geophys Res 89:3059-3073.
314. Klinowska, M. (1970) Lunar Rhythms in Activity, Urinary Volume and Acidity in the Golden Hamster (*Mesocricetus Auratus Waterhouse*). J Interdiscipl Cycle Res 1:317-322.
315. Klinowska, M. (1972) Comparison of the Lunar and Solar Activity Rhythms of the Golden Hamster (*Mesocricetus Auratus Waterhouse*). J Interdiscipl Cycle Res 3:145-150.
316. Ko-Chen, C. (1973) Preliminary Study on the Climatic Fluctuations During the Last 5,000 Years in China. Sci Sin 14(2):243-261.
317. Kockarts, G. (1981) Effects of Solar Variations on the Upper Atmosphere. Aeronomica Acta Inst D'Aeronomie Spat Belg A Nor 229:1-44.
318. Kokus, M.S. (1987) Earthquakes, Earth Expansion and Tidal Cycles. Cycles 38:192-197.
319. Kokus, M.S. (1988) Planetary Conjunctions and the Fundamental Laws of Nature. Cycles 39:160-161.
320. Kokus, M.S. (1988) Sunspots, Earthquakes, and the San Andreas Fault. Cycles 39:85-87.
321. Kokus, M.S. (1990) 18.6-Year and 19-Year Periods in Southern California and Upland Earthquakes. Cycles 41:76-79.
322. Kokus, M.S., Ritter, D. (1988) Earthquakes at the First and Last Quarter of the Lunar Cycle in the Southeastern United States. Cycles 39:56-57.
323. Kokus, M.S., Ritter, D. (1988) Earthquakes at the First and Last Quarter of the Moon. Cycles 39:28-29.
324. Kolisko, E., Kolisko, L. (1939) Planetary Influences Upon Crystallisation. p.73 in (ed) E. Kolisko, Agriculture of Tomorrow. Gloucester, UK, John Jennings Ltd.
325. Koppen, W. (1914) [Air Temperature, Solar Activity and Volcanic Eruptions.] Lufttemperaturen, Sonnenflecken und Vulkanausbruche. Meteorol Z 31:305.
326. Korneev, V.V., et al. (1983) On Soft Electron Beams in Solar Active and Flare Regions. Adv Space Res 2(11):139-144.

327. Kozyrev, N.A. (1972) On the Interaction Between Tectonic Processes of the Earth and the Moon. pp.220-225 in (eds) S.K. Runcorn, H. Urey. The Moon. Dordrecht, Reidel.

328. Krauss, L.M. (1990) Correlation of Solar Neutrino Modulation with Solar Cycle Variation in P-Mode Acoustic Spectra. Nature 348:403-410.

329. Krimigis, S.M., Venkatesan, D. (1988) In Situ Acceleration and Gradients of Charged Particles in the Outer Solar System Observed by the Voyager Spacecraft. pp.463-486 in (eds) C.-G. Falthammer, et al., Plasma and the Universe. Boston, Kluwer Acad. [and] Astrophys Space Phys 144:463-486.

330. Kuhn, J.R., et al. (1988) Surface Temperature of the Sun and Changes in the Solar Constant. Science 242:908-911.

331. Labitzke, K., van Loon, H. (1988) Associations Between the 11-Year Solar Cycle, the QBO, and the Atmosphere, Part I: The Troposphere and Stratosphere in the Northern Hemisphere in Winter. J Atmo Terres Phys 50:197-206.

332. Lada, C.J., Shu, F.H. (1990) Formation of Sun-like Stars. Science 248:564-572.

333. Laj, C., et al. (1987) Rapid Changes and Near Stationarity of the Geomagnetic Field During a Polar Reversal. Nature 330:145-148.

334. La Marche, Jr., V.C., Fritts, H.C. (1972) Tree Rings and Sunspot Numbers. Tree Ring Bull 32:19-33.

335. Lambeck, K., Hopgood, P. (1982) Earth's Rotation and Atmospheric Circulation: 1958-1980. Geophys J Roy Astron Soc 71:581-587.

336. Landsberg, H.E., Taylor, R.E. (1976) Spectral Analysis of Long Meteorological Series. J Interdiscipl Cycle Res 7:237-242.

337. Landsberg, H.E., Taylor, R.E. (1977) Statistical Analysis of Tokyo Winter Temperature Approximations, 1443-1970. Geophys Res Lett 4:105-107.

338. Landscheidt, T. (1981) Swinging Sun, 79-Year Cycle, and Climatic Change. J Interdiscipl Cycle Res 12:3-19.

339. Landscheidt, T. (1984) Cycles of Solar Flares and Weather. pp.473-481 in (eds) N.A. Morner, W. Karlen, Climatic Changes on a Yearly to Millennial Basis. Dordrect, Reidel.

340. Landscheidt, T. (1987) Long-Range Forecasts of Solar Cycles and Climate Change. pp.421-445 in (eds) M.R. Rampino, et al., Climate: History, Periodicity and Predictability. NY, Van Nostrand Reinhold.

341. Landscheidt, T. (1989) Sun-Earth-Man: A Mesh of Cosmic Oscillations. London, Urania Trust.

342. Landscheidt, T. (1989) Predictable Cycles in Geomagnetic Activity and Ozone Levels. Cycles 40:261-264.

343. Landscheidt, T. (1989) Mini-Crash in Tune with Cosmic Rhythms. Cycles 40:317-319.

344. Landscheidt, T. (1990) Relationship Between Rainfall in the Northern Hemisphere and Impulses of the Torque in the Sun's Motion. Cycles 41:128-133.

345. Lang, H.J. (1968) [Variations in the Lunar Period Upon Color Changes of the Guppy (Libistes Reticulatus)] Uber Lunarperiodische Schwankungen des Farbempfindlichkeit Beim Guppy (Libistes Reticulatus). Verh Dtsch Zool Ges 379-386.

346. Lang, H.J. (1977) Lunar Periodicity in Colour Sense of Fish. J Interdiscipl Cycle Res 8:317-321.

347. Lanzerotti, L.J., Gregori, G.P. (1986) Telluric Currents: The Natural Environment and Interactions with Man-Made Systems. pp.232-257 in The Earth's Electrical Environment. NRC Geophys Study Comm. Wash, DC, Natl Acad.

348. Larsen, M.F., Kelley, M.C. (1977) Study of an Observed and Forecasted Meteorological Index and its Relation to the Interplanetary Magnetic Field. Geophys Res Lett 4:337-340.

349. Larson, M.A. (1986) Do Planets Affect Our Weather? Sci Fair Rpt, Denver, Co.

350. Lastovicka, J. (1988) IMF Sector Boundary Effects in the Middle Atmosphere. Adv Space Res 8(7):201-204.

351. Lavrovskii, A.A. (1977) Periodic Activity of Natural Foci of Plague and Its Causes. pp.74-81 in (eds) M.N. Gnevyshev, A.I. Oi'. Effects of Solar Activity on the Earth's Atmosphere and Biosphere. Jerursalem, Israel Prog Sci Trans.

352. Le Mouel, J.L. (1984) Outer-Core Geostrophic Flow and Secular Variations of Earth's Geomagnetic Field. Nature 311:734-735.

353. Le Roy Ladurie, E. (1971) Times of Feast, Times of Famine. Garden City, NY, Doubleday.

354. Lethbridge, M.D. (1970) Relationship Between Thunderstorm Frequency and Lunar Phase and Declination. J Geophys Res 75:5149.

355. Lethbridge, M.D. (1981) Cosmic Rays and Thunderstorm Frequency. Geophys Res Lett 8:521.

356. Libby, L.M. (1983) Past Climates: Tree Thermometers, Commodities, and People. Austin, Univ Texas Press.

357. Link, F. (1964) [Manifestations of Solar Activity from Past History] Manifestations de L'Actvite Solaire Dans le Passe Historique. Planet Space Sci 12:333-348.

358. Link, F. (1981) Auroral Fluctuations A.D. 500-1600 Observed in China and in Europe. J Interdiscipl Cycle Res 12:129-32.

359. Lisitzin, E. (1974) Sea-Level Changes. NY, Elsevier.

360. Liritzis, Y., Petropoulus, B. (1987) Latitude Dependence of Auroral Frequency in Relation to Solar Terrestrial and Interplanetary Parameters. Earth, Moon & Planets 39:75-91.

361. Livshits, M.A., et al. (1979) On a Possible Connection Between the Irregularity of the Earth's Rotation and High-Latitude Magnetic Field of the Sun. Soviet Astron 23:328-331.

362. Loder, J.W., Garrett, C. (1978) 18.6-Year Cycle of Sea Surface Temperature in Shallow Seas Due to Variations in Tidal Mixing. J Geophys Res 83:1967-1970.

363. Lo-Pao-Yung, Li-Wei-Pao (1978) To Investigate the Stability of the Sunspot Period by Means of the Periodic Analysis of the Aurora and Earthquake Records in Ancient China. [in Chinese] Kexue Tongbao (China) 23(6):362-366.

364. Lund, I.A. (1965) Indications of a Lunar Synodical Period in Unites States Observations of Sunshine. Atmos Sci 22:24-39.

365. Luo-Shi-Fang, et al. (1977) Analysis of Periodicity in the Irregular Rotation of the Earth. Chinese Astron 1:221-227.

366. Machado, M.E. (1983) Energy Transfer in Solar Flares. Adv Space Res 2(11):115-133.

367. Machado, M.E., Somov, B.V. (1983) Flares of April 1980. Adv Space Res 2(11):101-104.

368. MacKenzie, T. (1886) Meteorological Phenomena. Nature 33:245.

369. Maksimov, A.A. (1977) Population Dynamics and Rhythms of Epizootics Among Rodents Correlated with Solar Activity Cycles. pp.58-72 in (eds) M.N. Gnevyshev, A.I. Oi', Effects of Solar Activity on the Atmosphere and Biosphere. Jersulaem, Israel Prog Sci Trans.

370. Malherbe, J.M., et al. (1983) Preflare Heating of Filaments. Adv Space Res 2(11):53-56.

371. Malin, S.R.C., Srivastava, B.J. (1979) Correlation Between Heart Attacks and Magnetic Activity. Nature 277:646-648.

372. Mannila, T. (1979) Lunar and Planetary Periodicity of Temperature and Rainfall in Helsinki, 1902-1977. Geophysica 16(1):97-107.

373. Mansinha, L., Smylie, D.E. (1967) Effect of Earthquakes on the Chandler Wobble and the Secular Polar Shift. J Geophys Res 72:4731-4743.

374. Mansurov, S.M., et al. (1975) Certain Regularities of Geomagnetic and Baric Field at High Latitudes. in (eds) W.R. Bandeen, S.P. Maran. Possible Relationship Between Solar Activity and Meteorological Phenomena. NASA SP-366, Wash, DC.

375. Markson, R. (1971) Considerations Regarding Solar and Lunar Modulation of Geophysical Parameters, Atmospheric Electricity and Thunderstorms. Pure Appl Geophys 84:161-202.

376. Markson, R. (1978) Solar Modulation of Atmospheric Electrification and Possible Implications for the Sun-Weather Relationship. Nature 273:103-109.

377. Markson, R. (1981) Modulation of the Earth's Electric Field by Cosmic Radiation. Nature 291:304.

378. Markson, R., Muir, R. (1980) Solar Wind Control of the Earth's Electric Field. Science 208:979-990.

379. Markson, R., Nelson, R. (1971) Mountain-Peak Potential-Gradient Measurements and the Andes Glow. Weather 25:350.

380. Marshall, E. (1980) Navy Lab Concludes the Vela Saw a Bomb. Science 209:996-997.

381. Martin, S.F., et al. (1983) Emerging Magnetic Flux, Flares and Filaments—FBS Interval 16-23 June 1980. Adv Space Res 2(11):39-51.

382. Martin, S.F., et al. (1984) Relationships of a Growing Magnetic Flux Region to Flares. Adv Space Res 4(7):61-70.

383. Masamura, S. (1977) Solar Activity as a Weighty Factor in Road Accidents. pp.238-240. in (eds) M.N. Gnevyshev, A.I. Oi'. Effects of Solar Activity on the Atmosphere and Biosphere. Jerusalem, Israel Prog Sci Trans.

384. Mason, B.J. (1976) Towards the Understanding and Prediction of Climatic Variations. Q J Roy Meteorol Soc 102:473-485.

385. Mass, C., Schneider, S.H. (1977) Statistical Evidence on the Influence of Sunspots and Volcanic Dust on Long-Term Temperature Records. J Atmos Sci 34:1095-2004.

386. Mauk, F.J. (1979) Triggering of Activity at Soufrierede St. Vincent, April 1979 by Solid Earth Tides. Eos 60:833.

387. Mauk, F.J., Johnston, M.J.S. (1973) On the Triggering of Volcanic Eruptions by Earth Tides. J Geophys Res 78:3356-3362.

388. Mauk, F.J., Kienle, J. (1973) Microearthquakes at St. Augustine Volcano, Alaska, Triggered by Earth Tides. Science 182:386-389.

389. Maunder, A.S.D. (1907) An Apparent Influence of the Earth on the Numbers and Areas of Sun-Spots in the Cycle 1889-1901. Mon Not Roy Astron Soc (May).

390. Maximov, I.V. (1952) On the 80-Year Cycle in Climatic Variations. Doklady Akademii Nauk 86(5):917-920.

391. Maximov, I.V. (1960) Nutational Phenomena in the High Latitudes and Their Role in the Formation of Climate. pp.103-123 in Problems of the North, No 1. Ottawa, Nat Res Council Canada.

392. Maximov, I.V., Karklin, V.P. (1970) Seasonal and Long-term Variations of the Depth and Geographical Position of the Aleutian Low in Air Pressure for 1899-1959. Izvestiya Vses Geograf Obsheh 102:422-431.

393. McClellan, P.H. (1984) Earthquake Seasonality Before the 1906 San Francisco Earthquake. Nature 307:153-6.

394. McDonald, K.L., Gunst, R.H. (1967) Analysis of the Earth's Magnetic Field from 1835 to 1965. Essa Tech Rept IER 46-1ES. Wash, DC, USGPO.

395. McKenna-Lawlor, S.M.P., Richter, A.K. (1983) Physical Interpretation of Interdisciplinary Solar/Interplanetary Observations Relevant to the 27-29 June 1980 SMY/STIP Event No 5. Adv Space Res 2(11):239-251.

396. Meeus, J. (1975) Comments on the Jupiter Effect. Icarus 26:257-267.

397. Meeus, J. (1975) Reply to Gribbin and Plagemann. Icarus 26:270.

398. Meeus, J. (1976) Solar Activity and Earthquakes. Ciel Terre (Belgium) 92(4):233-6.

399. Meeus, J. (1977) "Unexpected Anomaly" in the Annual Distribution of the Maximum of the 11-Year Sunspot Cycle. J Interdisciplinary Cycle Res 8:205-6.

400. Mein, N., et al. (1984) Activity in the Homologous Flare Site. Adv Space Res 4(7):33-35.

401. Mendillo, M., Forbes, J. (1977) Spatial-Temporal Development of Molecular Releases Capable of Creating Large-Scale F-Region Holes. Agard Conf Proc 192(14).

402. Michels, D.J., et al. (1984) Synoptic Observations of Coronal Transients and Their Interplanetary Consequences. Adv Space Res 4(7):311-321.

403. Mignard, F. (1986) Tidal and Non-Tidal Acceleration of the Earth's Rotation. pp.93-110 in (ed) A. Cazenave, Earth Rotation: Solved and Unsolved Problems. Boston, MA, Reidel.

404. Milgrom, M. (1989) Alternatives to Dark Matter. Comments Astrophys 13:215-230.

405. Minakata, H., Nunokawa, H. (1989) Hybrid Solution of the Solar Neutrino Problem in Anticorrelation with Sunspot Activity. Phys Rev Lett 63:121-124.

406. Mock, S.J., Hibler, III, W.D. (1976) 20-Yr Oscillation in Eastern North American Temperature Records. Nature 261:484-486.

407. Mogey, R. (1989) Introduction to Solar Phenomenon. Cycles 40:257-260.

408. Mogey, R. (1990) Cycles in War and Peace. Cycles 41:17-20.

409. Mogi, K. (1974) Active Periods in the World's Chief Seismic Belts. Tectonophysics 22:265-282.

410. Monastersky, R. (1987) Solar Cycle Linked to Weather. Sci News 132:388-389.

411. Monastersky, R. (1989) Depleted Ring Around Ozone Hole. Sci News 136:324.

412. Moore, P.D. (1989) Ancient Climate from Fossils. Nature 340:18-19.

413. Morgan, W.J., et al. (1961) Periodicity of Earthquakes and the Invariance of the Gravitational Constant. J Geophys Res 66:3831-3843.

414. Mori, Y. (1981) Evidence of an 11-Year Periodicity in Tree-Ring Series from Formosa Related to the Sunspot Cycle. J Climatol 1:345-353.

415. Morrison, L.V., Stephenson, F.R. (1986) Observations of Secular and Decade Changes in the Earth's Rotation. pp.69-78 in (ed) A. Cazenave, Earth Rotation: Solved and Unsolved Problems. Boston, MA, Reidel.

416. Morth, H.T., Schlamminger, L. (1979) Planetary Motion, Sunspots and Climate. pp.193-207 in (eds) B.M. McCormac, T.A. Seliga. Solar-Terrestrial Influences on Weather and Climate. Boston, MA, Reidel.

417. Murray, M. (1986) Why the Southeast is Sweltering. Sci News 130:68.

418. Mustel, E.R. (1977) Solar Activity and the Troposphere. pp.38-42 in (eds) M.N. Gnevyshev, A.I. Oi'. Effects of Solar Activity on the Atmosphere and Biosphere. Jerusalem, Israel Prog Sci Trans.

419. Muzalevskaya, N.I. (1977) Biological Activity of the Disturbed Geomagnetic Field. pp.128-137 in (eds) M.N. Gnevyshev, A.I. Oi'. Effects of Solar Activity on the Atmosphere and Biosphere. Jerusalem, Israel Prog Sci Trans.

420. Myerson, R.J. (1970) Long-Term Evidence for the Association of Earthquakes with the Excitation of the Chandler Wobble. J Geophys Res 75:6612-6617.

421. Nelson, J.H. (1979) Certain Planetary Angles Related to Magnetic Storms. Cycles 30(6):129-133.

422. Nelson, J.H. (1987) Planetary Effects on Shortwave Radio. Cycles 38(7):157-158.

423. Newton, R.R. (1970) Ancient Astronomical Observations and the Acceleration of the Earth and Moon. Baltimore, Johns Hopkins.

424. Newton, R.R. (1972) Astronomical Evidence Concerning Non-Gravitational Forces in the Earth-Moon System. Astrophys Space Sci 16:179-200.

425. Newton, R.R. (1972) Medieval Chronicles and the Rotation of the Earth. Baltimore, John Hopkins.

426. Newton, R.R. (1979) Moon's Acceleration and its Physical Origin. Vol.1. As Deduced from Solar Eclipses. Baltimore, John Hopkins.

427. Novikova, K.F., Ryvkin, B.A. (1977) Solar Activity and Cardiovascular Diseases. pp.184-199 in (eds) M.N. Gnevyshev, A.I., Oi'. Effects of Solar Activity on the Earth's Atmosphere and Biosphere. Jerusalem, Israel Prog Sci Trans.

428. Noyes, R.W., Rosner, R. (1981) Ring Around the Sun. Sciences 20(12):15-18,32.

429. Noyes, R.W., et al. (1984) Relation Between Stellar Rotation Rate and Activity Cycle Periods. Astrophys J 287:769-773.

430. O'Connell, R.J., Dziewonski, A.M. (1976) Excitation of the Chandler Wobble by Large Earthquakes. Nature 262:259-262.

431. Odinets, M.G. (1983) Statistical Analysis of a Sequence of Earthquakes of the Far East and Central Asia. Izv Acad Sci USSR Phys Sold Earth (USA) 19(8):597-602.

432. Oehmke, M.G. (1973) Lunar Periodicity in Flight Activity of Honey Bees. J Interdiscipl Cycle Res 4:319-335.

433. Ohl, A.I. (1966) About the Relationship Between Solar Activity and the Troposphere. Solnechyne Dannye 1:69-75.

434. Ohl, A.I. (1972) Solar Activity Cycles and Their Geophysical Manifestations. A Review. J Interdispl Cycle Res 3:395-408.

435. Okal, E., Anderson, D.L. (1975) On the Planetary Theory of Sunspots. Nature 253:511-13.

436. Olson, R.H. (1977) Sun and the Weather. Nature 270:11.

437. Olson, R.H., et al. (1977) Solar Plages and the Vorticity of the Earth's Atmosphere. Nature 274:140-143.

438. Ong, S.G. (1958) Cosmic Radiation and Tuberculosis I. Sci Rec 2:289-296.

439. Ong, S.G. (1959) Cosmic Radiation and Tuberculosis II. Action of Cosmic Radiation on Tubercle Bacilli. Sci Rec 3:40-45.

440. Ong, S.G. (1962) Cosmic Radiation and Tuberculosis III. Influence of Cosmic Radiation on Tuberculosis at High Altitude and at Sea Level. Sci Sinica 5:645-676.

441. Ong, S.G. (1963) Cosmic Radiation and Cancer I. Influence of Cosmic Radiation at Sea Level on Induced Cancer. Sci Sinica 12:1760-1761.

442. Ong, S.G. (1964) Cosmic Radiation and Tuberculosis IV. Influence of Cosmic Radiation on Tuberculosis at High Altitude (3130 m) and at Sea Level. Sci Sinica 13:61-68.

443. Ong, S.G. (1964) Cosmic Radiation and Tuberculosis V. Influence of Cosmic Radiation on Tuberculosis at High Altitude (2300 m) and at Sea Level. Sci Sinica 2:241-258.

444. Ong, S.G. (1964) Cosmic Radiation and Tuberculosis VI. Immunizing Property of Tubercle Bacilli Exposed to Cosmic Radiation. Sci Sinica 2:259-261.

445. Ong, S.G. (1964) Cosmic Radiation and Tuberculosis VII. Action of Cosmic Radiation on Tubercle Bacilli at 2300 m and at Sea Level. Sci Sinica 2:263-267.

446. Opik, E. (1972) Planetary Tides and Sunspots. Icarus 10:298-301.

447. Orville, R.E. (1990) Peak-Current Variations of Lightning Return Strokes as a Function of Latitude. Nature 343:149-151.

448. Osipov, A.I., Desyatov, V.P. (1977) Mechanism of the Influence of Solar-Activity Oscillations on the Human Organism. p.234 in (eds) M.N. Gnevyshev, A.I. Oi'. Effects of Solar Activity on the Earth's Atmosphere and Biosphere. Jerusalem, Israel Prog Sci Trans.

449. Ottestad, P. (1979) Sunspot Series and Biospheric Series Regarded as Results Due to a Common Cause. Sci Rpt Agric Univ Norway 58(9):26-29.

450. Palmer, T.N., Brankovic', C. (1989) 1988 US Drought Linked to Anomalous Sea Surface Temperature. Nature 338:54-56.

451. Palumbo, A. (1989) Gravitational and Geomagnetic Tidal Source of Earthquake Triggering. Il Nuovo Cimento 12C:685-694.

452. Papamarinopoulos, S.P. (1988) Geomagnetic Intensity Measurement from Northern Greece and Their Comparison with Other Data from Central and Northeastern Europe for the Period 0-2000 yr AD. pp.437-442 in (eds) F.R. Stephenson, A.W. Wolfendale. Secularand Geomagnetic Variations in the Last 10,000 Years. Boston, Kluwer.

453. Park, C.G. (1976) Solar Magnetic Sector Effects on the Vertical Electrical Field at Vostok, Antarctica. Geophys Res Lett 3:475-478.

454. Parker, D.E., Clark, J.B. (1980) Solar Influences and Tornadoes in Britain. Weather 35(1):26-29.

455. Parker, E.N. (1983) Magnetic Fields in the Cosmos. Sci Amer 249(2):44-65.

456. Pasichnyk, R. (1990) Solar and Lunar Cycles in Earthquakes: An Electrostatic Trigger. Cycles 41:321-328.

457. Pasichnyk, R. (1991) The Solar-Terrestrial Linkage in Climate and War. Part I: Climate. Cycles 42:125-133.

458. Pasichnyk, R. (1991) The Solar-Terrestrial Linkage in Climate and War. Part II: War. Cycles 42:316-322.

459. Pavlov, N. (1968) Variations of the Earth's Rotation, Deformations in the Earth's Crust and Solar Activity. Izv Glav Astron Obs Pulkove 183:1.

460. Pay, R. (1967) Position of Planets Linked to Solar Flare Prediction. Technol Week 15:35-8.

461. Payne, B. (1983) Prediction of Storms Based on Planetary Positions and Geomagnetic Field Disturbances. Geocosmic Res 8(2).

462. Payne, B. (1984) Cycles of Peace, Sunspots and Geomagnetic Activity. Cycles 35:101-105.

463. Pecker, J.-C. (1988) Difficulties of Standard Cosmologies. pp.295-312 in (eds) F. Bertola, et al., New Ideas in Astronomy. NY, Cambridge Univ.

464. Pepin, R.O., et al. (eds) (1980) Ancient Sun—Fossil Record in the Earth, Moon and Meteorites. NY, Pergamon.

465. Perez-Peraza, J., et al. (1983) Particle Charge Interchange During Acceleration in Flare Regions. Adv Space Res 2(11):197-200.

466. Perret, F.A. (1908) Some Conditions Affecting Volcanic Eruptions. Science 28:277-287.

467. Perret, F.A. (1940) Moon, Sun Help Release Pent-Up Volcanic Energy. Sci News Lett 37:233.

468. Pesses, M.E. (1983) Particle Acceleration by Coronal and Interplanetary Shock Waves. Adv Space Res 2(11):255-264.

469. Peterson, I. (1988) New Pictures of the Sun Reveal a Number of Surprising and Puzzling Solar Features. Sci News 134:8-11.

470. Petit, J.-R., et al. (1981) Ice Age Aerosol Content from East Antarctic Ice Core Samples and Past Wind Strength. Nature 293:391-394.

471. Petrov, V.N., et al (1966) Periodicity of Snow Accumulation in the Antarctic. [In Russian] Inform Bull Sovetskoi Anarkticheskoi Exped 57:97-106.

472. Philander, G.H. (1986) Unusual Conditions in the Tropical Atlantic Ocean in 1984. Nature 322:236-237.

473. Piccardi, G. (1962) Chemical Basis of Medical Climatology. Springfield, Ill, C Thomas.

474. Piccardi, G. (1977) Solar Activity and Chemical Tests. pp.155-163. in (eds) M.N. Gnevyshev, A.I. Oi', Effects of Solar Activity on the Atmosphere and Biosphere. Jerusalem, Israeli Prog Sci Trans.

475. Pierce, E.T. (1976) Winter Thunderstorms in Japan—A Hazard to Aviation. Nav Res Rev 29:12.

476. Pines, D., Shaham, J. (1973) Seismic Activity, Polar Tides and the Chandler Wobble. Nature 245:77.

477. Planel, H., et al. (1969) [The Biological Role of Natural Irradiation] Le Role Biologique de Lirradiation Naturelle. Bordeaux Med 1:131-135.

478. Planel, H., et al. (1971) [Research on the Effects of Ionizing Radiation on Human Cancer Cell Cultures, In Vitro] Recherches sur L'action des Radiations Ionisantes Naturelles sur des Lignees Cellulaires Cancereuses Humaines Cultivees in Vitro. C R Soc Biol 165:2189-2193.

479. Platnova, A.T. (1977) Solar Activity and the Variation of Blood Coagulation Between 1949 and 1966. pp.215-217 in (eds) M.N. Gnevyshev, A.I. Oi'. Effects of Solar Activity on the Earth's Atmosphere and Biosphere. Jerusalem, Israel Prog Sci Trans.

480. Pneuman, G.W. (1983) Diamagnetic Aspects of the Coronal Transient Phenomenon. Adv Space Res 2(11):233-236.

481. Pokrovskaya, T.V. (1977) Solar Activity and Climate. pp.1ff in (eds) M.N. Gnevyshev, A.I. Oi'. Effects of Solar Activity on the Atmosphere and Biosphere. Jerusalem, Israel, Prog Sci Trans.

482. Pool, R. (1988) Was Newton Wrong? Science 241:789-790.

483. Pontius, D.H., Jr., Wolf, R.A. (1990) Transient Flux Tubes in the Terrestrial Magnetosphere. Geophys Res Lett 17:49-52.

484. Poma, A., Proverbio, E. (1981) Random and Long Periodic Variations in the Earth's Motion. J Interdiscipl Cycle Res 12(3):237-246.

485. Porter, S.C. (1981) Recent Glacier Variations and Volcanic Eruptions. Nature 291:139-42.

486. Press, F., Briggs, P. (1975) Chandler Wobble, Earthquakes, Rotation and Geomagnetic Changes. Nature 256:270.

487. Prevot, M., et al. (1985) How the Geomagnetic Field Vector Reverses Polarity. Nature 316:230-234.

488. Priest, E.R. (1984) Role of Newly Emerging Flux in the Flare Process. Adv Space Res 4(7):37-48.

489. Proverbio, E., Poma, A. (1975) Astronomical Evidence of Change in the Rate of the Earth's Rotation and Continental Motion. pp.385-395 in (eds) G.D. Rosenberg, S.K. Runcorn, Growth Rhythms, And the History of the Earth's Rotation. NY, J Wiley.

490. Raisbeck, G.M., et al. (1985) Evidence for an Increase in Cosmogenic 10Be During a Geomagnetic Reversal. Nature 315:315-317.

491. Ramesh, R., Et al. (1990) Climatic Significance of D Variations in Tropical Tree Species from India. Nature 337:149-150.

492. Rawson, H.E. (1907) Anticyclones as Aides to Long Distance Forecasts. J Roy Meteorol Soc 33:309-310.

493. Rawson, H.E. (1908) Anticyclonic Belt of the Southern Hemisphere. Q J Roy Meteorol Soc 34:165-188.

494. Rawson, H.E. (1909) Anticyclonic Belt of the Northern Hemisphere. Q J Roy Meteorol Soc 35:233-248.

495. Reder, F., et al. (1967) Interfering VLF Radio Signals Observed on GBR-16-0 Kc/s Transmissions During November and December 1965. Nature 213:584.

496. Reinsel, G.C., et al. (1987) Statistical Analysis of Total Ozone and Stratospheric Umkehr Data for Trends and Solar Cycle Relationship. J Geophys Res 92D:2201-2209.

497. Reiter, R. (1969) Solar Flares and Their Impact on Potential Gradient and Air-Earth Current Characteristics at High Mountain Stations. Pure Appl Geophys 72:259-267.

498. Reiter, R. (1970) Increased Influx of Stratospheric Air Into the Lower Troposphere After Solar H[-alpha] and X-Ray Flares. J Geophys Res 78:6167-6172.

499. Reiter, R. (1971) Further Evidence for the Impact of Solar Flares on Potential Gradient and Air-Earth Current Characteristics at High Mountain Stations. Pure Appl Geophys 86(3):142-158.

500. Reiter, R. (1972) Case Study Concerning the Impact of Solar Activity Upon Potential Gradient and Air Earth Current in the Lower Troposphere. Pure Appl Geophys 94:218.

501. Reiter, R. (1973) Influx of Stratospheric Air Masses Into the Lower Troposphere After Solar Flares. Naturwissenschaften 60(3):52-153.

502. Reiter, R. (1973) Solar-Terrestrial Relationships of an Atmospheric Electrical and Meteorological Nature: New Findings. Riv Ital Geofis 22:(5/6):247-258.

503. Reiter, R. (1976) Increased Frequency of Stratospheric Injections Into the Tropopause as Triggered by Solar Events. J Atmos Terres Phys 38:503-510.

504. Reiter, R. (1977) Electric Potential of the Ionosphere as Controlled by the Solar Magnetic Sector Structure. Result of a Study Over the Period of a Solar Cycle. J Atmos Terr Phys 39:95-99.

505. Reiter, R. (1978) Influences of Solar Activity on the Electric Potential Between the Ionosphere and the Earth. In Proc Symp/Workshop Solar-Terrestrial Influences on Weather and Climate, 24-28 July. Ohio State Univ, Fawcett Ctr Tomorrow.

506. Ren, Z., Li, Z. (1980) Effects of Motions of Planets on Climatic Changes in China. Kexue Tongbao 25:417-422.

507. Robbins, R.W. (1985) The Case of the Missing Sunspot Peak. Cycles 36(2):52-55.

508. Robbins, R.W. (1985) Does the Sun Really Drive the Economy? Cycles 36(1):11-14.

509. Roberts, W.O. (1975) Relationships Between Solar Activity and Climate Change. In (eds) W.R. Bandeen, S.P. Maran, Possible Relationship Between Solar Activity and Meteorological Phenomena, NASA SP-366. Wash, DC, USGPO.

510. Roberts, W.O., Olson, R.H. (1973) Geomagnetic Storms and Wintertime 300-mb Trough Development in the North Pacific —North America Area. J Atmos Sci 30(1):135-140.

511. Roberts, W.O., Olson, R. (1973) New Evidence for Effects of Variable Solar Corpuscular Emission on the Weather. Rev Geophys Spac Phys 11:731-740.

512. Roble, R.G. (1986) Chemistry in the Thermosphere and Ionosphere. Chem Eng News 64(24):23-38.

513. Roeloffs, E.A., et al. (1989) Hydrologic Effects on Water Level Changes Associated with Episodic Fault Creep Near Parkfield California. J Geophys Res 94A:12387-402.

514. Rokityansky, I.I. (1982) Geoelectromagnetic Investigation of Earth's Crust and Mantle. Trans N. Chobotova. NY, Springer Verlag.

515. Rompolt, B. (1984) Eruption of Huge Magnetic Systems from the Sun. Adv Spac Res 4(7):357-361.

516. Rosenberg, R.L. (1970) Unified Theory of Interplanetary Magnetic Field. Solar Phys 15:72-78.

517. Rossignol-Strick, M., Planchais, N. (1989) Climate Patterns Revealed by Pollen and Oxygen Isoptope Records in a Tyrrhenian Sea Core. Nature 342:413-416.

518. Rotanova, N.M., Fillippov, S.V. (1987) Identification and Analysis of the 1969 Jerk in Secular Geomagnetic Variations. Geomagnetism Aeronomy 27:867-870.

519. Rothen, A., Kincaid, M. (1974) Influence of a Magnetic Field on Immunologic and Enzymatic Reactions Carried Out at a Solid-Liquid Interface. Physiol Chem Phys 6:417-27.

520. Rozhdestvenskaya, E.D., Novikova, K.F. (1977) Influence of Solar Activity on the Fibrinolytic System of the Blood. pp.218-223 in (eds) M.N. Gnevyshev, A.I. Oi'. Effects of Solar Activity on the Earth's Atmosphere and Biosphere. Jerusalem, Israeli Prog Sci Trans.

521. Rubashev, B.M. (1964) Problems of Solar Activity. Moscow-Leningrad, Nauka. NASA TT F-244. Wash, DC.

522. Rubashev, B.H. (1978) Effects of Long-Period Solar Activity Fluctuation on Temperature and Pressure of the Terrestrial Atmosphere. Wash, DC, NASA.

523. Ruelle, D. (1980) [Strange Attractors] Les Attracteurs Estranges. La Recherche 11:132.

524. Runcorn, S.K. (1975) Palaeontology and Astronomical Observations on the Rotational History of the Earth and Moon. pp.285-291 in (eds) G.D. Rosenberg, S.K. Runcorn, Growth Rhythms and the History of the Earth's Rotation. NY, J Wiley.

525. Russell, C.T. (1989) Universal Time Variation of Geomagnetic Activity. Geophys Res Lett 16:555-8.

526. Rust, D.M. (1983) Study of Energy Release in Flares. Adv Space Res 2(11):5.

527. Rymer, N., Brown, G. (1989) Gravity Changes as a Precursor to Volcanic Eruption at Poas Volcano, Costa Rica. Nature 432:902-905.

528. Sadeh, D. (1972) Possible Sidereal Period for the Seismic Lunar Activity. Nature 240:139-140.

529. Sadeh, D., Meidav, M. (1972) Periodicities in Seismic Response Caused by Pulsar CP1133. Nature 240:136-38.

530. Saiko, J. (1979) Influence of Solar Activity on the Physical Processes of the Atmosphere. Kosmische Einflusse 47:102-9.

531. Sanford, F. (1936) Influence of Planetary Configurations Upon the Frequency of Visible Sun Spots. (Pub 3391) Smithsonian Misc Coll 95(11):1-5.

532. Sastri, J.H. (1985) IMF Polarity Effects on the Equatorial Ionospheric F-Region. Adv Space Res 5(4):199-204.

533. Sauers, C.J. (1985) Will Mt. St. Helens Reawaken? Cycles 36:75-76.

534. Sawyer, C. (1983) 1980 April 12 Flare and Transient: Report on Progress in Interpretation. Adv Space Res 2(11):265-270.

535. Saxl, E.J., Allen, M. (1971) 1970 Solar Eclipse as "Seen" by a Torsion Pendulum. Phys Rev D 3:823-825.

536. Sazonov, B.I. (1964) High-Level Pressure Formations and Solar Activity. pp.112-125. Pub Glav Geofiz Obs [in Russian]. Leningrad, Gidrometeoizdat.

537. Schmahl, E.J. (1983) Flare Build-Up in X-Rays, UV, Microwaves and White Light. Adv Space Res 2(11):73-90.

538. Schneider, D.A., Kent, D.V. (1990) Ivory Coast Microtektites and Geomagnetic Reversals. Geophys Res Lett 17:163-166.

539. Schneider, S.H., Mass, C. (1975) Volcanic Dust, Sunspots and Temperature Trends. Science 190:741-46.

540. Schove, D.J. (ed) (1983) Sunspot Cycles. Stroudsberg, Pa, Hutchinson Ross.

541. Schuurmans, C.J.E. (1978) Influence of Solar Activity on Winter Temperatures: New Climatological Evidence. Climatic Change 1(3):231-237.

542. Schuurmans, C.J.E., Oort, A.H. (1969) Statistical Study of Pressure Changes in the Troposphere and Lower Stratosphere After Strong Solar Flares. Pure Appl Geophys 75(4):233-246.

543. Schwartz, S.J., et al. (1985) Active Current Sheet in the Solar Wind. Nature 318:269-271.

544. Shao-guang, C. (1989) Does Vacuum Polarization Influence Gravitation? Il Nuovo Cimento 104B:611-619.

545. Sheehan, R.G., Grieves, R. (1982) Sunspots and Cycles: A Test of Causation. So Economic J 48:775-7.

546. Sheeley, Jr., N.R. (1964) Polar Faculae During the Sunspot Cycle. Astrophys J 140:731-5.

547. Shimshoni, M. (1971) Evidence for Higher Seismic Activity During the Night. Geophys J 24:97-99.

548. Shirk, G. (1958) War. Cycles 9(3):102.

549. Shirley, J.H. (1985/86) Shallow Moonquakes and Large Shallow Earthquakes: A Temporal Correlation. Earth Planet Sci Lett 76:241-253.

550. Shirley, J.H. (1986) Lunar Periodicity in Great Earthquakes 1950-65. Gerlands Beitr Geophys (Germany) 95(6):509-15.

551. Shirley, J.H. (1989) When the Sun Goes Backward: Solar Motion, Volcanic Activity and Climate, 1990-2000. Cycles 40:70-76.

552. Siedentopf (1958) [Cosmic Phenomena and Biometeorology]. Wien Med Wochenschr 108:126.

553. Simpson, J.F. (1967) Solar Activity as a Triggering Mechanism for Earthquakes. Earth Planet Sci Lett 3:417.

554. Siscoe, G.L. (1978) Solar-Terrestrial Influences on Weather and Climate. Nature 276:348-352.

555. Siscoe, G.L. (1980) Evidence in the Auroral Record for Secular Solar Variability. Rev Geophys Space Phys 18:647-8.

556. Sleeper, Jr., H.P. (1972) Planetary Resonances, Bi-Stable Oscillation Modes and Solar Activity Cycles. NAS 1.26 2035.

557. Sleeper, Jr., H.P. (1974) Astrometeorology: The Influence of the Planets on Solar Activity and Climate Change. Explorer 5:4-6.

558. Smith, P.J. (1979) Magnetism, Climate and Eccentricity. Nature 277:354.

559. Snodgrass, H.B. (1988) Evidence for a Solar Cycle. Phys Today 41(1):S11-S12.

560. Sobakar', G.T., Dejneko, V.I. (1978) Space-Time Correlations Between Gravity Quasi-Periodic Variations, Solar Activity, Earthquake Energy and Earth's Crustal Structure [in Russian]. Geofiz SB, (USSR) (82):3-8.

561. Somov, B.V., Titov, V.S. (1984) Magnetic Reconnection in a High-Temperature Plasma of Solar Flares. Adv Space Res 4(7):183-185.

562. Sonett, C.P., Suess, H.E. (1984) Correlation of Bristle Cone Pine Ring Widths with Atmospheric 14C Variations: A Climate-Sun Relation. Nature 307:141-143.

563. Song, M.T., Wu, S.T. (1984) On the Heating Mechanism of Magnetic Flux Loops in the Solar Atmosphere. Adv Space Res 4(7):275-278.

564. Sonneman, G., et al. (1985) Do There Exist Effects in the Thermospheric Plasma, Arising from Dynamic Variations in the Middle Atmosphere? Adv Space Res 5(4):299-304.

565. Souriau, A. (1986) Influence of Earthquakes on the Polar Motion. pp.229-240 in (ed) A. Cazenave, Earth Rotation: Solved and Unsolved Problems. Boston, Reidel.

566. Southwood, D.J. (1980) Thunderstorms and Substorms: Any Connection? Nature 284:599.

567. Stacey, C.M. (1963) Cyclical Measures: Some Tidal Aspects Concerning Equinoctial Years. Ann NY Acad Sci 105:421-460.

568. Stadol'nik, V.E. (1977) Effect of Heliogeophysical Factors on the Evolution of Infectious Diseases of Man. p.107 in (eds) M.N. Gnevyshev, A.I. Oi', Effects of Solar Activity on the Earth's Atmosphere and Biosphere. Jerusalem, Isreal Prog Sci Trans.

569. Starr, V.P., Oort, A.H. (1973) Five-year Climate Trend for the Northern Hemisphere. Nature 242:310-313.

570. Stearns, H.T., MacDonald, G.A. (1946) Geology and Groundwater Resources of the Island of Hawaii. Div Hydrogeography, Territory of Hawaii, Bull 9:89-90.

571. Stening, R.J. (1989) Diurnal Modulation of the Lunar Tide in the Upper Atmosphere. Geophys Res Lett 16:307-310.

572. Stetson, H.T. (1935) Correlation of Deep-Focus Earthquakes with Lunar Hour Angle and Declination. Science 82:523-524.

573. Stetson, H.T. (1947) Sunspots in Action. NY, Ronald.

574. Stevens, C.O. (1908) Long-Lived Solar Halo. Nature 78:221.

575. Stothers, R. (1980) Giant Solar Flares in Antarctic Ice. Nature 287:365.

576. Stothers, R.B. (1989) Seasonal Variation of Volcanic Eruption Frequencies. Geophys Res Lett 16:453-456.

577. Stringfellow, M.F. (1974) Lightning Incidence in Britain and the Solar Cycle. Nature 249:332-333.

578. Stuiver, M. (1980) Solar Variability and Climatic Change During the Current Millennium. Nature 286:868-871.

579. Stuiver, M., Braziunas, T.F. (1988) Solar Component of the Atmosphere 14C Record. pp.245-266 in (eds) F.R. Stephenson, A.W. Wolfendale. Secular Solar and Geomagnetic Variations in the Last 10,000 Years. Boston, Kluwer.

580. Stuiver, M., Grootes, P.M. (1980) Trees and the Ancient Record of Heliomagnetic Cosmic Ray Flux Modulation. pp.165-174 in (eds) R.O. Pepin, et al., Ancient Sun: Fossil Record in the Earth, Moon and Meteorites. NY, Pergamon.

581. Stuiver, M., Quay, P.D. (1980) Changes in Atmospheric Carbon-14 Attributed to a Variable Sun. Science 207:11-19.

582. Stutz, A.M. (1973) Synodic Monthly Rhythms in the Mongolian Gerbil (*Meriones unguiculatus*). J Interdiscipl Cycle Res 4:229-236.

583. Sugimoto, Y., Uchida, E. (1978) Relation Between Solar Activity and Tropospheric Short-Term Meteorological Phenomena in the Polar Region. pp.279-285 in (eds) K. Takahasi, M.M. Yoshino, Climatic Change and Food Production. Intl Symp Recent Climatic Change Food Production, 4-8 Oct 1976. Tsukuba & Tokyo.

584. Svalgaard, L. (1968) Sector Structure of the Interplanetary Magnetic Field and Daily Variation of the Geomagnetic Field at High Latitudes. Geophys Paper R-6 (Aug) Danish Metero Instit. Charlottenlund, Denmark.

585. Svalgaard, L. (1972) Interplanetary Magnetic Sector Structure, 1926-1971. J Geophys Res 77:4027-4033.

586. Svenonius, B., Olausson, E. (1977) Solar Activity and Weather Conditions in Sweden for the Period 1756-1975. J Interdiscipl Cycle Res 8:222-225.

587. Svestka, Z. (1983) Flare Build-Up Study in the SMA Period. Adv Space Res 2(11):3-4.

588. Svestka, Z., et al. (eds) (1983) Solar Maximum Year. Adv Space Res 2 (11).

589. Szekeres (1968) Effect of Gravitation on Frequency. Nature 220:1116-1118.

590. Takeuchi, T., et al. (1973) On Lightning Discharges in Winter Thunderstorms. J Meteorol Soc Jap 51:494.

591. Takeuti, T., Nakano, M. (1978) Anomalous Winter Thunderstorms of the Hokuriku Coast. J Geophys Res 83:2385-2394.

592. Tamrazyan, G.P. (1957) Tbilisi Earthquakes and Cosmic Conditions of the Earth. Reports Acad Sci Georgian SSR 19(2).

593. Tamrazyan, G.P. (1968) Earthquakes of Nevada (USA) and the Tidal Forces. J Geophys Res 73:6013-6018.

594. Tarcsai, Gy. (1985) Ionosphere-Plasmasphere Electron Fluxes at Middle Latitudes Obtained from Whistlers. Adv Space Res 5(4):155-158.

595. Tarling, D.H. (1975) Geological Processes and the Earth's Rotation in the Past. pp.397-412 in (eds) G.D. Rosenberg, S.K. Runcorn, Growth Rhythms, and the History of the Earth's Rotation. NY, J Wiley.

596. Tarling, D.H. (1988) Secular Variations of the Geomagnetic Field—The Archaeomagnetic Record. pp.349-365 in (eds) F.R. Stephenson, A.W. Wolfendale. Secular Solar and Geomagnetic Variations in the Last 10,000 Years. Boston, Kluwer.

597. Taylor, H., et al. (1984) Corotating Interplanetary Streams and Associated Ionospheric Disturbances at Venus and Earth. Adv Space Res 4(7):343-346.

598. Thomas, R.J. (1988) What Makes Flares Recur at 155 or 51 Days? Phys Today 41(1):S12-S13.

599. Thompson, L. M. (1989) Sunspots and Lunar Cycles: Their Possible Relation to Weather Cycles. Cycles 40:265-268.

600. Thomsen, D.E. (1986) Efficient Wimps Would Rescue the Sun. Sci News 129:325.

601. Tomaschek, R. (1959) Great Earthquakes and the Astronomical Positions of Uranus. Nature 184:177-8.

602. Tomaschek, R. (1960) Earthquakes and Uranus: Misuse of a Statistical Test of Significance [A Reply]. Nature 186:337-338.

603. Toptygin, I.N. (1985) Cosmic Rays in Interplanetary Magnetic Fields. Boston, Reidel.

604. Tourtellott, R. (1987) Mavericks of History. Cycles 38:169-170.

605. Trenberth, K.E., et al. (1988) Origins of the 1988 North American Drought. Science 242:1640-1645.

606. Trimble, V. (1988) New- and Old-Ideas on the Universal J(M) Relationship. pp.239-242 in (eds) F. Bertola, et al., New Ideas in Astronomy. NY, Cambridge Univ.

607. Tripol'nikov, V.P. (1977) Results of an Experiment on the Forecasting of the Time of Occurrence of Strong Earthquakes. Geomagnet Aerono 17:765-767.

608. Triskova, L. (1988) Asymmetry of the Main Solar Dipole Field Resulting in a 12-month Wave in Geomagnetic Activity. Adv Space Res 8(7):195-198.

609. Triskova, L. (1989) Vernal-Autumnal Asymmetry in the Seasonal Variation of Geomagnetic Actvity. J Atmo Terres Phys 51:111-118.

610. Tromp, S.W. (1965) Influence of Weather and Climate on Blood Pressure, Blood Composition and Physico-Chemical State of Blood of Blood Donors at Leiden, the Netherlands (Period 1953-1963). Mongr Ser Biometeor res Centre, Leiden (7).

611. Tromp, S.W. (1972) Possible Effects of Extra-Terrestrial Stimuli on Colloidal Systems and Living Organisms. pp.239-248 in (eds) S.W. Tromp, J.J. Bouma, Biometeorological Aspects of Plants, Trees and Animals in Human Life. Amsterdam, Swets & Zeitlinger.

612. Tromp, S.W. (1975) Further Evidence for Possibly Extra-Terrestrial Influences on the Living and Non-Living World. Rpt Biometeorol Res Ctr (15).

613. Tromp, S.W. (1975) Possible Extra-Terrestrial Triggers of Interdisciplinary Cycles on Earth: A Review. J Interdiscipl Cycle Res 6:303-315.

614. Tromp, S.W. (1982) Studies Suggesting Extra-Terrestrial Influences (Apart from Solar Radiation on Biological Phenomena and Physio-Chemical Processes on Earth. Cycles 33:179-192 [and] pp.219-243 in (eds) S.W. Tromp, J.J. Bouma. Biometeorological Survey. Vol.1, 1973-1978. Part A. Human Biometerology. London, Heyden & Son.

615. Turman, B.N. (1977) Detection of Lightning Superbolts. J Geophys Res 82:2566-2568.

616. Tyson, P.D. (1980) Temporal and Spatial Variation of Rainfall Anomalies in Africa South of Latitude 22o During the Period of Meteorological Record. Climatic Change 2:363-371.

617. Tyson, P.D. (1981) Atmospheric Circulation Variation and the Occurrence of Extended Wet and Dry Spells Over Southern Africa. J Climatol 1:115-130.

618. Uffen, R.J. (1963) Influence of the Earth's Core on the Origin and Evolution of Life. Nature 198:143-4.

619. Valdimirov, O.A., et al. (1970) Solar Activity and Baric Field Oscillations in High Latitudes in the Northern Hemisphere. Trudy Arktich i Antarktich, Nauchno-Issledov Instit 296:160-167.

620. Valet, J.-P., et al. (1986) High-Resolution Sedimentary Record of a Geomagnetic Reversal. Nature 322:27-32.

621. Valet, J.-P., et al. (1986) Volcanic Record of Reversal. Nature 316:217-218.

622. Van Fossen, M.C., Channell, J.E.T. (1989) Paleomagnetic Evidence for Tertiary Anticlockwise Rotation in Southwest Puerto Rico. Geophys Res Lett 16:819-822.

623. Van Hoven, G., Hurford, G.J. (1984) Flare Precursors and Onset. Adv Space Res 4(7):95-103.

624. Van Loon, H., Labitzke, K. (1988) Association Between the 11-Year Solar Cycle, the QBO, and the Atmosphere, Part II: Surface and 700mb on the Northern Hemisphere in Winter. J Climat 1:905-920.

625. Van Loon, H., Labitzke, K. (1988) Association Between the 11-Year Solar Cycle, the QBO, and the Atmosphere, Part III: Aspects of the Association. J Climat 2:554-565.

626. Van Loon, H., Labitzke, K. (1988) When the Wind Blows. New Sci 119(1629):58-60.

627. Vaughan, O.H., Vonnegut, O.H. (1989) Recent Observations of Lightning Discharges from the Top of a Thundercloud Into the Clear Air Above. J Geophys Res 94D:13179-13182.

628. Venne, D.E., et al. (1983) Comment on "Evidence for a Solar Cycle Signal in Tropospheric Winds" by G.D. Nastrom and A.D. Melmont. J Geophys Res 88:11025-30.

629. Verfaillie, G.R.M. (1969) Correlation Between the Rate of Growth of Rice Seedling and the P-Indices of the Chemical Test of Piccardi: A Solar Hypothesis. Intl J Biometeor 13:113-121.

630. Vines, R.G. (1980) Analyses of South African Rainfall. So African J Sci 76:404-409.

631. Vines, R.G. (1982) Rainfall Patterns in the Western United States. J Geophys Res 87:7303-7311.

632. Visvanathan, T.R. (1966) Formation of Depressions in the Indian Seas and Lunar Phase. Nature 210:406-407.

633. Vladimirskii, B.M. (1977) Possible Solar-Activity Factors Affecting Processes in the Biosphere. pp.139-142. in (eds) M.N. Gnevyshev, A.I. Oi'. Effects of Solar Activity on the Atmosphere and Biosphere. Jerusalem, Tsrael Prog Sci Trans.

634. Vladimirskii, B.M., et al. (1977) An Experimental Study of the Effect of Ultra-Low-Frequency Electromagnetic Fields on Warm-blooded Animals and Microorganisms. pp.259-269 in (eds) M.N. Gnevyshev, A.I. Oi'. Effects of Solar Activity on the Atmosphere and Biosphere. Jerusalem, Tsrael Prog Sci Trans.

635. Vlahos, L. (1989) Particle Acceleration in Solar Flares. Solar Phys 121:431-447.

636. Volland, H. (1982) CRC Handbook of Atmospherics. Boca Raton, Florida, CRC Press.

637. Volland, H. (1984) Atmospheric Electrodynamics. Physics and Chemistry in Space. Vol.11. NY, Springer-Verlag.

638. Vondrak, J, (1977) Rotation of the Earth Between 1955.5 and 1976.5. Studia Geoph et Geod 21:107-117.

639. Vonnegut, B., et al. (1989) Nocturnal Photographs Taken from A U-2 Airplane Looking Down on Tops Of Clouds Illuminated by Lightning. Amer Meteor Soc Bull 70:1263-1271.

640. Vorpahl, J.A., et al. (1970) Satellite Observations of Lightning. Science 169:860-862.

641. Wagner, W.J. (1983) SERF Studies of Mass Motions Arising in Solar Flares. Adv Space Res 2(11):203-19.

642. Wang, P.-K. (1980) On the Relationship Between Winter Thunder and the Climatic Change in China in the Past 2200 Years. Climatic Change 3:37-46.

643. Wang, P.-K., Siscoe, G.L. (1980) Ancient Chinese Observations of Physical Phenomena Attending Solar Eclipses. Solar Phys 66:187-193.

644. Wang, Y.-M., et al. (1989) Magnetic Flux Transport on the Sun. Science 245:712-718.

645. Weber, J. (1969) Evidence for the Discovery of Gravitational Radiation. Phys Rev Lett 22:1320-1324.

646. Weisburd, S. (1986) Stalking the Weather Bombs. Sci News 129:314-317.

647. Wesson, P.S. (1978) Cosmology and Geophysics. NY, Oxford Univ.

648. Wesson, P.S. (1979) Self-Similarity and the Angular Momenta of Astronomical Systems: A Basic Rule in Astronomy. Astron Astrophys 80:296-300.

649. Wesson, P.S. (1981) Clue to the Unification of Gravitation and Particle Physics. Phys Rev 23D:1730-1734.

650. Wigley, T.M.L. (1988) Climate of the Past 10,000 Years and the Role of the Sun. Climatic Res Unit, UEA, Norwich, UK.

651. Wilcox, J.M. (1968) Interplanetary Magnetic Field. Solar Origin and Terrestrial Effects. Space Sci Rev 8:250-328.

652. Wilcox, J.M. (1972) Inferring the Interplanetary Magnetic Field by Observing the Polar Magnetic Field. Rev Geophys Space Phys 10:1003-1015.

653. Wilcox, J.M. (1975) In (eds) W.R. Bandeen, S.P. Maran, Possible Relationship Between Solar Activity and Meteorological Phenomena. NASA Rept SP-366. Wash, DC, USGPO.

654. Wilcox, J.M. (1979) Solar Activity and Changes in Atmospheric Circulation. J Atmos Terr Phys 41:753-763.

655. Wilcox, J.M., et al. (1974) Influence of Solar Magnetic Sector Structure on Terrestrial Atmospheric Vorticity. J Atmos Sci 31:581-588.

656. Wilcox, J.M., et al. (1976) On the Reality of a Sun-Weather Effect. J Atmos Sci 33:1113-1116.

657. Wilcox, J.M., et al. (1979) Interplanetary Magnetic Field Polarity and the Size of Low-Pressure Troughs Near 180o W Longitude. Science 204:60.

658. Wilcox, J.M., Scherer, P. (1972) Annual and Solar Magnetic-Cycle Variations in the Interplanetary Magnetic Field, 1926-1971. J Geophys Res 77:5385-8.

659. Williams, D. (1973) Theories of the Cause of Sunspots. Cycles 24:205-214.

660. Williams, G.E. (1986) Solar Cycle in Precambrian Times. Sci Amer 255(8):88-96.

661. Williams, J. (1873) Chinese Observations of Solar Spots. Mon Not Roy Astron Soc 33:370-375.

662. Williams, R.G. (1978) Study of the Energetics of a Particular Sun-Weather Relation. Geophys Res Lett 5:519-522.

663. Williams, R.G. (1979) Comments on 'Large Amplitude Standing Planetary Waves Induced in the Troposphere by the Sun' by J.W. King, et al. J Atmos Terr Phys 41:643-645.

664. Willis, D.M. (1976) Energetics of Sun-Weather Relationships Magnetospheric Processes. J Atmos Terr Phys 38:685-689.

665. Willis, D.M., et al. (1988) Seasonal and Secular Variations of the Oriental Sunspot Sightings. pp.187-202 in (eds) F.R. Stephenson, A.W. Wolfendale. Secular Solar and Geomagnetic Variations in the Last 10,000 Years. Boston, Kluwer Acad.

666. Winstanley, D. (1973) Recent Rainfall Trends in Africa, The Middle East and India. Nature 243:464-5.

667. Wollin, G., et al. (1973) Magnetic Intensity and Climate Changes, 1925-1970. Nature 242:34-37.

668. Wollin, G., et al. (1987) Abrupt Geomagnetic Variations—Predictive Signals for Temperature Changes 3-7yr in Advance. pp.241-255 in (eds) M.R. Rampino, et al., Climate: History, Periodicity, and Predictability. NY, Van Nostrand Reinhold.

669. Wood, C.A., Lovett, R.R. (1974) Rainfall, Drought and the Solar Cycle. Nature 251:594-596.

670. Wood, H.O. (1917) On Cyclical Variations in Eruption at Kilauea. Second Rpt Hawaiian Volcano Observatory. Cambridge, Mass, MIT.

671. Wood, K.D. (1972) Sunspots and Planets. Nature 240:91-93.

672. Wood, R.M. (1975) Comparison of Sunspot Periods with Planetary Synodic Period Resonances. Nature 255:312-313.

673. Wood, R.M., Wood, K.D. (1965) Solar Motion and Sunspot Comparison. Nature 208:129-131.

674. Woodgate, B.E. (1983) Flare Build-Up Study—Homologous. Flares Group Interim Report. Adv Space Res 2(11):61-64.

675. Woodworth, P.L. (1985) A Worldwide Search for the 11-Year Solar Cycle in Mean Sea Level Records. Geophys J Roy Astron Soc 80:743-755.

676. World Data Center A for Solar-Terrestrial Physics (1977) Solar-Terrestrial Physics and Meteorology: Working Document II. Boulder, Co, World Data Ctr for Solar-Terrestrial Phys, Environmental Data Service.

677. Xanthakis, J., et al. (1981) Study of the Inferred Interplanetary Magnetic Field Polarity Periodicities. J Interdiscipl Cycle Res 12(3):205-215.

678. Yagodinskii, V.N., et al. (1977) Epidemic Process as a Function of Solar Activity. pp.82-103 in (eds) M.N. Gnevyshev, A.I. Oi', Effects of Solar Activity on the Atmosphere and Biosphere. Jerusalem, Israel Prog Sci Trans.

679. Yeager, D.M., Frank, L.A. (1976) Low-Energy Electron Intensities at Large Distances Over the Earth's Polar Cap. J Geophys Res 81:3966-3976.

680. Yukutake, T. (1965) Solar Cycle Contribution to the Secular Change in the Geomagnetic Field. J Geomagn Geoelectr 17:287-309.

681. Yunnan Observatory. Ancient Sunspot Records Research Group (1976) Re-Compilation of Our Country's Records of Sunspots Through the Ages and an Inquiry Into Possible Periodicities in Their Activity. Acta Astron Sinica 17:217-227 [and] Chinese Astron 1:347-359 (1977).

682. Zahnle, K.J., Walker, J.C. (1987) Climatic Oscillations During the Precambian Era. Climatic Change 10:269-284.

683. Zerefos, C.S. (1981) Short Period Oscillations in Wintertime Atmospheric Vorticity Patterns and the Geomagnetic Field. J Interdiscipl Cycle Res 12(2):181-185.

684. Zhang, G. et al. (1989) Quasi-Steady Corotating Structure of the Interplanetary Geomagnetic Disturbances: A Survey of Solar Cycles. J Geophys Res 94A:1235-1245.

685. Zhenqui, R., Zhisen, L. (1980) Effects of Motions of Planets on Climatic Changes in China. Kexue Toughao (Beijing) 25(5):417-422.

686. Zhokhov, V.P., Indeken, E.N. (1978) Correlation Between Acute Attacks of Glaucoma and Oscillations of the Geomagnetic Field. pp.241-243 in (eds) M.N. Gnevyshev, A.I. Oi'. Effects of Solar Activity on the Earth's Atmosphere and Biosphere. Jerusalem, Israel Prog Sci Trans.

687. Zook, H.A. (1980) On Lunar Evidence for a Possible Large Increase in Solar Flare Activity [approximately] 2 x 104 Years Ago. pp.245-266 in (eds) R.O. Pepin, et al., Ancient Sun: Fossil Record in the Earth, Moon and Meteorites. NY, Pergammon.

688. Zook, H.A., et al. (1977) Solar Flare Activity: Evidence for Large-Scale Changes in the Past. Icarus 32:106-126.

Conclusions

1.   Abramowicz, M., Ellis, G. (1989) Elusive Anthropic Principle. Nature 337:411-412.

2.   Alfven, H. (1981) Cosmic Plasma. Boston, D Reidel.

3.   Allais, M.F.C. (1959) Should the Laws of Gravitation Be Reconsidered? Part I. Aerospace Eng 18(9):46-52.

4.   Allais, M.F.C. (1959) Should the Laws of Gravitation Be Reconsidered? Part II. Aerospace Eng 18(10):51-55.

5.   Allais, M.F.C. (1959) Should the Laws of Gravitation Be Reconsidered? Commentary Note. Aerospace Eng 18(11):55.

6.   Altarelli, G., Barbieri, R. (1991) Vacuum Polarization Effects of New Physics on Electroweak Processes. Phys Lett 253B:161-167.

7.   Bagby, J.P. (1990) Review of Some Recent Anomalous Gravity Experiments. Cycles 41:134-139.

8.   Babosunov, M.B., Gurovich, V.Ts. (1991) Conformal Correspondence and Quantization of Scalar Excitations in Cosmological Models with Vacuum Polarization. Sov Astron Lett 16:413-415.

9.   Balashov, Yu.V. (1990) Multifaced Anthropic Principle. Comments Astrophys 15:19-28.

10. Barklow, T., Martin, P. (1990) Elementary Particle Physics. pp.171-204 in (ed) R.A. Meyers. Encyclopedia of Modern Physics. NY, Academic.

11. Bekenstein, J., Milgrom, M. (1984) Does the Missing Mass Problem Signal the Breakdown of Newtonian Gravity? Astophys J 286:7-14.

12. Berman, M.S. (1991) Inflation in the Einstein-Cartan Model. Gen Relativ Gravit 23:1083-88

13. Blackmore, S.J. (1985) Rupert Sheldrake's New Science of Life: Science, Parascience or Pseudo-Science? Parapsych Rev 16(3):6-9.

14. Bohm, D. (1980) Wholeness and the Implicate Order. London, Routledge, Kegan Paul.

15. Bohm, D., et al. (1987) An Ontological Basis for the Quantum Theory. Phys Rep 144:322-75

16. Bostick, W.H. (1987) Similarities Between the Plasma Vortex Filaments (Relativistic and Nonrelativitic) Observed in the Plasma Focus and in Conventional Relativistic Electron-Beam Machines. Fusion Tech 12:92-103.

17. Bouchiat, M.-A., Pottier, L. (1984) An Atomic Preference Between Left and Right. Sci Amer 250(6):100-105.

18. Braude, S.E. (1983) Radical Provincialism in the Life Sciences: A Review of Rupert Sheldrake's *A New Science of Life*. J Amer Soc Psychical Res 77(1):63-78.

19. Brown, J.D., et al. (1989) Scalar Field Wormholes. Nuc Phys 328B:213-222.

20. Brush, S.G. (1990) Theories of the Origin of the Solar System 1956-1985. Rev Mod Phys 62:43-112.

21. Bunyard, P. (1987) Gaia: Its Implications for Industrialized Society. pp.217-236 in (eds) P. Bunyard, E. Goldsmith. Gaia: The Thesis, the Mechanisms and the Implications. Worthyvale Manor, Camelford, Cornwall, Wadebridge Ecological Center [and] Ecologist 18:196-211.

22. Bunyard, P., Goldsmith, E. (1989) Gaia and Evolution. Proc 2nd Annual Camelford Conference on the Implications of the Gaia Thesis. Worthyvale Manor, Camelford, Cornwall, Wadebridge Ecological Centre.

23. Burgess, C.P., Cloutier, J. (1988) Astrophysical Evidence for a Weak New Force? Phys Rev 38D:2944-2950.

24. Carr, B.J., Rees, M.J. (1979) The Anthropic Principle and the Structure of the Physical World. Nature 278:605-612.

25. Caves, C.M. (1990) Quantitative Limits on the Ability of a Maxwell Demon to Extract Work from Heat. Phys Rev Lett 64:2111-2113.

26. Charney, E. (1979) Molecular Basis of Optical Activity. Malabar, Fl, Krieger.

27. Clement, B.E.P. (1991) Monitoring Environmental Pollution in Real Time and on a Global Basis by Means of Differential Computation. pp.82-89 in Proc Intl Conf Environ Pollution (Lisbon, Portugal). Interscience Enterprises, Geneva, Switz.

28. Coisson, R. (1973) On the Vector Potential of Coriolis Forces. Amer J Phys 41:585.

29. Coles, P. (1992) Primordial Magnetic Fields and the Large-Scale Structure of the Universe. Comments Astrophys 16:45-60.

30. Cordero, S., Busse, F.H. (1992) Experiments on Convection in Rotating Hemispherical Shells: Transition to a Quasi-periodic State. Geophys Res Lett 19:733-736.

31. Costa de Beauregard, O. (1978) S-Matrix, Feynman Zigzag and Einstein Correlation. Phys Lett 67A:171-173.

32. Coveney, P.V. (1988) The Second Law of Thermodynamics: Entropy, Irreversibility and Dynamics. Nature 333:409-415.

33. Craik, A.D.D. (1989) Stability of Unbounded Two- and Three-Dimensional Flows Subject to Body Forces: Some Exact Solutions. J Fluid Mech 198:275-292.

34. Davies, P.C.W. (1984) Superforce: The Search for a Grand Unified Theory of Nature. London, Heinemann.

35. Davies, P.C.W. (1987) Cosmic Blueprint. London, Heinemann.

36. D'Espagnat, B. (1979) Quantum Theory and Reality. Sci Amer 241(11):158-181.

37. De Rujula, A. (1990) Dark Matter. pp.23-32 in (ed) A. Zichichi. The Challenging Questions. NY, Plenum.

38. Desloge, E.A. (1989) Further Comments on the Horizontal Deflection of a Falling Object. Amer J Phys 57:282-284.

39. DeVaucouleur, G. (1970) The Case for a Hierarchal Cosmology. Science 167:1203-1213.

40.   Dolan, L. (1990) Unified Theories. pp.737-747 in (ed) R.A. Meyers. Encyclopedia of Modern Physics. NY, Academic.

41.   Dragilev, V.M. (1990) Vacuum Polarization of a Scalar Field in Anistropic Multidimensional Cosmology. Theoret Math Phys 84:887-893.

42.   Dragone, L.R. (1985) Gravitational Lift via the Coriolis Force. Hadronic J 8:85-92.

43.   Ehrlich, P.R., Ehrlich, A.H. (1991) Healing the Planet: Strategies for Resolving the Environmental Crisis. NY, Addison-Wesley.

44.   Engheta, N., et al. (1989) Effect of Chirality on the Doppler Shift and Aberration of Light Waves. J Appl Phys 66:2274-2276.

45.   Evans, M.W. (1990) Axial Birefringence Due to Electromagnetic Fields: Spin Chiral Effects. Phys Lett 146A:475-477.

46.   Ferencz, Cs., Tarcsai, Gy. (1971) Interaction of Gravitational and Electromagnetic Fields or Another Effect? Nature 233:404-406.

47.   Ferguson, M. (1977) Special Issue: A New Perspective on Reality. Brain/Mind Bull 2(16).

48.   Fischbach, E. (1986) Reanalysis of the Eotvos Experiment. Phys Rev Lett 56:3-6.

49.   Gal'tsov, D.V., Morozov, M., Yu. (1989) Vacuum Polarization and Dynamical Symmetry Breaking in de Sitter Space. Sov J Nuc Phys 49:1090-1097.

50.   Gardner, M. (1979) Ambidextrous Universe. NY, Scribner.

51.   Giorgi, A. (1976) Phenomenology and the Foundations of Psychology. In (ed) W.J. Arnold, Symposium on Motivation: Conceptual Foundations of Psychology, vol.24. Lincoln, Univ Nebraska.

52.   Globus, G., Franklin, S. (1980) Prospects for the Scientific Observer of Perceptual Consciousness. pp.465-482 in (eds) J.M. Davidson, R.M. Davidson, Psychobiology of Consciousness. NY, Plenum.

53.   Goodwin, G.L., Hobson, G.J. (1978) Atmospheric Gravity Waves Generated During a Solar Eclipse. Nature 275:109-111.

54.   Greenstein, G. (1988) Symbiotic Universe: Life and Mind in the Cosmos. NY, William Morrow & Co.

55.   Gunzig, E., et al. (1987) Entropy and Cosmology. Nature 330:621-624.

56.   Harman, W.W. (1981) Broader Implications of Recent Findings in Psychological and Psychic Research. pp.113-132 in (ed) R.G. Jahn, Role of Consciousness in the Physical World. AAAS. Boulder, Co, Westview.

57.   Harrison, E.R. (1970) Generation of Magnetic Fields in the Radiation Era. Mon Not Roy Astron Soc 147:279-286.

58.   Hartshone, C. (1978) Panpsychism: Mind as Sole Reality. Ultimate Reality and Meaning 1:115-129.

59.   Hegstrom, R.A., Kondepudi, D.K. (1990) The Handedness of the Universe. Sci Amer 262(1):108-115.

60.   Hegstrom, R.A., et al. (1988) Mapping the Weak Chirality of Atoms. Amer J Phys 56:1086-1092.

61.   Hogan, C.J. (1983) Magnetospheric Effects of a First-Order Cosmological Phase Transition. Phys Rev Lett 51:1488-1491.

62.   Holding, S.C., et al (1986) Gravity in Mines—An Investigation of Newton's Law. Phys Rev 33D:3487-3494.

63. Holding, S.C., Tuck, G.J. (1984) A New Mine Determination of the Newtonian Gravitational Constant. Nature 307:714-716.

64. Honorton, C. (1981) Psycho physical Interaction. pp.19-36 in (ed) R.G. Jahn, Role of Consciousness in the Physical World. AAAS. Boulder, Co, Westview.

65. Hsui, A. T. (1987) Borehole Measurement of the Newtonian Gravitational Constant. Science 237:881-883.

66. Jahn, R.G., Dunne, B.J. (1986) On the Quantum Mechanics of Consciousness, with Implications to Anomalous Phenomena. Found Phys 16:721-772.

67. Jahn, R.G., Dunne, B.J. (1987) Margins of Reality. Orlando, Florida, Harcourt Brace Jovanovich.

68. Jahn, R.G. (ed) (1981) Role of Consciousness in the Physical World. AAAS. Boulder, Co, Westview.

69. James, D.F.V., Wolf, E. (1990) Doppler-like Frequency Shifts Generated by Dynamic Scattering. Phys Lett 146A:167-170.

70. Jaynes, J. (1976) Origin of the Consciousness in the Breakdown of the Bicameral Mind. Boston, Houghton Millfin.

71. Johnson, A.R., Morris, W.D. (1992) Experimental Investigation into the Effects of Rotation on the Isothermal Flow Resistence in Circular Tubes Rotating about a Parallel Axis. Intl J Heat Fluid Flow 13:132-140.

72. Jones, A.K. (1990) Social Symbiosis: Gaian Theory and Sociology. Ecologist 20:108-112.

73. Joseph, L.E. (1990) Gaia: Growth of an Idea. NY, St Martin's.

74. Josephson, B.D. (1988) Limits to the Universality of Quantum Mechanics. Found Phys 18:1195-1204.

75. Josephson, B.D., Pallikari-Viras, F. (1991) Biological Utilization of Quantum Nonlocality. Found Phys 21:197-207.

76. Jung, C.G. (1952) Synchronicity: An Acausal Connecting Principle. Coll Works, vol.8. NJ, Princeton: Bollingen Series.

77. Jung, C.G. (1975) Archetypes and the Collective Unconscious. Trans R.F.C. Hall. NJ, Princeton Univ.

78. Kawati, S., Kokado, A. (1987) Dynamical Symmetry Breaking in a de Sitter-Invariant Vacuum. Phys Rev 35D:3092-3099.

79. Kawati, S., Kokado, A. (1989) Magnetic Field Induced by Polarization in de Sitter Spacetime. Phys Rev 39D:3612-18.

80. Keutzer, C.S. (1982) Archetypes, Synchronicity and the Theory of Formative Causation. J Analyt Psych 27(3):255-262.

81. Keutzer, C.S. (1983) Theory of "Formative Causation" and Its Implications for Archetypes, Parallel Inventions and the "Hundredth Monkey Phenomenon." J Mind Behav 4(3):353-367.

82. Keyes, K. (1982) Hundredth Monkey. NY, Vision Books.

83. Kikuyama, K., et al. (1991) Effects of Channel Rotation on Turbulent Boundary Layer Along a Convex Surface. Trans Japan Soc Mech Eng 57:2261-2268.

84. Kim, Y.E. (1989) Apparent Anomalies in Borehole and Seafloor Gravity Measurements. Phys Lett 216B:212-216.

85. Klinger, E. (1978) Modes of Normal Conscious Flow. p.225 in (eds) K.S. Pope, J.L. Singer, Waking Stream of Consciousness. NY, Plenum.

86. Konoplich, R.V. (1989) The Casmir Effect with Nontrivial Field Configurations. Hadronic J 12:19-24.

87. Koppar, S.S., Patel, L.K. (1989) Interior Fields of Charged Fluid Sphere in the Einstein-Cartan Theory. Acta Physica Polonica 20B:165-175.

88. Kubler-Ross, E. (1970) On Death and Dying. NY, Macmillian.

89. Lakhtakia, A., et al. (1989) Time-Harmonic Electromagnetic Fields in Chiral Media. NY, Springer Verlag.

90. Langacker, P., Mann, A.K. (1989) Unification of Electro-magnetism with the Weak Force. Phys Today 42(12):22-31.

91. Lawrance, W.D., Knight, A.E.W. (1988) Rotational Mechanisms in Intramolecular Vibrational Energy Redistribution: An Examination of Centrifugal and Coriolis Coupling and the States Contributing to Time Dynamical Measurements. J Phys Chem 92:5900-5908.

92. Lerner, E. (1991) The Big Bang Never Happened. NY, Time Books/Random House.

93. Lerner, E. (1990) Radio Absorption by the Intergalactic Medium. Astrophys J 361:63-68.

94. Lerner, E. (1989) Galactic Model of Element Formation. IEEE Trans Plasma Phys 17:259-263.

95. Lerner, E. (1988) Plasma Model of the Microwave Background and Primordial Elements. Laser & Particle Beams 6:457-469.

96. Lerner, E. (1986) Magnetic Self-Compression in Laboratory Plasma, Quasars and Radio Galaxies. Laser & Particle Beams 4:193-222.

97. Lloyd, S. (1989) Use of Mutual Information to Decrease Entropy: Implications For the Second Law of Thermodynamics. Phys Rev 39A:5378-5386.

98. Lovelock, J.E. (1987) The Gaia Hypothesis. pp.35-45 in (eds) Bunyard, P., Goldsmith, E., Gaia: The Thesis, the Mechanisms and the Implications. Worthyvale Manor, Camelford, Cornwall, Wadebridge Ecological Centre.

99. Lovelock, J.E. (1989) Geophysiology. Trans Roy Soc Edinburgh 80:169-176.

100. Lovelock, J.E. (1989) Geophysiology, the Science of Gaia. Rev Geophys 27:215-222.

101. Maddox, J. (1989) Down With the Big Bang. Nature 340:425.

102. Marino, E.C. (1988) Quantum Theory of Nonlocal Vortex Fields. Phys Rev 38D:3194-96.

103. Maslow, A.H. (1948) Some Theoretical Consequences of Basic Need-Gratification. J Personal 16:402-416.

104. Maslow, A.H. (1950) Self-Actualizing People; A Study in Psychological Health. Personal 1:11-34.

105. Maslow, A.H. (1970) Motivation and Personality. NY, Harper & Row.

106. Maslow, A.H. (1975) Love in Healthy People. In (ed) A. Montagu, The Practice of Love. Englewood Cliffs, NJ, Prentice-Hall.

107. Miller, A.S. (1991) Gaia Connections: An Introduction to Ecology, Ecoethics, and Economics. Savage, Md, Rowman & Littlefield.

108. Morris, M.S., et al. (1989) Wormholes, Time Machines, and the Weak Energy Condition. Phys Rev Lett 61:1446-1448.

109. Mostepanenko, V.M., Sokolov, I.Yu. (1989) Restrictions on the Parameters of the Sin-1 Antigraviton and the Dilation Resulting from the Casmir Effect and from the Eotvos and Cavendish Experiments. Sov J Nucl Phys 49:1118-1120.

110. Newton, R.R. (1972) Astronomical Evidence Concerning Non-Gravitational Forces in the Earth-Moon System. Astrophys Space Sci 16:179-200.

111. Nguyen, D., Odagaki, T. (1987) Quantum and Classical Electrons in a Potential Well with Uniform Electric Field. Amer J Phys 55:466-469.

112. Noguchi, T., et al (1989) Coriolis Coupling Effects on the Angular Distribution of Photoelectrons from Time-Resolved Two-Photon Ionization of Polyatomic Molecules. Chem Phys Lett 155:177-182.

113. O'Hara, M.M. (1984) Reflections on Sheldrake, Wilber, and "New Science." J Humanistic Psych 24(2):116-120.

114. Oman, H. (1992) Extracting Earth Power for a Northeast-Southwest Train. Proc Intersoc Energy Conv Eng 6:171-176.

115. Opat, G.I. (1990) Coriolis and Magnetic Forces: The Gyrocompass and Magnetic Compass as Analogs. Amer J Phys 58:1173-1176.

116. O'Shaughnessy, B. (1972) Mental Structure and Self-Consciousness. Inquiry 15:30-63.

117. Park, Y., et al. (1990) Wormhole Induced Supersymmetry Breaking in String Theory. Phys Lett 244B:393-395.

118. Pasini, A. (1990) Precessions of Opposite Chirality for the Spin Vector in a Riemann-Cartan Framework. Phys Lett 151A:459-463.

119. Paul, S.N., et al. (1992) Nonlinear Effects on the Propagation of Waves in a Magnetized Rotating Plasma. IEEE Trans Plasma Sci 20:481-486.

120. Pelet, P., Nader, E. (1990) Theory of Chirowaveguides. IEEE Trans Ant Prop 38:90-97.

121. Peratt, A.L. (1986) Evolution of the Plasma Universe. IEEE Trans Plasma Sci 14:639-660 & 763-778.

122. Peratt, A.L., et al. (1980) Evolution of Colliding Plasmas. Phys Rev Lett 44:1767-1770.

123. Peratt, A.L., Green, J. (1983) On the Evolution of Interacting, Magnetized, Galactic Plasmas. Astrophys Space Sci 91:19-33.

124. Pope, K.S., Singer, J.L. (1980) Waking Stream of Consciousness. pp.169-192 in (eds) J.M. Davidson, R.J. Davidson, Psychobiology of Consciousness. NY, Plenum.

125. Popper, K., Eccles, J.C. (1977) Self and Its Brain. Berlin, Springer Intl.

126. Prigogine, I. (1978) Time, Structure and Fluctuations. Science 201:777-785.

127. Prigogine, I., Stengers, I. (1984) Order Out of Chaos. NY, Bantam Books.

128. Puchov, I.V., Chalenko, N.S. (1989) Analysis of Force-Reproduction Methods. Measurement Techniques 32:150-152.

129. Puthoff, H.E., et al. (1981) Experimental Psi Research: Implications for Physics. pp.37-86 in (ed) R.G. Jahn, Role of Consciousness in the Physical World. AAAS, Boulder, Co, Westview.

130. Radin, D.I., Nelson, R.D. (1989) Evidence for Consciousness-Related Anomalies in Random Physical Systems. Found Phys 19:1499-1514.

131. Ravetz, J. (1987) Gaia and the Philosophy of Science. pp.133-144 in (eds) Bunyard, P., Goldsmith, E., Gaia: The Thesis, the Mechanisms and the Implications. Worthyvale Manor, Camelford, Cornwall, Wadebridge Ecological Centre.

132. Riscalla, L.M. (1974) An Electromagnetic Field Theory of Consciousness. J Amer Soc Psychosom Dentistry Med 21(2):40-51.

133. Rothman, M.A. (1988) A Physicist's Guide to Skepticism. Buffalo, NY, Prometheus.

134. Roukens De Lange, A. (1982) Evolution and Formative Causation. Parapsych J So Africa 3(2):84-105.

135. Roukens De Lange, A. (1982) Matter, Life, Mind and Psi. Parapsych J So Africa 3(1):28-49.

136. Sagan, D., Margulis, L. (1990) Biospheres: Metamorphosis of Planet Earth. NY, McGraw Hill.

137. Sailendra Nath, P., et al. (1992) Nonlinear Effects on the Propagation of Waves in a Magnetized Rotating Plasma. IEEE Trans Plasma Sci 20:481-486.

138. Scherk, J. (1980) Gravitation at Short Range and Supergravity. pp.381-410 in (eds) S. Ferrara, et al. Unification of the Fundamental Particle Interactions. NY, Plenum.

139. Schiffer, I. (1981) Role of Illusion in Mental Life. Union Seminary Q Rev 36(2/3):83-93.

140. Schmid, P. (1986) Holistic and Bio-Logical Light Weight Structures for Healthy and Individual Housing, Including a Methodical Approach to Details. pp.176-181 in LSA 86: Lightweight Structures in Architecture, Proc 1st Intl Conf. (Sydney, Australia), Vol.1. Kensington, Australia.

141. Schmidt, A.G. (1986) Coriolis Acceleration and Conservation of Angular Momentum. Amer J Phys 54:755-757.

142. Schneider, S.H., Boston, P.J. (eds) (1991) Scientists on Gaia. Cambridge, Mass, MIT.

143. Schwarzschild, B. (1989) Why is the Cosmological Constant So Small? Phys Today 42(3):21-24.

144. Semon, M.D., Schmeig, G.M. (1981) Note on the Analogy Between Inertial and Electromagnetic Forces. Amer J Phys 49:689-690.

145. Shao-guang, C. (1989) Does Vacuum Polarization Influence Gravitation? Il Nuovo Cimento 104:611-619.

146. Sheldrake, R. (1982) Morphic Resonance, Memory and Psychical Research. Parapsych J So Africa 3(2):70-76.

147. Sheldrake, R. (1985) A New Science of Life: The Hypothesis of Formative Causation. London, Anthony Blond.

148. Silverman, M.P. (1990) Effect of the Earth's Rotation on the Optical Properties of Atoms. Phys Lett 146A:175-180.

149. Soleng, H.H. (1990) Spin-polarized Cylinder in Einstein-Cartan Theory. Classical & Quantum Gravity 7:999-1007.

150. Soleng, H.H. (1991) Torsion Vector and Variable G. Gen Relativ Gravit 23:1089-1111.

151. Sperry, R.W. (1964) Problems Outstanding in the Evolution of Brain Function. J. Arthur Lecture. NY, Amer Museum Nat Hist.

152. Sperry, R.W. (1972) Science and the Problem of Values. Perspect Biol Med 16:115-130.

153. Sperry, R.W. (1980) Mind-Brain Interaction, Mentalism, Yes; Dualism, No. Neuroscience 5:195-206.

154. Sperry, R.W. (1981) Changing Priorities. Annu Rev Neurosci 4:1-15.

155. Squires, E.J. (1987) Many Views of One World—An Interpretation of Quantum Theory. Eur J Phys 8:173.

156. Stacey, F.D., et al. (1987) Geophysics and the Law of Gravity. Rev Mod Phys 59:157-174.

157. Stacey, F.D., Tuck, G.J. (1981) Geophysical Evidence for Non-Newtonian Gravity. Nature 292:230-232.

158. Subramanian, C.S., et al. (1992) Surface Heat Transfer and Flow Properties of Vortex Arrays Induced Artificially and from Centrifugal Instabilities. Intl J Heat Fluid Flow 13:210- 223.

159. Sun, X., Jaggard, D.L. (1991) Accelerated Particle Radiation in Chiral Media. J Appl Phys 69:34-38.

160. Talbot, M. (1991) Holographic Universe. NY, Harper Collins.

161. Tieng, S.M., Yan, A.C. (1992) Investigation of Mixed Convection about a Rotating Sphere by Holographic Interferometry. J Thermophys Heat Transfer 6:727-732.

162. Tudge, C. (1989) The Rise and Fall of *Homo sapiens sapiens*. Phil Trans Roy London 325B:479-488.

163. Vainshtein, A.I., Zakharov, V.I. (1989) In Search of Symmetry Behind Topological Chiral Current. Nuc Phys 324B:495-502.

164. van Ballegooijen, A.A., Choudhuri, A.R. (1988) Possible Role of Meridional Flows in Suppressing Magnetic Buoyancy. Astrophys J 333:965-977.

165. Vilenkin, A., Leahy, D.A. (1982) Parity Nonconservation and the Origin of Cosmic Magnetic Fields. 254:77-81.

166. Weinberg, S. (1974) Unified Theories of Elementary-Particle Interaction. Sci Amer 231(1):50-59.

167. Weldon, H.A. (1989) Particles and Holes. Physica 158A:169-177.

168. Westbroek, P. (1991) Life as a Geological Force: Dynamics of the Earth. NY, Norton.

169. Westbroek, P., de Bruyn, G.-J. (1987) Geological Impact of Life. pp.99-111 in (eds) Bunyard, P., Goldsmith, E., Gaia: The Thesis, the Mechanisms and the Implications. Worthyvale Manor, Camelford, Cornwall, Wadebridge Ecological Centre.

170. Wheeler, J.A. (1974) Universe as Home for Man. Amer Sci 62:683-691.

171. Wilber, K. (1984) Sheldrake's Theory of Morphogensis. J Humanistic Psych 24(2):107-115.

172. Will, C.M. (1984) Confrontation Between General Relativity and Experiment: An Update. Phys Reports 113:345-422.

173. Wolpert, D. (1988) Chaos of the Brussels School May Not Be Irreversible. Nature 335:595.

174. Xu, J.-Z., Chen, Y.-H. (1991) Intermediate-Range Force and Gravitation. Gen Relativ Gravit 23:169-175.

175. Zweibel, E.G. (1988) Growth of Magnetic Fields Prior to Galaxy Formation. Astophys J 329:L1-L4.

INDEX TO VOLUMES ONE AND TWO

A

time-varying electrostatic auroral conductivity, 86

Azores High, 922, 933

B

C

E

F

H

I

J

L

M

N

O

Q

R

S

T

U

V

W

X

0-595-21086-4